Statistics and Computing

Series editor

Wolfgang Karl Härdle, Humboldt-Universität zu Berlin, Berlin, Germany

Statistics and Computing (SC) includes monographs and advanced texts on statistical computing and statistical packages.

More information about this series at http://www.springer.com/series/3022

Wolfgang Karl Härdle · Cathy Yi-Hsuan Chen
Ludger Overbeck
Editors

Applied Quantitative Finance

Third Edition

Springer

Editors
Wolfgang Karl Härdle
C.A.S.E.—Center for Applied Statistics
and Economics
Humboldt-Universität zu Berlin
Berlin
Germany

Ludger Overbeck
Department of Mathematics
University of Giessen
Giessen
Germany

Cathy Yi-Hsuan Chen
School of Business and Economics
Humboldt-Universität zu Berlin
Berlin
Germany

ISSN 1431-8784 ISSN 2197-1706 (electronic)
Statistics and Computing
ISBN 978-3-662-57199-6 ISBN 978-3-662-54486-0 (eBook)
DOI 10.1007/978-3-662-54486-0

Printed on acid-free paper

This Springer imprint is published by Springer Nature
The registered company is Springer-Verlag GmbH Germany
The registered company address is: Heidelberger Platz 3, 14197 Berlin, Germany

With the kind permission of MFO: Ludger Overbeck, Cathy Chen, Wolfgang Karl Härdle

Preface to the Third Edition

The third edition of Applied Quantitative Finance moves the focus to risk management. As a consequence, we changed the basic structure from four to three chapters with many more contributions to market and credit risk. We revisit important market risk issues in Chap. 1. Chapter 2 introduces novel concepts in credit risk along with renewed quantitative methods being proposed accordingly. A wider range of coverage in recent development of credit risk and its management is presented in this version. The last chapter is on dynamics of risk management and includes risk analysis of energy markets and for cryptocurrencies. Digital assets, such as block chain-based currencies, become popular but are theoretically challenging when based on conventional methods. A modern text mining method called Dynamic Topic Modelling is introduced in detail and applied to the message board of Bitcoins. A time-varying LASSO technique for tail events is at the heart of a new financial risk meter. This third edition brings together modern risk analysis based on quantitative methods and textual analytics for the need of the new challenges in banking and finance.

Berlin/Giessen, Germany
April 2017

Wolfgang Karl Härdle
Cathy Yi-Hsuan Chen
Ludger Overbeck

Contents

Part I
Market Risk

Chapter 1
VaR in High Dimensional Systems-A Conditional Correlation Approach

H. Herwartz, B. Pedrinha and F.H.C. Raters

Abstract In empirical finance, multivariate volatility models are widely used to capture both volatility clustering and contemporaneous correlation of asset return vectors. In higher dimensional systems, parametric specifications often become intractable for empirical analysis owing to large parameter spaces. On the contrary, feasible specifications impose strong restrictions that may not be met by financial data as, for instance, constant conditional correlation (CCC). Recently, dynamic conditional correlation (DCC) models have been introduced as a means to solve the trade off between model feasibility and flexibility. Here, we employ alternatively the CCC and the DCC modeling framework to evaluate the Value-at-Risk associated with portfolios comprising major U.S. stocks. In addition, we compare their performances with corresponding results obtained from modeling portfolio returns directly via univariate volatility models.

H. Herwartz · F.H.C. Raters (✉)
Department of Economics, University of Göttingen, Humboldtallee 3, 37073 Göttingen, Germany
e-mail: raters@uni-goettingen.de

H. Herwartz
e-mail: hherwartz@uni-goettingen.de

B. Pedrinha
WINGAS GmbH, Königstor 20, 34117 Kassel, Germany
e-mail: bruno.pedrinha@wingas.de

W.K. Härdle et al. (eds.), *Applied Quantitative Finance*, Statistics and Computing,
DOI 10.1007/978-3-662-54486-0_1

1.1 Introduction

Volatility clustering, i.e. positive correlation of price variations observed on speculative markets, motivated the introduction of autoregressive conditionally heteroskedastic (ARCH) processes by Engle (1982) and its popular generalizations by Bollerslev (1986) (Generalized ARCH, GARCH) and Nelson (1991) (Exponential GARCH). Being univariate in nature, however, these models neglect a further stylized feature of empirical price variations, namely contemporaneous correlation over a cross section of assets, stock or foreign exchange markets (Engle et al. 1990a; Hamao et al. 1990; Hafner and Herwartz 1998; Lee and Long 2009).

The covariance between asset returns is of essential importance in finance. Effectively, many problems in financial theory and practice, such as asset allocation, hedging strategies or Value-at-Risk (VaR) evaluation, require some formalization not merely of univariate risk measures but rather of the entire covariance matrix (Bollerslev et al. 1988; Cecchetti et al. 1988). Similarly, pricing of options with more than one underlying asset will require some (dynamic) forecasting scheme for time varying variances and covariances as well (Duan 1995).

When modeling time dependent second order moments, a multivariate model is a natural framework to take cross sectional information into account. Over recent years, multivariate volatility models have been attracting high interest in econometric research and practice. Popular examples of multivariate volatility models comprise the GARCH model class recently reviewed by Bauwens et al. (2006). Numerous versions of the multivariate GARCH (MGARCH) model suffer from huge parameter spaces. Thus, their scope in empirical finance is limited since the dimension of vector valued systems of asset returns should not exceed five (Ding and Engle 2001). Factor structures (Engle et al. 1990b) and so-called correlation models (Bollerslev 1990) have been introduced to cope with the curse of dimensionality in higher dimensional systems. The latter start from univariate GARCH specifications to describe volatility patterns and formalize in a second step the conditional covariances implicitly via some model for the systems' conditional correlations. Recently, dynamic conditional correlation models have been put forth by Engle (2002), Engle and Sheppard (2001) and Tse and Tsui (2002) that overcome the restrictive CCC pattern (Bollerslev 1990) while retaining its computational feasibility.

Here, we will briefly review two competing classes of MGARCH models, namely the half-vec model family and correlation models. The latter will be applied to evaluate the VaR associated with portfolios comprised by stocks listed in the Dow Jones Industrial Average (DJIA) index. We motivate the idea for VaR backtesting and reference the recent literature on (un)conditional VaR coverage tests. We compare the performance of models building on constant and dynamic conditional correlation. Moreover, it is illustrated how a univariate volatility model performs in comparison with both correlation models.

The remainder of this paper is organized as follows. The next section introduces the MGARCH model and briefly mentions some specifications that fall within the class of so-called half-vec MGARCH models. Correlation models are the focus of Sect. 1.3 where issues like estimation or inference within this model family are discussed in some detail. In Sect. 1.4, we motivate and discuss VaR backtesting by means of (un)conditional coverage. An empirical application of basic correlation models to evaluate the VaR for portfolios comprising U.S. stocks is provided in Sect. 1.5.

1.2 Half-Vec Multivariate GARCH Models

Let $\varepsilon_t = (\varepsilon_{1t}, \varepsilon_{2t}, \ldots, \varepsilon_{Nt})^\top$ denote an N-dimensional vector of serially uncorrelated components with mean zero. The latter could be directly observed or estimated from a multivariate regression model. The process ε_t follows a multivariate GARCH process if it has the representation

$$\varepsilon_t|\mathcal{F}_{t-1} \sim N(0, \Sigma_t), \Sigma_t = [\sigma_{ij,t}], \tag{1.1}$$

where Σ_t is measurable with respect to information generated up to time $t-1$, formalized by means of the filtration $\mathcal{F}_{t-1}$. The $N \times N$ conditional covariance matrix, $\Sigma_t = \mathsf{E}[\varepsilon_t\varepsilon_t^\top|\mathcal{F}_{t-1}]$, has typical elements $\sigma_{ij,t}$ with $i = j$ $(i \neq j)$ indexing conditional variances (covariances). In a multivariate setting, potential dependencies of the second order moments in Σ_t on $\mathcal{F}_{t-1}$ become easily intractable for practical purposes.

The assumption of conditional normality in (1.1) allows to specify the likelihood function for observed processes ε_t, $t = 1, 2, \ldots, T$. In empirical applications of GARCH models, it turned out that conditional normality of speculative returns is more an exception than the rule. Maximizing the misspecified Gaussian log-likelihood function is justified by quasi maximum likelihood (QML) theory. Asymptotic theory on properties of the QML estimator in univariate GARCH models is well developed (Bollerslev and Wooldridge 1992; Lee and Hansen 1994; Lumsdaine 1996 and a few results on consistency Jeantheau 1998) and asymptotic normality Comte and Lieberman (2003); Ling and McAleer (2003) have been derived for multivariate processes.

The so-called half-vec specification encompasses all MGARCH variants that are linear in (lagged) second order moments or squares and cross products of elements in (lagged) ε_t. Let vech(B) denote the half-vectorization operator stacking the elements of a $(m \times m)$ matrix B from the main diagonal downwards in a $m(m+1)/2$ dimensional column vector. We concentrate the formalization of MGARCH models on the MGARCH(1,1) case which is, by far, the dominating model order used in the empirical literature (Bollerslev et al. 1994). Within the half-vec representation of the GARCH(1, 1) model Σ_t is specified as follows:

$$\text{vech}(\Sigma_t) = c + A\,\text{vech}(\varepsilon_{t-1}\varepsilon_{t-1}^\top) + G\,\text{vech}(\Sigma_{t-1}). \tag{1.2}$$

In (1.2), the matrices A and G each contain $\{N(N+1)/2\}^2$ elements. Deterministic covariance components are collected in c, a column vector of dimension $N(N+1)/2$. On the one hand, the half-vec model in (1.2) allows a very general dynamic structure of the multivariate volatility process. On the other hand, this specification suffers from huge dimensionality of the relevant parameter space which is of order $\mathcal{O}(N^4)$. In addition, it might be cumbersome or even impossible in applied work to restrict the admissible parameter space such that the time path of implied matrices Σ_t is positive definite.

To reduce the dimensionality of MGARCH models, numerous avenues have been followed that can be nested in the general class of half-vec models. Prominent examples in this vein of research are the Diagonal model (Bollerslev et al. 1988), the BEKK model (Baba et al. 1990; Engle and Kroner 1995), the Factor GARCH (Engle et al. 1990b), the orthogonal GARCH (OGARCH) (Alexander 1998, 2001) or the generalized OGARCH model put forth by Van der Weide (2002). Evaluating the merits of these proposals requires to weight model parsimony and computational issues against the implied loss of generality. For instance, the BEKK model is convenient to allow for cross sectional dynamics of conditional covariances, and weak restrictions have been formalized keeping Σ_t positive definite over time (Engle and Kroner 1995). Implementing the model will, however, involve simultaneous estimation of $\mathcal{O}(N^2)$ parameters such that the BEKK model has been rarely applied in higher dimensional systems ($N > 4$). Factor models build upon univariate factors, such as an observed stock market index (Engle et al. 1990b) or underlying principal components (Alexander 1998, 2001). The latter are assumed to exhibit volatility dynamics which are suitably modeled by univariate GARCH-type models. Thereby, factor models drastically reduce the number of model parameters undergoing simultaneous estimation. Model feasibility is, however, paid with restrictive correlation dynamics implied by the (time invariant) loading coefficients. Moreover, it is worthwhile mentioning that in case of factor specifications still $\mathcal{O}(N)$ parameters have to be estimated jointly when maximizing the Gaussian (quasi) likelihood function.

1.3 Correlation Models

1.3.1 Motivation

Correlation models comprise a class of multivariate volatility models that is not nested within the half-vec specification. Similar to factor models, correlation models circumvent the curse of dimensionality by separating the empirical analysis in two steps. First, univariate volatility models are employed to estimate volatility dynamics of each asset specific return process ε_{it}, $i = 1, \ldots, N$. In a second step Σ_t is obtained imposing some parsimonious structure on the correlation matrix (Bollerslev 1990). Thus, in the framework of correlation models we have

$$\Sigma_t = V_t(\theta)R_t(\phi)V_t(\theta), \tag{1.3}$$

where $V_t = \text{diag}(\sqrt{\sigma_{11,t}}, \ldots, \sqrt{\sigma_{NN,t}})$ is a diagonal matrix having as typical elements the square roots of the conditional variances estimates $\sigma_{ii,t}$. The latter could be obtained from some univariate volatility model specified with parameter vectors θ_i stacked in $\theta = (\theta_1^\top, \ldots, \theta_N^\top)^\top$. If univariate GARCH(1,1) models are used for the conditional volatilities $\sigma_{ii,t}$, θ_i will contain 3 parameters such that θ is of length $3N$. Owing to its interpretation of a correlation matrix, the diagonal elements in $R(\phi)$ are unity ($r_{ii} = 1,\ i = 1, \ldots, N$). From the general representation in (1.3) it is apparent that alternative correlation models particularly differ with regard to the formalization of the correlation matrix $R_t(\phi)$ specified with parameter vector ϕ.

In this section, we will highlight a few aspects of correlation models. First, a log-likelihood decomposition is given that motivates the stepwise empirical analysis. Then, two major variants of correlation models are outlined, the early CCC model (Bollerslev 1990) and the DCC approach introduced by Engle (2002) and Engle and Sheppard (2001). Tools for inference in correlation models that have been applied in the empirical part of the paper are collected in an own subsection. Also, a few remarks on recent generalizations of the basic DCC specification are provided.

1.3.2 Log-Likelihood Decomposition

The adopted separation of volatility and correlation analysis is motivated by a decomposition of the Gaussian log-likelihood function (Engle 2002) applying to the model in (1.1) and (1.3):

$$\begin{aligned} l(\theta, \phi) &= -\frac{1}{2}\left\{\sum_{t=1}^{T} N\log(2\pi) + \log(|\Sigma_t|) + \varepsilon_t^\top \Sigma_t^{-1}\varepsilon_t\right\} \\ &= -\frac{1}{2}\left\{\sum_{t=1}^{T} N\log(2\pi) + 2\log(|V_t|) + \log(|R_t|) + \varepsilon_t^\top \Sigma_t^{-1}\varepsilon_t\right\} \\ &= \sum_{t=1}^{T} l_t(\theta, \varphi), \end{aligned}$$

$$\begin{aligned} l_t(\theta, \phi) &= l_t^V(\theta) + l_t^C(\theta, \phi), &(1.4)\\ l_t^V(\theta) &= -\frac{1}{2}\left\{N\log 2\pi + 2\log(|V_t(\theta)|) + \varepsilon_t' V_t(\theta)^{-2}\varepsilon_t\right\} &(1.5)\\ l_t^C(\theta, \phi) &= -\frac{1}{2}\left(\log|R_t(\phi)| + v_t^\top R_t(\phi)^{-1}v_t - v_t^\top v_t\right). &(1.6) \end{aligned}$$

According to (1.5) and (1.6), the maximization of the log-likelihood function may proceed in two steps. First, univariate volatility models are used to maximize the volatility component, $l_t^V(\theta)$, and conditional on first step estimates $\hat{\theta}$, the correlation part $l_t^C(\theta, \phi)$ is maximized in a second step. To perform a sequential estimation procedure efficiently, it is required that the volatility and correlation parameters are variation free (Engle et al. 1983) meaning that there are no cross relationships linking single parameters in θ and ϕ when maximizing the Gaussian log-likelihood function. In the present case, the parameters in θ will impact on $v_t = V_t^{-1}\varepsilon_t$, $v_t = (v_{1t}, v_{2t}, \ldots, v_{Nt})^\top$, and, thus, the condition necessary to have full information and limited information estimation equivalent is violated. Note, however, that univariate GARCH estimates $(\hat{\theta})$ will be consistent. Thus, owing to the huge number of available observations which is typical for empirical analyses of financial data, the efficiency loss involved with a sequential procedure is likely to be smaller in comparison with the gain in estimation feasibility.

1.3.3 Constant Conditional Correlation Model

Bollerslev (1990) proposes a constant conditional correlation (CCC) model

$$\sigma_{ij,t} = r_{ij}\sqrt{\sigma_{ii,t}\sigma_{jj,t}},\ i, j = 1, \ldots, N,\ i \neq j. \tag{1.7}$$

Given positive time paths of the systems' volatilities, positive definiteness of Σ_t is easily guaranteed for the CCC model ($|r_{ij}| < 1,\ i \neq j$). As an additional objective of this specification, it is important to notice that the estimation of the correlation pattern may avoid iterative QML estimation of the $\{N(N-1)/2\}$ correlation parameters r_{ij} comprising $R_t(\phi) = R$. Instead, one may generalize the idea of variance targeting (Engle and Mezrich 1996) towards the case of correlation targeting. Then, $D = E[v_t v_t^\top]$ is estimated as the unconditional covariance matrix of standardized returns, $v_t = V_t^{-1}\varepsilon_t$, and R is the correlation matrix implied by D. With $'\odot'$ denoting matrix multiplication by element, we have formally

$$\hat{R} = \hat{D}^{*-1/2}\hat{D}\hat{D}^{*-1/2},\ \hat{D} = \frac{1}{T}\sum_{t=1}^{T} v_t v_t^\top,\ \hat{D}^* = \hat{D} \odot I_N. \tag{1.8}$$

The price paid for the feasibility of CCC is, however, the assumption of a rather restrictive conditional correlation pattern which is likely at odds with empirical systems of speculative returns. Applying this model in practice therefore requires at least some pretest for constant correlation (Tse 2000; Engle 2002).

1.3.4 Dynamic Conditional Correlation Model

The dynamic conditional correlation model introduced by Engle (2002) and Engle and Sheppard (2001) preserves the analytic separability of the models' volatilities and correlations, but allows a richer dynamic structure for the latter. For convenience, we focus the representation of the DCC model again on the DCC(1,1) case formalizing the conditional correlation matrix $R_t(\phi)$ as follows:

$$R_t(\phi) = \{Q_t^*(\phi)\}^{-1/2} Q_t(\phi) \{Q_t^*(\phi)\}^{-1/2}, \ Q_t^*(\phi) = Q_t(\phi) \odot I_N, \tag{1.9}$$

with

$$Q_t(\phi) = R(1 - \alpha - \beta) + \alpha v_{t-1} v_{t-1}^\top + \beta Q_{t-1}(\phi) \tag{1.10}$$

and R is a positive definite (unconditional) correlation matrix of v_t.

Sufficient conditions guaranteeing positive definiteness of the time path of conditional covariance matrices Σ_t implied by (1.3), (1.9) and (1.10) are given in Engle and Sheppard (2001). Apart from well known positivity constraints to hold for the univariate GARCH components, the DCC(1,1) model will deliver positive definite covariances if $\alpha > 0$, $\beta > 0$ while $\alpha + \beta < 1$ and λ_{min}, the smallest eigenvalue of R, is strictly positive, i.e. $\lambda_{min} > \delta > 0$. It is worthwhile to point out that the DCC framework not only preserves the separability of volatility and correlation estimation, but also allows to estimate the nontrivial parameters in R via correlation targeting described in (1.8).

Given consistent estimates of unconditional correlations r_{ij}, $i \neq j$, the remaining parameters describing the correlation dynamics are collected in the two-dimensional vector $\varphi = (\alpha, \beta)^\top$. Note that making use of correlation targeting the number of parameters undergoing nonlinear iterative estimation in the DCC model is constant (= 2), and, thus, avoids the curse of dimensionality even in case of very large systems of asset returns.

Instead of estimating the model in three steps, one could alternatively estimate the unconditional correlation parameters in R and the coefficients in φ jointly. Note that the number of unknown parameters in R is $\mathcal{O}(N^2)$. Formal representations of first and second order derivatives to implement the two step estimation and inference can be found in Hafner and Herwartz (2008). We prefer the three step approach here, since it avoids iterative estimation procedures in large parameter spaces.

1.3.5 Inference in the Correlation Models

QML-inference on significance of univariate GARCH parameter estimates is discussed in Bollerslev and Wooldridge (1992). Analytical expressions necessary to evaluate the asymptotic covariance matrix are given in Bollerslev (1986). In the empirical part of the chapter, we will not provide univariate GARCH parameter estimates at all to economize on space. Two issues of evaluating parameter significance remain, inference for the correlation estimates given in (1.8) and for the estimated DCC parameters $\hat{\varphi}$. We consider these two issues in turn:

1. Inference for unconditional correlations

   Conditional on estimates $\hat{\theta}$, we estimate R from standardized univariate GARCH residuals as formalized in (1.8). The elements in $\hat{R}$ are obtained as a nonlinear and continuous transformation of the elements in $\hat{D}$, i.e. $\hat{R} = \hat{D}^{*^{-1/2}} \hat{D} \hat{D}^{*^{-1/2}}$. Denote with vechl$(B)$ an operator stacking the elements below the diagonal of a symmetric $(m \times m)$ matrix B in a $\{m(m-1)/2\}$ dimensional column vector $b_l =$ vechl(B). Thus, $\hat{r}_l =$ vechl$(\hat{R})$ collects the nontrivial elements in $\hat{R}$. Standard errors for the estimates in $\hat{r}_l$ can be obtained from a robust estimator of the covariance of the (nontrivial) elements in $\hat{D}$, $\hat{d} =$ vech$(\hat{D})$, via the delta method. To be precise, we estimate the covariance of $\hat{r}_l$ by means of the following result (Ruud 2000):

$$\sqrt{T}(\hat{r}_l - r_l) \xrightarrow{\mathcal{L}} N\left(0, \mathcal{H}(\hat{r})\mathcal{G}\mathcal{H}(\hat{r})^\top\right), \tag{1.11}$$

   where $\mathcal{G}$ is an estimate of the covariance matrix of the elements in d, $\mathcal{G} = \widehat{\text{Cov}(\hat{d})}$, and $\mathcal{H}(\hat{r})$ is a $\{N(N-1)/2 \times (N(N+1)/2)\}$ dimensional matrix collecting the first order derivatives $\partial r_l / \partial d^\top$ evaluated at $\hat{d}$. We determine $\mathcal{G}$ by means of the covariance estimator

$$\mathcal{G} = \frac{1}{T} \sum_{t=1}^{T} (vv)_t (vv)_t^\top, \quad (vv)_t = \text{vech}(v_t v_t^\top) - \hat{d}. \tag{1.12}$$

   The derivatives in $\mathcal{H}(r)$ are derived from a result in Hafner and Herwartz (2008) as

$$\frac{\partial r_l}{\partial d^\top} = P_{N,-}^\top (D^* \otimes D^*) P_N + P_{N,-}^\top (DD^* \otimes I_N + I_N \otimes DD^*) P_N \frac{\partial \text{vech}(D^*)}{\partial \text{vech}(D)^\top}$$

   and

$$\frac{\partial \text{vech}(D^*)}{\partial \text{vech}(D)^\top} = -\frac{1}{2} \text{diag}\left[\text{vech}\left\{(I_N \odot D)^{-3/2}\right\}\right],$$

   where the matrices $P_{N,-}$ and P_N serve as duplication matrices (Lütkepohl 1996) such that $(B) = P_{N,-}$vechl(B) and $(B) = P_N$vech(B).
2. Inference for correlation parameters
   The correlation parameters are estimated by maximizing the correlation part, $l^C(\theta, \phi)$, of the Gaussian (quasi) log-likelihood function. When evaluating the estimation uncertainty associated with $\hat{\varphi} = (\hat{\alpha}, \hat{\beta})^\top$, the sequential character of the estimation procedure has to be taken into account. To provide standard errors

for QML estimates $\hat{\varphi}$, we follow a GMM approach introduced in Newey and McFadden (1994), which works in case of sequential GMM estimation under typical regularity conditions. In particular, it is assumed that all steps of a sequential estimation procedure are consistent. The following result on the asymptotic behavior of $\hat{\gamma} = (\hat{\theta}^\top, \hat{\varphi}^\top)^\top$ applies:

$$\sqrt{T}(\hat{\gamma} - \gamma) \xrightarrow{\mathcal{L}} N(0, \mathcal{N}^{-1}\mathcal{M}(\mathcal{N}^{-1})^\top). \tag{1.13}$$

In (1.13), $\mathcal{M}$ is the (estimated) expectation of the outer product of the scores of the log-likelihood function evaluated at $\hat{\gamma}$,

$$\mathcal{M} = \frac{1}{T}\sum_{t=1}^{T}\left(\frac{\partial l_t}{\partial \gamma}\right)\left(\frac{\partial l_t}{\partial \gamma}\right)^\top, \quad \frac{\partial l_t}{\partial \gamma} = \left(\frac{\partial l_t^V}{\partial \theta^\top}, \frac{\partial l_t^C}{\partial \varphi^\top}\right)^\top. \tag{1.14}$$

Compact formal representations for the derivatives in (1.14) can be found in Hafner and Herwartz (2008) and Bollerslev (1986). The matrix $\mathcal{N}$ in (1.13) has a lower block diagonal structure containing (estimates) of expected second order derivatives, i.e.

$$\mathcal{N} = \begin{pmatrix} \mathcal{N}_{11} & 0 \\ \mathcal{N}_{21} & \mathcal{N}_{22} \end{pmatrix},$$

with

$$\mathcal{N}_{11} = \frac{1}{T}\sum_{t=1}^{T}\frac{\partial^2 l_t^V}{\partial\theta\partial\theta^\top}, \; \mathcal{N}_{21} = \frac{1}{T}\sum_{t=1}^{T}\frac{\partial^2 l_t^C}{\partial\varphi\partial\theta^\top}, \; \mathcal{N}_{22} = \frac{1}{T}\sum_{t=1}^{T}\frac{\partial^2 l_t^C}{\partial\varphi\partial\varphi^\top}.$$

Formal representations of the latter second order quantities are provided in Hafner and Herwartz (2008).

1.3.6 Generalizations of the DCC Model

Generalizing the basic DCC(1,1) model in (1.9) and (1.10) towards higher model orders is straightforward and in analogy to the common GARCH volatility model. In fact, it turns out that the DCC(1,1) model is often sufficient to capture empirical correlation dynamics (Engle and Sheppard 2001). Tse and Tsui (2002) propose a direct formalization of the dynamic correlation matrix R_t as a weighted average of unconditional correlation, lagged correlation and a local correlation matrix estimated over a time window comprising the M most recent GARCH innovation vectors ξ_{t-i}, $i = 1, ..., M$, $M \geq N$. As discussed so far, dynamic correlation models are restrictive in the sense that asset specific dynamics are excluded. Hafner and Franses (2003) discuss a generalized DCC model where the parameters α and β in (1.10) are replaced by outer products of N-dimensional vectors, e.g. $\tilde{\alpha} = (\alpha_1, \alpha_2, \ldots, \alpha_N)^\top$, obtaining

$$Q_t = R(1 - \tilde{\alpha}\tilde{\alpha}^\top - \tilde{\beta}\tilde{\beta}^\top) + \tilde{\alpha}\tilde{\alpha}^\top \odot v_{t-1}v_{t-1}^\top + \tilde{\beta}\tilde{\beta}^\top \odot Q_{t-1}. \tag{1.15}$$

From (1.15) it is apparent that implied time paths of conditional correlations show asset specific characteristics. Similar to the generalization of the basic GARCH volatility model towards threshold specifications (Glosten et al. 1993), one may also introduce asymmetric dependencies of Q_t on $\text{vech}(v_t v_t^\top)$ as in Cappiello et al. (2006). A semiparametric conditional correlation model is provided by Hafner et al. (2006). In this model, the elements in Q_t are determined via local averaging where the weights entering the nonparametric estimates depend on a univariate factor as, for instance, market volatility or market returns.

1.4 Value-at-Risk

Financial institutions and corporations can suffer financial losses in their portfolios or treasury department due to unpredictable and sometimes extreme movements in the financial markets. The recent increase in volatility in financial markets and the surge in corporate failures are driving investors, management and regulators to search for ways to quantify and measure risk exposure. One answer came in the form of Value-at-Risk (VaR) being the minimum loss a portfolio will not exceed with a given probability over a specific time horizon (Jorion 2007; Christoffersen et al. 2001). For a critical review of the VaR approach see Acerbi and Tasche (2002). They also discuss the merits of an important and closely related risk measure, the expected shortfall. It is defined as the expected tail return conditional on a specific VaR level and provides further sensitive insights into the loss distribution, i.e. the expected portfolio loss when the portfolio value exceeds the VaR.

The VaR of some portfolio (.) may be defined as a one-sided confidence interval of expected h-periods ahead losses:

$$\text{VaR}_{t+h,\varsigma}^{(.)} = \Xi_t^{(.)}(1 + \bar{\xi}_{t+h,\varsigma}), \tag{1.16}$$

where $\Xi_t^{(.)}$ is the value of a portfolio in time t and $\bar{\xi}_{t+h,\varsigma}$ is a time dependent quantile of the conditional distribution of portfolio returns $\xi_{t+h}^{(.)}$ such that

$$\text{P}[\xi_{t+h}^{(.)} < \bar{\xi}_{t+h,\varsigma}] = \varsigma, \; \bar{\xi}_{t+h,\varsigma} = \sigma_{t+h} z_\varsigma, \tag{1.17}$$

and z_ς is a quantile from an unconditional distribution with unit variance. In the light of the assumption of conditional normality in (1.1), we will take the quantiles z_ς from the Gaussian distribution. As outlined in (1.16) and (1.17), the quantities $\bar{\xi}_{t+h,\varsigma}$ and σ_{t+h} generally depend on the portfolio composition. For convenience, however, our notation does not indicate this relationship. Depending on the risk averseness of the agent, the parameter ζ is typically chosen as some small probability, for instance, $\zeta = 0.005, 0.01, 0.05$.

In order to assess the performance of distinct VaR models in-sample and out-of-sample, one can employ VaR backtesting methods. Several contributions in the recent literature exploit the statistical properties of the empirical hit series. A literature review and a comparative simulation study can be found in Campbell (2006). Given ζ, a so-called hit in time $t+h$ is defined by

$$\text{hit}_{t+h}(\zeta) = \mathbf{1}\left(\Xi_{t+h}^{(.)} < \text{VaR}_{t+h,\zeta}^{(.)}\right).$$

The indicator function $\mathbf{1}$ becomes unity if the portfolio value falls below its computed VaR and is zero otherwise. If the model is correctly specified the empirical hit rate, $\hat{\zeta} = 1/T \sum_{t=1}^{T} \text{hit}_{t+h}(\zeta)$, for $T \to \infty$ periods converges to ζ. In the empirical part, we will exploit this fact and compare the unconditional coverage of the estimated VaR series for the discussed volatility models.

Secondly, if the model is correctly specified, the observed hits do not provide any serial information and they are assumed to be independent. To validate the unconditional and conditional VaR coverage, Christoffersen (1998) suggests two likelihood ratio tests. These tests have been widely employed in the literature on multivariate volatility (Chib et al. 2006). A similar idea on testing the conditional coverage, Engle and Manganelli (2004) propose a dynamic quantile test assessing an autoregressive model on the series of centered hits by a Wald test for joint significance of the coefficients. A linear dependency of the hits in time contradicts the VaR model specification. Ready to use software implementations for VaR backtesting are briefly exposed in Chap. 1 Appendix.

1.5 An Empirical Illustration

1.5.1 Equal and Value Weighted Portfolios

We analyze portfolios comprised by all 30 stocks listed in the Dow Jones Industrial Average (DJIA) over the period Jan, 2nd, 1990 to Jan, 31st, 2005. The asset returns were computed using historical closing prices provided by Yahoo Finance. Measured at the daily frequency, 3803 observations are used for the empirical analysis. Two alternative portfolio compositions are considered. In the first place, we analyze a portfolio weighting each asset equally. Returns of this equal weight portfolio (EWP) are obtained from asset specific returns (ε_{it}, $i = 1, \ldots, N$) as

$$\xi_t^{(e)} = \sum_{i=1}^{N} w_{it}^{(e)} \varepsilon_{it}, \quad w_{it}^{(e)} = N^{-1}.$$

Secondly, we consider value weighted portfolios (VWP) determined as:

$$\xi_t^{(v)} = \sum_{i=1}^{N} w_{it}^{(v)} \varepsilon_{it}, \; w_{it}^{(v)} = w_{it-1}(1+\varepsilon_{it-1})/w_t^{(v)}, \; w_t^{(v)} = \sum_i w_{it-1}(1+\varepsilon_{it-1}).$$

Complementary to an analysis of EWP and VWP, dynamics of minimum variance portfolios (MVP) could also be of interest. The MVP, however, will typically depend on some measure of the assets' volatilities and covariances. The latter, in turn, depend on the particular volatility model used for the analysis. Since the comparison of alternative measures of volatility in determining VaR is a key issue of this investigation, we will not consider MVP to immunize our empirical results from impacts of volatility specific portfolio compositions.

Our empirical comparison of alternative approaches to implement VaR concentrates on the relative performance of one step ahead ex-ante evaluations of VaR ($h = 1$). Note, that the (M)GARCH model specifies covariance matrices Σ_t or univariate volatilities σ_t^2 conditional on $\mathcal{F}_{t-1}$. Therefore, we practically consider the issue of two step ahead forecasting when specifying

$$\mathrm{VaR}_{t+1,\varsigma}^{(.)}|\mathcal{F}_{t-1} = \mathrm{VaR}^{(.)}(\hat{\sigma}_{t+1}^2), \; \hat{\sigma}_{t+1}^2|\mathcal{F}_{t-1} = E[(\xi_{t+1}^{(.)})^2|\mathcal{F}_{t-1}].$$

The performance of alternative approaches to forecast VaR is assessed by means of the relative frequency of actual hits observed over the entire sample period, i.e.

$$\mathrm{hf}_{\varsigma}^{(.)} = \frac{1}{3800}\sum_{t=3}^{3802} \mathbf{1}(\xi_t^{(.)} < \bar{\xi}_{t,\varsigma}), \tag{1.18}$$

where $\mathbf{1}(.)$ is an indicator function. To determine the forecasted conditional standard deviation entering the VaR, we adopt three alternative strategies. As a benchmark, we consider standard deviation forecasts obtained from univariate GARCH processes fitted directly to the series of portfolio returns $\xi_t^{(.)}$. For the two remaining strategies, we exploit forecasts of the covariance matrix, $\hat{\Sigma}_{t+1} = E[\varepsilon_{t+1}\varepsilon_{t+1}^\top|\mathcal{F}_{t-1}]$, to determine VaR. Note that given portfolio weights $w_t = (w_{1t}, w_{2t}, \ldots, w_{Nt})^\top$, the expected conditional variance of the portfolio is $\hat{\sigma}_{t+1}^2 = w^\top \hat{\Sigma}_{t+1} w$. Feasible estimates for the expected covariance matrix are determined alternatively by means of the CCC and DCC model.

The empirical exercises first cover a joint analysis of all assets comprising the DJIA. Moreover, we consider 1000 portfolios composed of 5 securities randomly drawn from all assets listed in the DJIA. Implementing the volatility parts of both the CCC and the DCC model, we employ alternatively the symmetric GARCH(1,1) and the threshold GARCH(1,1) model as introduced by Glosten et al. (1993). Opposite to the symmetric GARCH model, the latter accounts for a potential leverage effect (Black 1976) stating that volatility is larger in the sequel of bad news (negative returns) in comparison with good news (positive returns).

Table 1.1 Estimation results and performance of VaR estimates. G and TG are short for GARCH(1,1) and TGARCH(1,1) models for asset specific volatilities, respectively. D, C and U indicate empirical results obtained from DCC, CCC and univariate GARCH(1,1) models applied to evaluate forecasts of conditional variances of equal weight (EWP) and value weighted portfolios (VWP). Entries in hf and s(hf) are relative frequencies of extreme losses and corresponding standard errors, respectively

$\zeta \cdot 1000$		$N = 30$		$N = 5$			
		G	TG	G		TG	
		hf	hf	$\bar{\text{hf}}$	s($\bar{\text{hf}}$)	$\bar{\text{hf}}$	s($\bar{\text{hf}}$)
EWP							
D	5.00	8.15	7.36	7.56	.033	7.13	.034
	10.0	13.2	12.4	11.7	.041	11.2	.042
	50.0	41.6	41.8	40.4	.075	40.3	.078
C	5.00	10.8	9.73	7.78	.034	7.36	.035
	10.0	14.2	14.2	11.9	.040	11.5	.042
	50.0	42.6	41.8	40.8	.074	40.7	.077
U	5.00	11.6	11.6	8.70	.036	8.36	.037
	10.0	14.7	14.7	13.2	.045	12.9	.045
	50.0	47.3	47.3	43.5	.076	44.0	.077
VWP							
D	5.00	6.58	7.10	7.86	.033	7.55	.033
	10.0	12.9	11.8	11.9	.043	11.6	.041
	50.0	41.6	40.5	40.3	.076	40.4	.078
C	5.00	9.21	9.21	8.18	.036	7.90	.035
	10.0	14.5	13.4	12.3	.043	12.1	.043
	50.0	42.6	41.8	41.1	.072	41.3	.071
U	5.00	9.99	9.99	8.71	.037	8.62	.035
	10.0	15.5	15.5	13.0	.048	12.9	.048
	50.0	43.7	43.7	42.6	.095	43.2	.098
Estimation results							
D	$\hat{\alpha}$	2.8e-03	2.8e-03	6.6e-03	4.5e-05	6.7e-03	4.8e-05
	t_α	17.5	17.3				
	$\hat{\beta}$	.992	.992	.989	8.3e-05	.989	9.5e-05
	t_β	1.8e+03	1.8e+03				

1.5.2 Estimation Results

A few selected estimation results are given in Table 1.1. Since we investigate 30 assets or 1000 random portfolios each containing $N = 5$ securities, we refrain from providing detailed results on univariate GARCH(1,1) or TGARCH(1,1) estimates. Moreover, we leave estimates of the unconditional correlation matrix R undocumented since the number of possible correlations in our sample is $N(N-1)/2 = 435$.

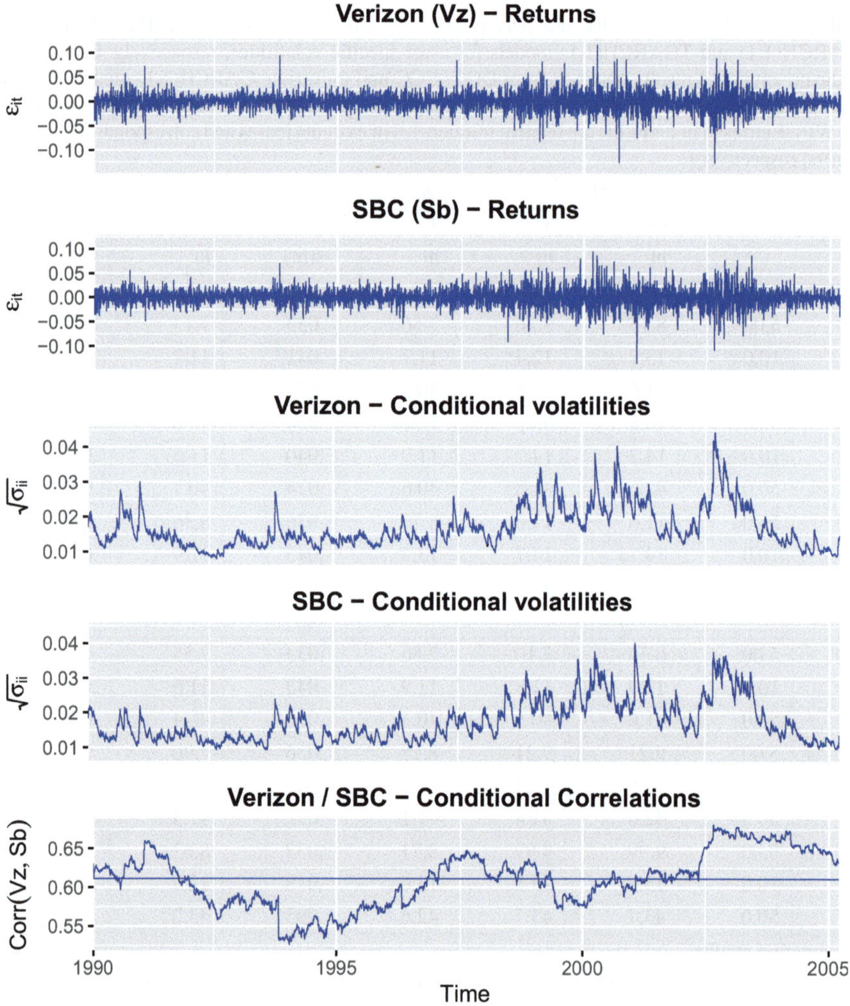

Fig. 1.1 Returns, conditional volatilities and correlations for Verizon and SBC communications

The lower left part of Table 1.1 provides estimates of the DCC parameters α and β and corresponding t-ratios for the analysis of all assets comprising the DJIA. Although the estimated α parameter governing the impact of lagged GARCH innovations on the conditional correlation matrix is very small (around $2.8 \cdot 10^{-3}$ for both implementations of the DCC model), it is significant at any reasonable significance level. The relative performance of the CCC and DCC model may also be evaluated in terms of the models' log-likelihood difference. Using symmetric and asymmetric volatility models for the diagonal elements of Σ_t, the log-likelihood difference between DCC and CCC is 645.66 and 622.00, respectively. Since the DCC specification has only two additional parameters, it apparently provides a substantial improvement of fitting multivariate returns. It is also instructive to compare, for the DCC case say, the log-likelihood improvement achieved when employing univariate TGARCH instead of a symmetric GARCH. Interestingly, implementing the DCC model with asymmetric GARCH the improvement of the log-likelihood is only 236.27, which is to be related to the number of $N = 30$ additional model parameters. Reviewing the latter two results, one may conclude that dynamic correlation is a more striking feature of U.S. stock market returns than leverage.

The sum of both DCC parameter estimates, $\hat{\alpha} + \hat{\beta}$, is slightly below unity and, thus, the estimated model of dynamic covariances is stationary. The lower right part of Table 1.1 gives average estimates obtained for the DCC parameters when modeling 1000 portfolios randomly composed of five securities contained in the DJIA. We also provide an estimator of the empirical standard error associated with the latter average. Irrespective of using a symmetric or asymmetric specification of univariate volatility models, estimates for α are small throughout. According to the reported standard error estimates, however, the true α parameter is apparently different from zero at any reasonable significance level.

The maximum over all 435 unconditional correlations is obtained for two firms operating on the telecommunication market, namely Verizon Communications and SBC Communications. To illustrate the performance of the DCC model and compare it with the more restrictive CCC counterpart, Fig. 1.1 provides the return processes for these two assets, the corresponding time paths of conditional standard deviations as implied by TGARCH(1,1) models and the estimated time paths of conditional correlations implied by the DCC model fitted over all assets contained in the DJIA. Facilitating the interpretation of the results, we also give the level of unconditional correlation.

Apparently, the univariate volatility models provide accurate descriptions of the return variability for both assets. Not surprisingly, estimated volatility turns out to be larger over the last third of the sample period in comparison with the first half. Although conditional correlation estimates vary around their unconditional level, the time path of correlation estimates exhibits only rather slow mean reversion. Interestingly, over the last part of the sample period, the conditional correlation measured between Verizon and SBC increases with the volatilities of both securities.

As mentioned, Verizon and SBC provide the largest measure of unconditional correlation within the DJIA over the considered sample period. To illustrate that time varying conditional correlation with slow mean reversion is also an issue for bivariate returns exhibiting medium or small correlation, we provide the conditional correlation estimates for Verizon and General Electric (medium unconditional correlation) and Verizon and Boeing (small unconditional correlation) in Fig. 1.2. For completeness, Fig. 1.3 provides empirical return processes for General Electric and Boeing.

The upper part of Table 1.1 shows relative frequencies of realized losses exceeding the one step ahead ex-ante VaR forecasts. We provide average relative frequencies when summarizing the outcome for 1000 portfolios with random composition. To

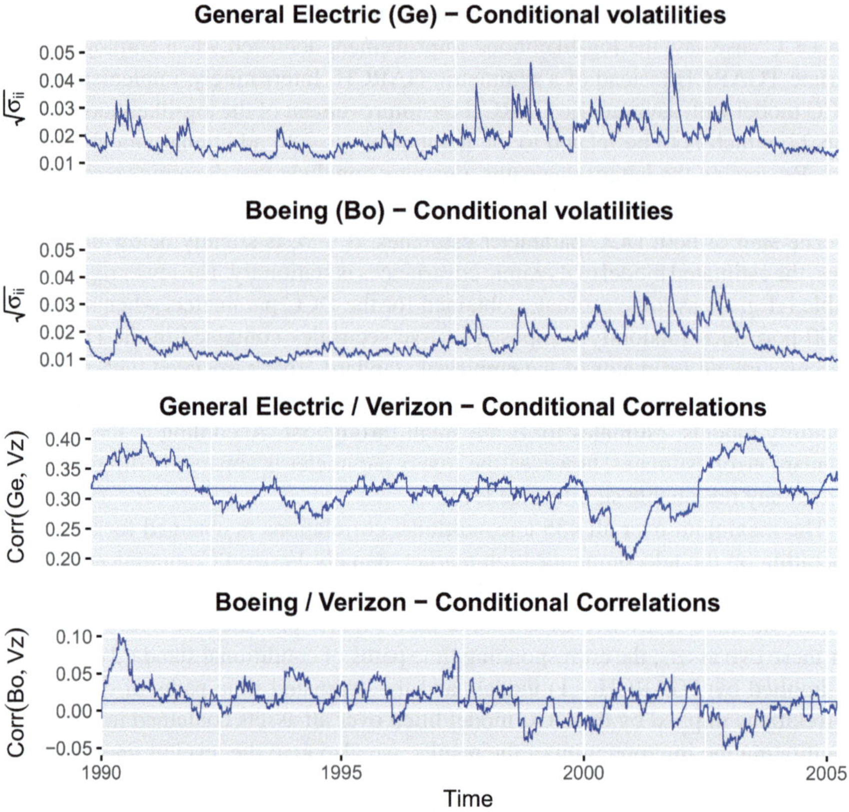

Fig. 1.2 Conditional volatilities for General Electric and Boeing and conditional correlations with Verizon

facilitate the discussion of the latter results, all frequencies given are multiplied with a factor of 1000.

The relative frequency of empirical hits of dynamic VaR estimates at the 5% level is uniformly below the nominal probability, indicating that dynamic VaR estimates are too conservative on average. For the remaining probability levels $\zeta = 0.5\%$ and $\zeta = 1\%$, the empirical frequencies of hitting the VaR exceed the nominal probability. We concentrate the discussion of empirical results on the latter cases. With regard to the performance of alternative implementations of VaR it is worthwhile to mention that the basic results are qualitatively similar for EWP in comparison with VWP. Similarly, employing an asymmetric GARCH model instead of symmetric GARCH has only minor impacts on the model comparison between the univariate benchmark and the CCC and DCC model, respectively. For the latter reason, we focus our discussion of the relative model performance on VaR modeling for EWP with symmetric GARCH(1,1) applied to estimate conditional variances.

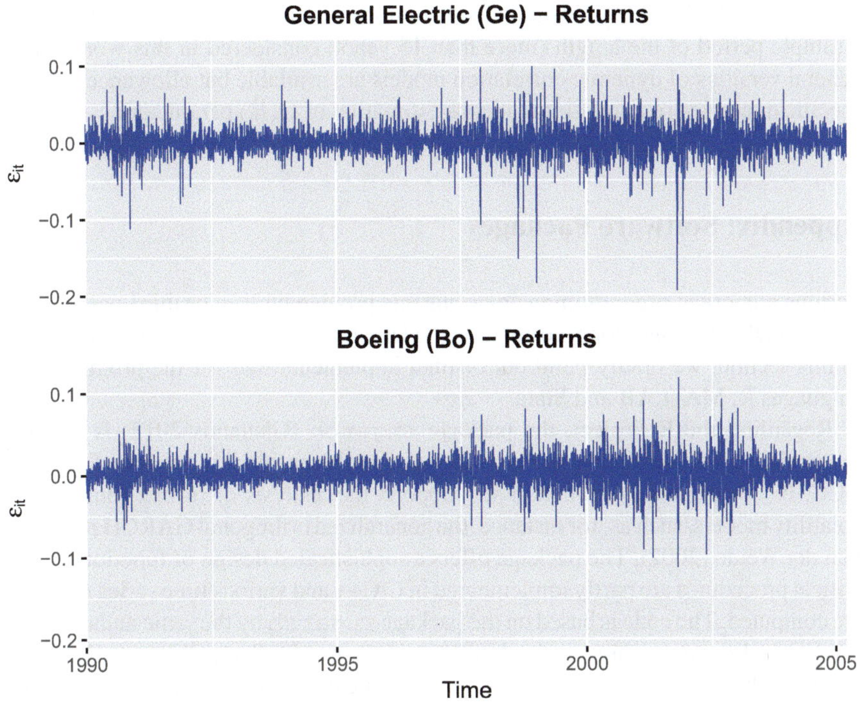

Fig. 1.3 Returns for General Electric and Boeing

Regarding portfolios composed of 30 securities, it turns out that for both probability levels, $\zeta = 1\%$ and $\zeta = 0.5\%$, the empirical frequencies of hitting the dynamic VaR estimates are closest to the nominal level for the DCC model and worst for modeling portfolio returns directly via univariate GARCH. Although it provides the best empirical frequencies of hitting the VaR, the DCC model still underestimates (in absolute value) on average the true quantile. For instance, the 0.5% VaR shows an empirical hit frequency of 0.82% (EWP) and 0.66% (VWP), respectively. Drawing randomly 5 out of 30 assets to form portfolios, and regarding the average empirical frequencies of hitting the VaR estimates, we obtain almost analogous results in comparison with the case $N = 30$. The reported standard errors of average frequencies, however, indicate that the discussed differences of nominal and empirical probabilities are significant at a 5% significance level since the difference between both exceeds twice the standard error estimates.

In summary, using the CCC and DCC model and, alternatively, univariate GARCH specifications to determine VaR, it turns out that the former outperform the univariate GARCH as empirical loss frequencies are closer to the nominal VaR coverage. DCC based VaR estimates in turn outperform corresponding quantities derived under the CCC assumption. Empirical frequencies of large losses, however, exceed the corresponding nominal levels if the latter are rather small, i.e. 0.5 and 1%. This might indicate that the DCC framework is likely to restrictive to hold homogeneously over a sample period of the length (more than 15 years) considered in this work. More general versions of dynamic correlation models are available but allowance of asset specific dynamics requires simultaneous estimation of $O(N)$ parameters.

Appendix: Software Packages

Various numerical programming environments provide built-in or third-party methods for analyzing conditional correlation models and Value-at-Risk backtesting tools. In this section, we briefly point out distinct implementations for the programming languages R, MATLAB and Stata.

Regarding the R Project, the package `rmgarch` (Ghalanos 2015) is suitable for modeling and analyzing the conditional correlation models, such as CCC and DCC. Its comprehensive function set supports the analysis of further multivariate volatility models, such as, for instance, the generalized orthogonal GARCH model by Van der Weide (2002). The package offers a sophisticated design of functions, time-critical procedures are partly implemented in C/C++ and various time series statistics are computed. The code is based on the package `rugarch` by the same author which can be used to study univariate volatility models in a similar sophisticated way. In addition, the latter package includes an implementation of the unconditional and conditional coverage VaR tests according to Christoffersen (1998). As an alternative, the package `ccgarch` Nakatani (2014) might be used for the evaluation of CCC and DCC models. Its functions were used to compute estimates and statistics quickly and correctly in several test applications. In comparison with `rmgarch`, its design and capabilities are less complex and it is restricted to conditional correlation models. Currently, there are no efforts by the authors of both packages to support the BEKK model.

Working with MATLAB, MathWorks' Econometrics Toolbox supports the simulation, estimation, and forecasting of different variants of univariate GARCH-type models. Its Risk Management Toolbox comprises an entire set of functions for assessing market risk, i.e. implementations of common approaches for VaR backtesting, which include the (un)conditional coverage tests described before. However, evaluations of multivariate volatility models including CCC or DCC can be carried out by means of the non-official MFE Toolbox. [1] It is the successor of the UCSD Toolbox by Kevin Sheppard. [2] The MFE project implements various univariate and multivariate volatility models and metrics. Its open source codebase is maintained and augmented by volunteers and particularly well suited as a starting point to study the programming of multivariate time series algorithms. Despite its wide range of functions, the user should always critically question the numerical results because the MFE project is still under development.

[1] Project website: https://www.kevinsheppard.com/MFE_Toolbox.

[2] Project website: https://www.kevinsheppard.com/UCSD_GARCH.

The Stata software package provides the user with comfortable fitting algorithms for conditional correlation models and diagonal half-vec models by means of the function `mgarch`. Its optimized program code proceeds rapidly and, at the same time, computes common metrics. The Stata documentation of the implemented methods is exemplary and might be a good complement while studying publicly available code examples of the volatility model implementations which are investigated in this chapter.

References

Acerbi, C., & Tasche, D. (2002). On the coherence of expected shortfall. *Journal of Banking and Finance*, *26*, 1487–1503.

Alexander, C. O. (1998). Volatility and correlation: Measurement, models and applications. In Alexander (Ed.), *Risk management and analysis: Measuring and modelling financial risk* (pp. 125–171). New York: Wiley.

Alexander, C. O. (2001). Orthogonal GARCH. In Alexander (Ed.), *Mastering risk* (Vol. II, pp. 21–38). Upper Saddle River: Prentice Hall.

Baba, Y., Engle, R. F., Kraft, D. F., & Kroner, K. F. (1990). Multivariate Simultaneous Generalized ARCH, mimeo, Department of Economics. San Diego: University of California.

Bauwens, L., Laurent, S., & Rombouts, J. V. K. (2006). Multivariate GARCH models: A survey. *Journal of Applied Econometrics*, *21*, 79–109.

Black, F. (1976). Studies of stock price volatility changes. In *Proceedings of the American Statistical Association, Business and Economic Statistics Section* (pp. 177–191).

Bollerslev, T. (1986). Generalized autoregressive conditional heteroscedasticity. *Journal of Econometrics*, *31*, 307–327.

Bollerslev, T. (1990). Modelling the coherence in short-run nominal exchange rates: A multivariate generalized ARCH model. *The Review of Economics and Statistics*, *72*, 498–505.

Bollerslev, T., & Wooldridge, J. M. (1992). Quasi-maximum likelihood estimation and inference in dynamic models with time-varying covariances. *Econometric Reviews*, *11*, 143–172.

Bollerslev, T., Engle, R. F., & Nelson, D. B. (1994). ARCH models. In R. F. Engle & D. L. McFadden (Eds.), *Handbook of econometrics* (Vol. 4, pp. 2959–3038). Amsterdam: Elsevier.

Bollerslev, T., Engle, R. F., & Wooldridge, J. M. (1988). A capital asset pricing model with time-varying covariances. *Journal of Political Economy*, *96*, 116–131.

Campbell, S. D. (2006). A review of backtesting and backtesting procedures. *The Journal of Risk*, *9*, 1–17.

Cappiello, L., Engle, R. F., & Sheppard, K. (2006). Asymmetric dynamics in the correlations of global equity and bond returns. *Journal of Financial Econometrics*, *4*, 537–572.

Cecchetti, S. G., Cumby, R. E., & Figlewski, S. (1988). Estimation of the optimal futures hedge. *The Review of Economics and Statistics*, *70*, 623–630.

Chib, S., Nardari, F., & Shephard, N. (2006). Analysis of high dimensional multivariate stochastic volatility models. *Journal of Econometrics*, *134*, 341–371.

Christoffersen, P. (1998). Evaluating interval forecasts. *International Economic Review*, *39*, 841–862.

Christoffersen, P., Hahn, J., & Inoue, A. (2001). Testing and comparing value-at-risk measures. *Journal of Empirical Finance*, *8*, 325–342.

Comte, F., & Lieberman, O. (2003). Asymptotic theory for multivariate GARCH processes. *Journal of Multivariate Analysis*, *84*, 61–84.

Ding, Z. & Engle, R. F. (2001). Large Scale Conditional Covariance Matrix Modeling, Estimation and Testing, mimeo, NYU.

Duan, J. C. (1995). The GARCH option pricing model. *Mathematical Finance*, *5*, 13–32.

Engle, R. F. (1982). Autoregressive conditional heteroscedasticity with estimates of the variance of United Kingdom inflation. *Econometrica*, *50*, 987–1007.

Engle, R. F. (2002). Dynamic conditional correlation: A simple class of multivariate generalized autoregressive conditional heteroskedasticity models. *Journal of Business and Economic Statistics*, *20*, 339–350.

Engle, R. F., & Kroner, K. F. (1995). Multivariate simultaneous generalized ARCH. *Econometric Theory*, *11*, 122–150.

Engle, R. F., & Mezrich, J. (1996). GARCH for groups. *Risk*, *9*, 36–40.

Engle, R. F. & Sheppard, K. (2001). Theoretical and Empirical Properties of Dynamic Conditional Correlation Multivariate GARCH, mimeo, National Bureau of Economic Research.

Engle, R. F., & Manganelli, S. (2004). CAViaR: Conditional autoregressive value at risk by regression quantiles. *Journal of Business and Economic Statistics*, *22*, 367–381.

Engle, R. F., Hendry, D. F., & Richard, J. F. (1983). Exogeneity. *Econometrica*, *51*, 277–304.

Engle, R. F., Ito, T., & Lin, W. L. (1990a). Meteor showers or heat waves? heteroskedastic intra-daily volatility in the foreign exchange market. *Econometrica*, *58*, 525–542.

Engle, R. F., Ng, V. M., & Rothschild, M. (1990b). Asset pricing with a factor-ARCH covariance structure: Empirical estimates for treasury bills. *Journal of Econometrics*, *45*, 213–237.

Ghalanos, A. (2015). rmgarch: Multivariate GARCH models, R package version 1.3-0.

Glosten, L. R., Jagannathan, R., & Runkle, D. E. (1993). On the relation between the expected value and the volatility of the nominal excess return on stocks. *Journal of Finance*, *48*, 1779–1801.

Hafner, C. M., & Herwartz, H. (1998). Structural analysis of portfolio risk using beta impulse response functions. *Statistica Neerlandica*, *52*, 336–355.

Hafner, C. M. & Franses, P. H. (2003). A Generalized Dynamic Conditional Correlation Model for Many Assets, Econometric Institute Report 18, Erasmus University Rotterdam.

Hafner, C. M., & Herwartz, H. (2008). Analytical quasi maximum likelihood inference in multivariate volatility models. *Metrika*, *67*, 219–239.

Hafner, C. M., van Dijk, D., & Franses, P. H. (2006). Semi-parametric modelling of correlation dynamics. In Fomby & Hill (Eds.), *Advances in econometrics part A* (Vol. 20, pp. 59–103). New York: Elsevier Science.

Hamao, Y., Masulis, R. W., & Ng, V. K. (1990). Correlations in price changes and volatility across international stock markets. *Review of Financial Studies*, *3*, 281–307.

Jeantheau, T. (1998). Strong consistency of estimators for multivariate ARCH models. *Econometric Theory*, *14*, 70–86.

Jorion, P. (2007). *Value at risk: The new benchmark for managing financial risk* (2nd ed.). New York: McGraw Hill.

Lee, S. W., & Hansen, B. E. (1994). Asymptotic theory for the GARCH(1,1) quasi-maximum likelihood estimator. *Econometric Theory*, *10*, 29–52.

Lee, T. H., & Long, X. (2009). Copula-based multivariate GARCH model with uncorrelated dependent errors. *Journal of Econometrics*, *150*, 207–218.

Ling, S., & McAleer, M. (2003). Asymptotic theory for a vector ARMA-GARCH model. *Econometric Theory*, *19*, 280–310.

Lütkepohl, H. (1996). *Handbook of matrices*. Chichester: Wiley.

Lumsdaine, R. L. (1996). Consistency and asymptotic normality of the quasi-maximum likelihood estimator in IGARCH(1,1) and covariance stationary GARCH(1,1) models. *Econometrica*, *64*, 575–596.

Nakatani, T. (2014). ccgarch: An R Package for Modelling Multivariate GARCH Models with Conditional Correlations. R package version 0.2.3.

Nelson, D. B. (1991). Conditional heteroskedasticity in asset returns: A new approach. *Econometrica*, *59*, 347–370.

Newey, W. K., & McFadden, D. (1994). Large sample estimation and hypothesis testing. In Engle & McFadden (Eds.), *Handbook of econometrics* (Vol. 4, pp. 2111–2245). Amsterdam: Elsevier.

Ruud, P. A. (2000). *An introduction to classical econometric theory*. Berkeley: University of California.

Tse, Y. K. (2000). A test for constant correlations in a multivariate GARCH model. *Journal of Econometrics*, *98*, 107–127.

Tse, Y. K., & Tsui, A. K. C. (2002). A multivariate generalized autoregressive conditional heteroscedasticity model with time-varying correlations. *Journal of Business and Economic Statistics*, *20*, 351–362.

Van der Weide, R. (2002). GO-GARCH: A multivariate generalized orthogonal GARCH model. *Journal of Applied Econometrics*, *17*, 549–564.

Chapter 2
Multivariate Volatility Models

M.R. Fengler, H. Herwartz and F.H.C. Raters

Abstract Multivariate volatility models are widely used in finance to capture both volatility clustering and contemporaneous correlation of asset return vectors. Here, we focus on multivariate GARCH models. In this common model class, it is assumed that the covariance of the error distribution follows a time dependent process conditional on information which is generated by the history of the process. To provide a particular example, we consider a system of exchange rates of two currencies measured against the US Dollar (USD), namely the Deutsche Mark (DEM) and the British Pound Sterling (GBP). For this process, we compare the dynamic properties of the bivariate model with univariate GARCH specifications where cross sectional dependencies are ignored. Moreover, we illustrate the scope of the bivariate model by ex-ante forecasts of bivariate exchange rate densities.

2.1 Introduction

Volatility clustering, i.e. positive correlation of price variations observed on speculative markets, motivated the introduction of autoregressive conditionally heteroskedastic (ARCH) processes by Engle (1982) and its popular generalizations

M.R. Fengler
Faculty of Mathematics and Statistics, University of St. Gallen, Bodanstrasse 6,
9000 St. Gallen, Switzerland
e-mail: matthias.fengler@unisg.ch

H. Herwartz · F.H.C. Raters (✉)
Department of Economics, University of Göttingen, Humboldtallee 3,
37073 Göttingen, Germany
e-mail: hherwartz@uni-goettingen.de

F.H.C. Raters
e-mail: raters@uni-goettingen.de

W.K. Härdle et al. (eds.), *Applied Quantitative Finance*, Statistics and Computing,
DOI 10.1007/978-3-662-54486-0_2

by Bollerslev (1986) (Generalized ARCH, GARCH) and Nelson (1991) (exponential GARCH, EGARCH). Being univariate in nature, however, such models neglect a further stylized fact of empirical price variations, namely contemporaneous cross correlation e.g. over a set of assets, stock market indices, or exchange rates.

Cross section relationships are often implied by economic theory. Interest rate parities, for instance, provide a close relation between domestic and foreign bond rates. Assuming absence of arbitrage, the so-called triangular equation formalizes the equality of an exchange rate between two currencies on the one hand and an implied rate constructed via exchange rates measured towards a third currency. Furthermore, stock prices of firms acting on the same market often show similar patterns in the sequel of news that are important for the entire market (Hafner and Herwartz 1998). Similarly, analyzing global volatility transmission Engle et al. (1990) and Hamao et al.(1990) found evidence in favor of volatility spillovers between the world's major trading areas occurring in the sequel of floor trading hours. From this point of view, when modeling time varying volatilities, a multivariate model appears to be a natural framework to take cross sectional information into account. Moreover, the covariance between financial assets is of essential importance in finance. Effectively, many problems in financial practice like portfolio optimization, hedging strategies, or Value-at-Risk evaluation require multivariate volatility measures (Bollerslev et al. 1988; Cecchetti et al. 1988).

2.1.1 Model Specifications

Let $\varepsilon_t = (\varepsilon_{1t}, \varepsilon_{2t}, \dots, \varepsilon_{Nt})^\top$ denote an N-dimensional error process, which is either directly observed or estimated from a multivariate regression model. The process ε_t follows a multivariate GARCH process if it has the representation

$$\varepsilon_t = \Sigma_t^{1/2} \xi_t, \tag{2.1}$$

where Σ_t is measurable with respect to information generated up to time $t-1$, denoted by the filtration $\mathcal{F}_{t-1}$. By assumption, the N components of ξ_t follow a multivariate Gaussian distribution with mean zero and a covariance matrix equal to the identity matrix.

The conditional covariance matrix, $\Sigma_t = \mathsf{E}[\varepsilon_t \varepsilon_t^\top | \mathcal{F}_{t-1}]$, has typical elements σ_{ij} with σ_{ii}, $i = 1, \dots, N$, denoting conditional variances and off-diagonal elements σ_{ij}, $i, j = 1, \dots, N$, $i \neq j$, denoting conditional covariances. To make the specification in (2.1) feasible, a parametric description relating Σ_t to $\mathcal{F}_{t-1}$ is necessary. In a multivariate setting, however, dependencies of the second order moments in Σ_t on $\mathcal{F}_{t-1}$ become easily computationally intractable for practical purposes.

Let vech(A) denote the half-vectorization operator stacking the elements of a quadratic $(N \times N)$-matrix A from the main diagonal downwards in a $\frac{1}{2}N(N+1)$ dimensional column vector. Within the so-called half-vec representation of the GARCH(p, q) model Σ_t is specified as follows:

$$\text{vech}(\Sigma_t) = c + \sum_{i=1}^{q} \widetilde{A}_i \text{vech}(\varepsilon_{t-i}\varepsilon_{t-i}^{\top}) + \sum_{i=1}^{p} \widetilde{G}_i \text{vech}(\Sigma_{t-i}). \tag{2.2}$$

In (2.2), the matrices $\widetilde{A}_i$ and $\widetilde{G}_i$ each contain $\{N(N+1)/2\}^2$ elements. Deterministic covariance components are collected in c, a column vector of dimension $N(N+1)/2$. We consider in the following the case $p = q = 1$ since in applied work the GARCH(1,1) model has turned out to be particularly useful to describe a wide variety of financial market data (Bollerslev et al., 1994).

On the one hand, the half-vec model in (2.2) allows for a very general dynamic structure of the multivariate volatility process. On the other hand, this specification suffers from high dimensionality of the relevant parameter space, which makes it almost intractable for empirical work. In addition, it might be cumbersome in applied work to restrict the admissible parameter space such that the implied matrices Σ_t, $t = 1, \ldots, T$, are positive definite. These issues motivated a considerable variety of competing multivariate GARCH specifications.

Prominent proposals reducing the dimensionality of (2.2) are the constant correlation model (Bollerslev et al. 1988) and the diagonal model (Bollerslev et al. 1988). Specifying diagonal elements of Σ_t both of these approaches assume the absence of cross equation dynamics, i.e. the only dynamics are

$$\sigma_{ii,t} = c_{ii} + a_i \varepsilon_{i,t-1}^2 + g_i \sigma_{ii,t-1}, \; i = 1, \ldots, N. \tag{2.3}$$

To determine off-diagonal elements of Σ_t, Bollerslev (1990) proposes a constant contemporaneous correlation,

$$\sigma_{ij,t} = \rho_{ij} \sqrt{\sigma_{ii}\sigma_{jj}}, \; i, j = 1, \ldots, N, \tag{2.4}$$

whereas Bollerslev et al. (1988) introduce an ARMA-type dynamic structure as in (2.3) for $\sigma_{ij,t}$ as well, i.e.

$$\sigma_{ij,t} = c_{ij} + a_{ij} \varepsilon_{i,t-1} \varepsilon_{j,t-1} + g_{ij} \sigma_{ij,t-1}, \; i, j = 1, \ldots, N. \tag{2.5}$$

For the bivariate case ($N = 2$) with $p = q = 1$, the constant correlation model contains only 7 parameters compared to 21 parameters encountered in the full model (2.2). The diagonal model is specified with 9 parameters. The price that both models pay for parsimony is in ruling out cross equation dynamics as allowed in the general half-vec model. Positive definiteness of Σ_t is easily guaranteed for the constant correlation model ($|\rho_{ij}| < 1$), whereas the diagonal model requires more complicated restrictions to provide positive definite covariance matrices.

The so-called BEKK model (Baba et al. 1990) provides a richer dynamic structure compared to both restricted processes mentioned before. Defining $N \times N$ matrices A_{ik} and G_{ik} and an upper triangular matrix C_0, the BEKK model reads in a general version as follows (see Engle and Kroner 1995):

$$\Sigma_t = C_0^\top C_0 + \sum_{k=1}^{K}\sum_{i=1}^{q} A_{ik}^\top \varepsilon_{t-i}\varepsilon_{t-i}^\top A_{ik} + \sum_{k=1}^{K}\sum_{i=1}^{p} G_{ik}^\top \Sigma_{t-i} G_{ik}. \tag{2.6}$$

If $K = q = p = 1$ and $N = 2$, the model in (2.6) contains 11 parameters and implies the following dynamic model for typical elements of Σ_t:

$$\begin{aligned}
\sigma_{11,t} &= c_{11} + a_{11}^2\varepsilon_{1,t-1}^2 + 2a_{11}a_{21}\varepsilon_{1,t-1}\varepsilon_{2,t-1} + a_{21}^2\varepsilon_{2,t-1}^2 \\
&\quad + g_{11}^2\sigma_{11,t-1} + 2g_{11}g_{21}\sigma_{21,t-1} + g_{21}^2\sigma_{22,t-1}, \\
\sigma_{21,t} &= c_{21} + a_{11}a_{22}\varepsilon_{1,t-1}^2 + (a_{21}a_{12} + a_{11}a_{22})\varepsilon_{1,t-1}\varepsilon_{2,t-1} + a_{21}a_{22}\varepsilon_{2,t-1}^2 \\
&\quad + g_{11}g_{22}\sigma_{11,t-1} + (g_{21}g_{12} + g_{11}g_{22})\sigma_{12,t-1} + g_{21}g_{22}\sigma_{22,t-1}, \\
\sigma_{22,t} &= c_{22} + a_{12}^2\varepsilon_{1,t-1}^2 + 2a_{12}a_{22}\varepsilon_{1,t-1}\varepsilon_{2,t-1} + a_{22}^2\varepsilon_{2,t-1}^2 \\
&\quad + g_{12}^2\sigma_{11,t-1} + 2g_{12}g_{22}\sigma_{21,t-1} + g_{22}^2\sigma_{22,t-1}.
\end{aligned}$$

Compared to the diagonal model, the BEKK–specification economizes on the number of parameters by restricting the half-vec model within and across equations. Since A_{ik} and G_{ik} are not required to be diagonal, the BEKK model is convenient to allow for cross dynamics of conditional covariances. The parameter K governs to which extent the general representation in (2.2) can be approximated by a BEKK-type model. In the following we assume $K = 1$. Note that in the bivariate case with $K = p = q = 1$ the BEKK model contains 11 parameters. If $K = 1$, the matrices A_{11} and $-A_{11}$ imply the same conditional covariances. Thus, for uniqueness of the BEKK-representation $a_{11} > 0$ and $g_{11} > 0$ is assumed. Note that the right hand side of (2.6) involves only quadratic terms and, hence, given convenient initial conditions, Σ_t is positive definite under the weak (sufficient) condition that at least one of the matrices C_0 or G_{ik} has full rank (Engle and Kroner 1995). It is worthwhile to mention that in a similar way the univariate GARCH volatility model can be augmented by threshold specifications (Glosten et al. 1993), a generalization for asymmetric effects in a BEKK-type model is discussed in Kroner and Ng (1998).

2.1.2 Estimation of the BEKK Model

As in the univariate case, the parameters of a multivariate GARCH model are estimated by maximum likelihood (ML) optimizing numerically the Gaussian log-likelihood function.

With f denoting the multivariate normal density, the contribution of a single observation, l_t, to the log-likelihood of a sample is given as:

$$\begin{aligned}
l_t &= \ln\{f(\varepsilon_t|\mathcal{F}_{t-1})\} \\
&= -\frac{N}{2}\ln(2\pi) - \frac{1}{2}\ln(|\Sigma_t|) - \frac{1}{2}\varepsilon_t^\top \Sigma_t^{-1}\varepsilon_t.
\end{aligned}$$

Maximizing the log-likelihood, $l = \sum_{t=1}^{T} l_t$, requires nonlinear maximization methods. Involving only first order derivatives, the BHHH algorithm introduced by Berndt et al. (1974) is easily implemented and particularly useful for the estimation of multivariate GARCH processes.

If the actual error distribution differs from the multivariate normal, maximizing the Gaussian log-likelihood has become popular as Quasi ML (QML) estimation. In the multivariate framework, results for the asymptotic properties of the (Q)ML-estimator have been derived by Jeantheau (1998) who proves the QML-estimator to be consistent under the main assumption that the considered multivariate process is strictly stationary and ergodic. Further assuming finiteness of moments of ε_t up to order eight, Comte and Lieberman (2003) derive asymptotic normality of the QML-estimator. The asymptotic distribution of the rescaled QML-estimator is analogous to the univariate case and discussed in Bollerslev and Wooldridge (1992).

2.2 An Empirical Illustration

2.2.1 *Data Description*

We analyze daily quotes of two European currencies measured against the USD, namely the DEM and the GBP. The sample period is December 31, 1979 to April 1, 1994, covering $T = 3720$ observations. Note that a subperiod of our sample has already been investigated by Bollerslev and Engle (1993) discussing common features of volatility processes.

Let the bivariate vector R_t denote the exchange rates (DEM/USD and GBP/USD) at time t. Before inspecting the sample statistics (`XFGmvol01.R`), we take the first differences of the log exchange rates, $\varepsilon_t = \ln(R_t) - \ln(R_{t-1})$. These log-differences are shown in Fig. 2.1. Evidently, the empirical means of both processes are very close to zero (-4.72e-06 and 1.10e-04, respectively). Also minimum, maximum and standard errors are of similar size. As is apparent from Fig. 2.1, variations of exchange rate log-differences exhibit an autoregressive pattern: Large log-differences of foreign exchange rates are followed by large log-differences of either sign. This is most obvious in periods of excessive log-differences. Note that these volatility clusters tend to coincide in both series. It is precisely this observation that justifies a multivariate GARCH specification.

2.2.2 *Estimating Bivariate GARCH*

A fast algorithm is used to estimate the BEKK representation of a bivariate GARCH (1,1) model: QML-estimation is implemented by means of the BHHH-algorithm which minimizes the negative Gaussian log-likelihood function. The algorithm

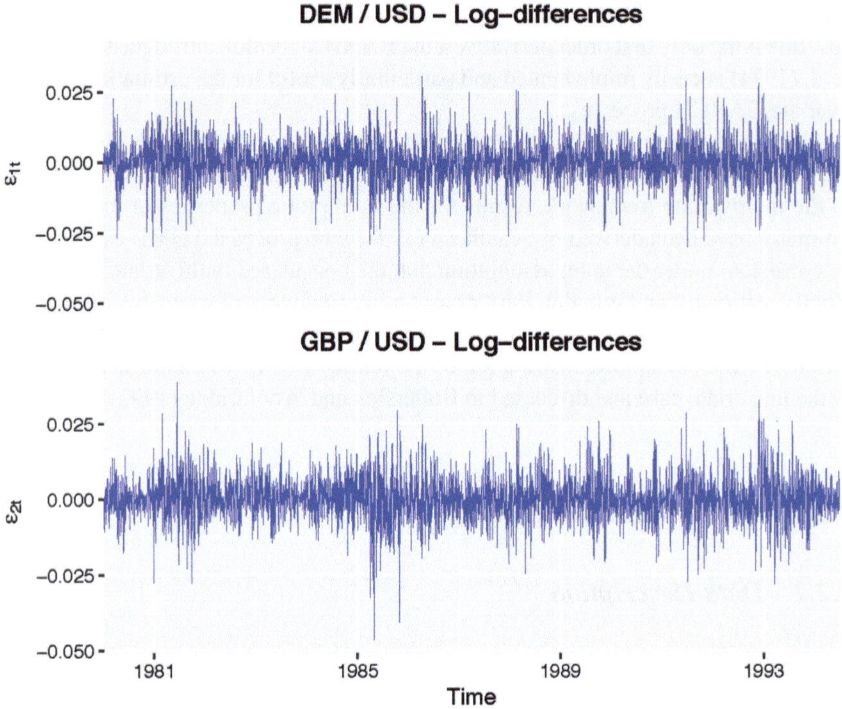

Fig. 2.1 Foreign exchange rate data: log-differences. XFGmvol01

employs analytical first order derivatives of the log-likelihood function Lütkepohl (1996) with respect to the 11-dimensional vector of parameters containing the elements of C_0, A_{11} and G_{11} as given in (2.6). Alternatively, the R package `mgarchBEKK` Schmidbauer et al. (2016) might be considered when estimating this model in *R*. Section 2.3 contains further references for implementations of the BEKK model in widely used numerical programming environments.

The estimation output contains the stacked elements of the parameter matrices C_0, A_{11} and G_{11} in (2.6) after numerical optimization of the Gaussian log-likelihood function. Being an iterative procedure, the algorithm requires to determine suitable initial parameters. For the diagonal elements of the matrices A_{11} and G_{11} values around 0.3 and 0.9 appear reasonable, since in univariate GARCH(1,1) models parameter estimates for a_1 and g_1 in (2.3) often take values around $0.3^2 = 0.09$ and $0.81 = 0.9^2$. There is no clear guidance how to determine initial values for off diagonal elements of A_{11} or G_{11}. Therefore, it might be reasonable to try alternative initializations of these parameters. Given an initialization of A_{11} and G_{11}, the starting values for the elements in C_0 are determined by the algorithm assuming the unconditional covariance of ε_t to exist (Engle and Kroner 1995).

Given our example under investigation, the bivariate GARCH estimation yields a vector of coefficient estimates,

$$\hat{\theta} = (.00115, .00031, .00076, .2819, -.0572, -.0504, .2934, .9389, .0251, .0275, .9391),$$

and a corresponding log-likelihood value $\hat{l} = 28599$ at the optimum. The first three estimates are the parameters of the upper triangular matrix C_0, the following four belong to the ARCH (A_{11}) and the last four to the GARCH parameters (G_{11}), i.e. for our model,

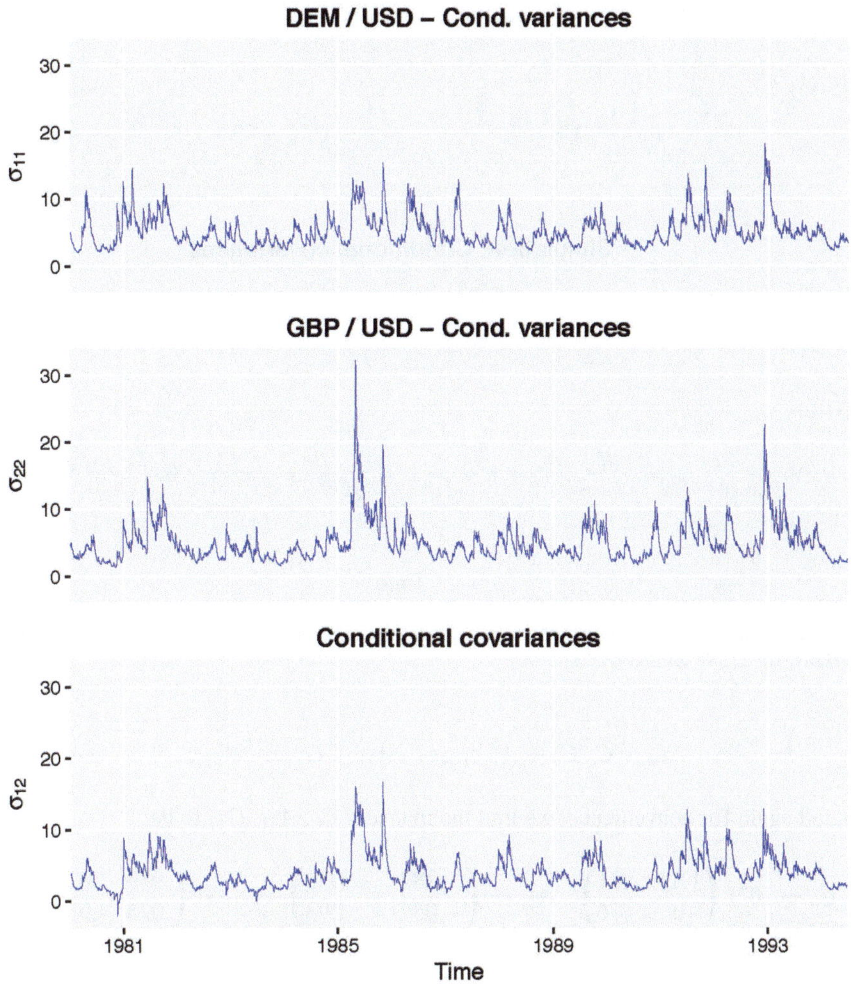

Fig. 2.2 Estimated variance and covariance processes, $10^5 \hat{\Sigma}_t$. XFGmvol02

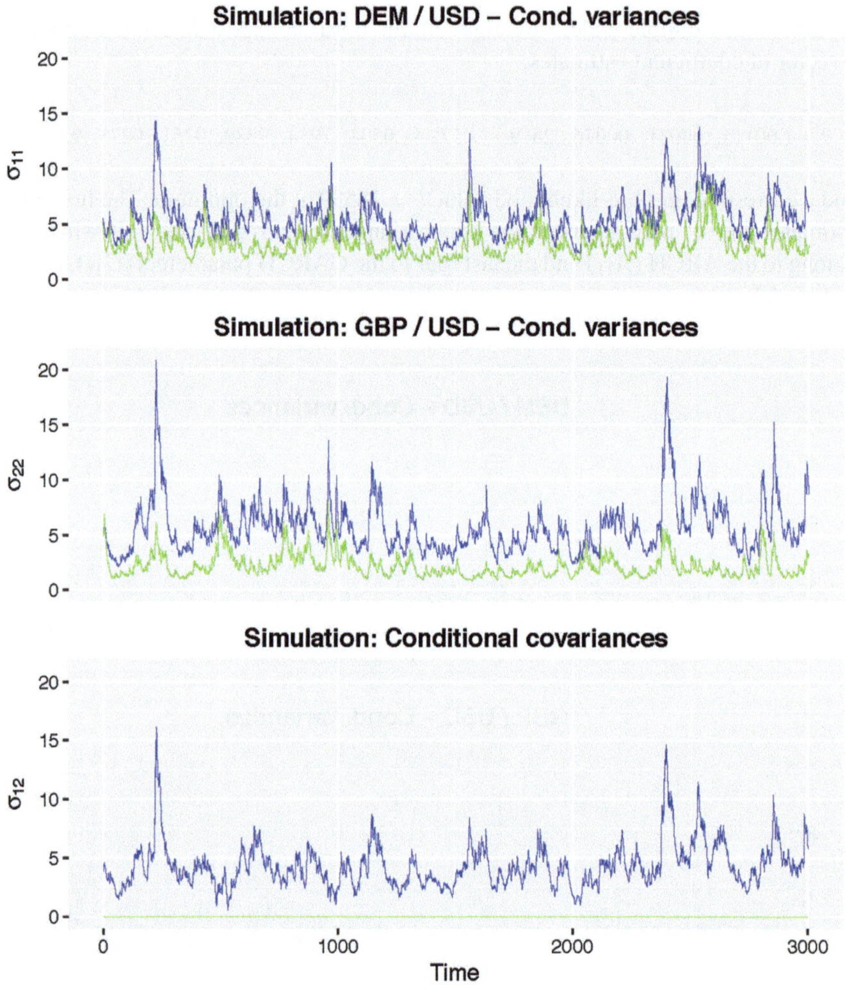

Fig. 2.3 Simulated variance and covariance processes, both bivariate (*blue*) and univariate case (*green*), $10^5\hat{\Sigma}_t$. XFGmvol03

$$\Sigma_t = C_0^\top C_0 + A_{11}^\top \varepsilon_{t-1}\varepsilon_{t-1}^\top A_{11} + G_{11}^\top \Sigma_{t-1} G_{11}, \tag{2.7}$$

stated again for convenience, we find the matrices C_0, A_{11}, G_{11} to be:

$$C_0 = 10^{-3}\begin{pmatrix} 1.15 & .31 \\ 0 & .76 \end{pmatrix}, \quad A_{11} = \begin{pmatrix} .282 & -.050 \\ -.057 & .293 \end{pmatrix}, \quad G_{11} = \begin{pmatrix} .939 & .028 \\ .025 & .939 \end{pmatrix}. \tag{2.8}$$

2.2.3 *Estimating the (co)variance Processes*

The (co)variance is obtained by sequentially calculating the difference equation (2.7) where we use the estimator for the unconditional covariance matrix as initial value ($\Sigma_0 = \frac{E^\top E}{T}$). Here, the $T \times 2$ matrix E contains log-differences of our foreign exchange rate data.

We display the estimated variance and covariance processes in Fig. 2.2. The quantlet XFGmvol02.R ss contains the code. The two upper panels of Fig. 2.2 show the variances of the DEM/USD and GBP/USD log-differences respectively, whereas in the lower panel we see the covariance process. Except for a very short period in the beginning of our sample, the covariance is positive and of non-negligible size throughout. This is evidence for cross sectional dependencies in currency markets which we mentioned earlier to motivate multivariate GARCH models.

Instead of estimating the realized path of variances as shown above, we could also use the estimated parameters to *simulate* volatility paths (XFGmvol03.R). For this, at each point in time an observation ε_t is drawn from a multivariate normal distribution with variance Σ_t. Given these observations, Σ_t is updated according to (2.7). Then, a new residual is drawn with covariance Σ_{t+1}. We apply this procedure for $T = 3000$. The results, displayed in the three panels of Fig. 2.3, show a similar pattern as the original process given in Fig. 2.2. For the upper two panels, we generate two variance processes from the *same* set of simulated residuals ξ_t. In this case, however, we set off-diagonal parameters in $C_0^\top C_0$, A_{11} and G_{11} to zero to illustrate how the unrestricted BEKK model incorporates cross equation dynamics. As can be seen, both approaches are convenient to capture volatility clustering. Depending on the particular state of the system, spillover effects operating through conditional covariances, however, have a considerable impact on the magnitude of conditional volatility.

2.3 Forecasting Exchange Rate Densities

The preceding section illustrated how the GARCH model may be employed effectively to describe empirical price variations of foreign exchange rates. For practical purposes, as for instance scenario analysis, Value-at-Risk estimation (Chap. 1), option pricing (see the corresponding chapter), one is often interested in the future joint density of a set of asset prices. Continuing the comparison of the univariate and bivariate approach to model volatility dynamics of exchange rates, it is thus natural to investigate the properties of these specifications in terms of forecasting performance.

We implement an iterative forecasting scheme along the following lines: Given the estimated univariate and bivariate volatility models and the corresponding information sets $\mathcal{F}_{t-1}, t = 1, \dots, T-5$ (Fig. 2.2), we employ the identified data generating processes to simulate one-week-ahead forecasts of both exchange rates. To get a reliable estimate of the future density, we set the number of simulations to 5000 for each initial scenario. This procedure yields two bivariate samples of future exchange rates, one simulated under bivariate, the other one simulated under univariate GARCH assumptions.

A review of evaluating competing density forecasts is offered by Tay and Wallis (2000). Adopting a Bayesian perspective the common approach is to compare the expected loss of actions evaluated under alternative density forecasts. In our pure time series framework, however, a particular action is hardly available for forecast density comparisons. Alternatively, one could concentrate on statistics directly derived from the simulated densities, such as first and second order moments or even quantiles. Due to the multivariate nature of the time series under consideration, it is a nontrivial issue to rank alternative density forecasts in terms of these statistics. Therefore, we regard a particular volatility model to be superior to another if it provides a higher simulated density estimate of the actual bivariate future exchange rate. This is accomplished by evaluating both densities at the actually realized exchange rate obtained from a bivariate kernel estimation. Since the latter comparison might suffer from different unconditional variances under univariate and multivariate volatility, the two simulated densities were rescaled to have identical variance. Performing the latter forecasting exercises iteratively over 3714 time points, we can test if the bivariate volatility model outperforms the univariate one.

To formalize the latter ideas, we define a success ratio SR_J as

$$SR_J = \frac{1}{|J|} \sum_{t \in J} \mathbf{1}\{\hat{f}_{biv}(R_{t+5}) > \hat{f}_{uni}(R_{t+5})\}, \tag{2.9}$$

where J denotes a time window containing $|J|$ observations and $\mathbf{1}$ an indicator function. $\hat{f}_{biv}(R_{t+5})$ and $\hat{f}_{uni}(R_{t+5})$ are the estimated densities of future exchange rates which are simulated by the bivariate and univariate GARCH processes, respectively, and which are evaluated at the actual exchange rate levels R_{t+5}. The simulations are performed in XFGmvol04.

Our results show that the bivariate model indeed outperforms the univariate one when both likelihoods are compared under the actual realizations of the exchange rate process. In 82.3% of all cases across the sample period, $SR_J = 0.823$, $J = \{t : t = 1, ..., T - 5\}$, the bivariate model provides a better forecast. This is highly significant. In Table 2.1, we show that the overall superiority of the bivariate volatility

Table 2.1 Time varying frequencies of the bivariate GARCH model outperforming the univariate one in terms of one-week-ahead forecasts (success ratio)

Time window J		Success ratio SR_J
1980	1981	0.762
1982	1983	0.786
1984	1985	0.868
1986	1987	0.780
1988	1989	0.872
1990	1991	0.835
1992	04/1994	0.854

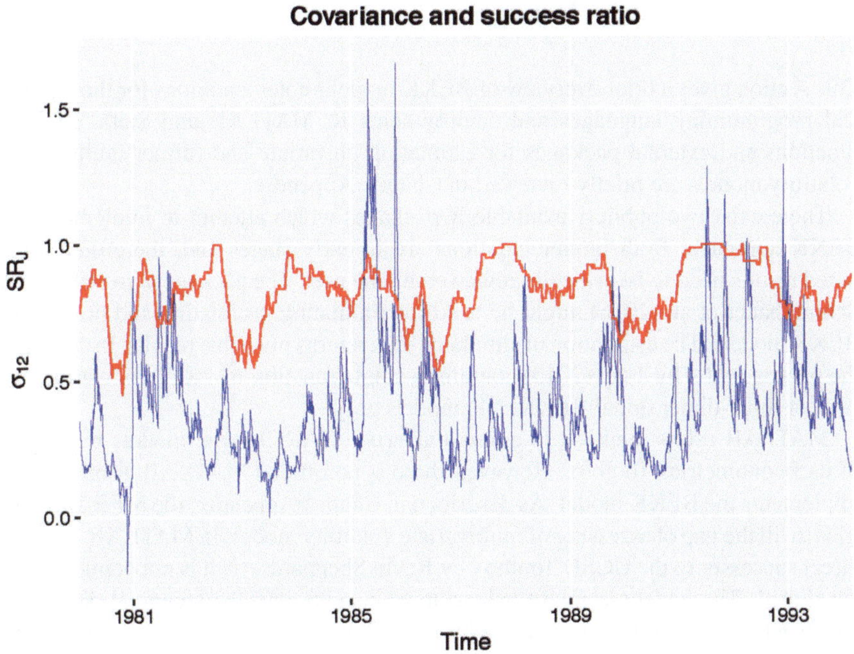

Fig. 2.4 Estimated covariance process from the bivariate GARCH model ($10^4\hat{\sigma}_{12}$, *blue*) and success ratio over overlapping time intervals with window length 80 days (*red*). XFGmvol04

approach is confirmed when considering subsamples of two-years length. A-priori, one may expect the bivariate model to outperform the univariate one the larger (in absolute value) the covariance between both log-difference processes is. To verify this argument, we display in Fig. 2.4 the empirical covariance estimates from Fig. 2.2 jointly with the success ratio evaluated over overlapping time intervals of length $|J| = 80$.

As is apparent from Fig. 2.4, there is a close co-movement between the success ratio and the general trend of the covariance process, which confirms our expectations: the forecasting power of the bivariate GARCH model is particularly strong in periods where the DEM/USD and GBP/USD exchange rate log-differences exhibit a high covariance. For completeness, it is worthwhile to mention that similar results are obtained if the window width is varied over reasonable choices of $|J|$ ranging from 40 to 150.

With respect to financial practice and research we take our results as strong support for a multivariate approach towards asset price modeling. Whenever contemporaneous correlation across markets matters, the system approach offers essential advantages. To name a few areas of interest, multivariate volatility models are supposed to yield useful insights for risk management, scenario analysis and option pricing.

Appendix: Software Packages

This section gives a brief overview of BEKK model implementations for the numerical programming languages and environments R, MATLAB and Stata. Built-in functions and external packages for estimating univariate and further multivariate volatility models are briefly reviewed in Chap. 1 Appendix.

There exist two publicly available R packages which attempt to implement the BEKK approach. Both implementations are in early stages and, therefore, computed results need to be critically reviewed by the user. The package `mgarchBEKK` Schmidbauer et al. (2016) might be used for simulating, estimating and predicting BEKK models. The estimation of simulated data returns plausible results. In contrast, the package `MTS` by Tsay (2015) contains a single function `BEKK11` for estimating two- or three-dimensional BEKK(1,1) models only.

MATLAB offers methods to assess univariate GARCH-type models by means of its Econometrics Toolbox. However, there is no official MATLAB Toolbox that implements the BEKK model. As described in Chap. 1 Appendix, the MFE Toolbox tries to fill the gap of assessing of multivariate volatility models in MATLAB. It is the direct successor to the UCSD Toolbox by Kevin Sheppard which is not being further developed. The codebase might help getting insights into the technical details of the BEKK approach. Because the toolbox is still under development, an optimized, error-free use can not be guaranteed.

Currently, Stata supports only the analysis of univariate volatility models, diagonal half-vec models, which are restricted versions of the half-vec model in (2.2), and conditional correlation models. It seems that there exists no publicly available extension to estimate a BEKK model. As an alternative, users might employ the tools of the independent software package JMulTi,[1] which is closely related to Lütkepohl and Krätzig (2004), for BEKK model estimation and investigation in combination with Stata.

References

Baba, Y., Engle, R. F., Kraft, D. F., & Kroner, K. F. (1990). *Multivariate simultaneous generalized ARCH, mimeo*. Department of Economics: University of California, San Diego.

Berndt, E. K., Hall, B. H., Hall, R. E., & Hausman, J. A. (1974). Estimation and inference in nonlinear structural models. *Annals of Economic and Social Measurement, 3*, 653–665.

Bollerslev, T. (1986). Generalized autoregressive conditional heteroskedasticity. *Journal of Econometrics, 31*, 307–327.

Bollerslev, T. (1990). Modelling the coherence in short-run nominal exchange rates: a multivariate generalized ARCH model. *The Review of Economics and Statistics, 72*, 498–505.

Bollerslev, T., & Engle, R. F. (1993). Common persistence in conditional variances. *Econometrica, 61*, 167–186.

Bollerslev, T., Engle, R.F. and Nelson, D.B. (1994). ARCH Models. In R.F. Engle, D.L. McFadden (Eds.), Handbook of Econometrics, Vol. 4, (pp. 2959–3038). Elsevier.

[1]Project website: http://www.jmulti.de.

Bollerslev, T., Engle, R. F., & Wooldridge, J. M. (1988). A capital asset pricing model with time-varying covariances. *Journal of Political Economy*, *96*, 116–131.
Bollerslev, T., & Wooldridge, J. M. (1992). Quasi-maximum likelihood estimation and inference in dynamic models with time-varying covariances. *Econometric Reviews*, *11*, 143–172.
Cecchetti, S. G., Cumby, R. E., & Figlewski, S. (1988). Estimation of the optimal futures hedge. *The Review of Economics and Statistics*, *70*, 623–630.
Comte, F., & Lieberman, O. (2003). Asymptotic theory for multivariate GARCH processes. *Journal of Multivariate Analysis*, *84*, 61–84.
Engle, R. F. (1982). Autoregressive conditional heteroscedasticity with estimates of the variance of United Kingdom inflation. *Econometrica*, *50*, 987–1007.
Engle, R. F., Ito, T., & Lin, W. L. (1990). Meteor showers or heat waves? *Heteroskedastic Intra-Daily Volatility in the Foreign Exchange Market, Econometrica*, *58*, 525–542.
Engle, R. F., & Kroner, K. F. (1995). Multivariate simultaneous generalized ARCH. *Econometric Theory*, *11*, 122–150.
Glosten, L. R., Jagannathan, R., & Runkle, D. E. (1993). On the relation between the expected value and the volatility of the nominal excess return on stocks, *Journal of Finance*, *48*, 1779–1801.
Hafner, C. M., & Herwartz, H. (1998). Structural analysis of portfolio risk using beta impulse response functions. *Statistica Neerlandica*, *52*, 336–355.
Hamao, Y., Masulis, R. W., & Ng, V. K. (1990). Correlations in price changes and volatility across international stock markets. *Review of Financial Studies*, *3*, 281–307.
Jeantheau, T. (1998). Strong consistency of estimators for multivariate ARCH models. *Econometric Theory*, *14*, 70–86.
Kroner, K. F., & Ng, V. K. (1998). Modeling asymmetric comovements of asset returns. *Review of Financial Studies*, *11*, 817–844.
Lütkepohl, H. (1996). *Handbook of matrices*. Chichester: Wiley.
Lütkepohl, H., & Krätzig, M. (2004). *Applied time series econometrics*. Cambridge: Cambridge University Press.
Nelson, D. B. (1991). Conditional heteroskedasticity in asset returns: a new approach. *Econometrica*, *59*, 347–370.
Schmidbauer, H., Roesch, A., & Tunalioglu, V. S. (2016). mgarchBEKK: Simulating, Estimating and Diagnosing MGARCH (BEKK and mGJR) Processes. *R package version*, 2.
Tay, A. S., & Wallis, K. F. (2000). Density forecasting: a survey. *Journal of Forecasting*, *19*, 235–254.
Tsay, R. S. (2015). MTS: All-Purpose Toolkit for Analyzing Multivariate Time Series (MTS) and Estimating Multivariate Volatility Models. *R package version*, 33.

Chapter 3
Portfolio Selection with Spectral Risk Measures

S.F. Huang, H.C. Lin and T.Y. Lin

Abstract In this chapter, a portfolio selection problem with spectral risk measure is considered. The spectral risk measure is a general family of coherent risk measures and is capable of reflecting investor's risk preference. A multivariate conditional heteroscedastic model with vine copulae is employed to describe the dynamics and dependence of the underlying asset returns. The technique of linear programming is used to accurately and quickly determine the optimal asset allocations. Simulation studies are conducted for investigating the impacts of the magnitude of tail dependence among the underlying assets and the degrees of risk aversion on the performance of the optimal portfolio. An empirical study is conducted by using the stock prices included in the FTSE TWSE Taiwan 100 Index. Numerical results indicate that the optimal portfolios have different reactions to different economic situations.

3.1 Introduction

In modern portfolio selection theory, the mean-variance (MV) portfolio optimization procedure introduced by Markowitz (1952; 1959) plays a crucial role in optimal asset allocations and investment diversification. In the MV procedure, investors attempt to maximize their portfolio expected return for a given level of portfolio risk, or equivalently to minimize the risk of investment with achieving a given amount of expected return, by determining the investment proportions of various securities (Markowitz 1952, 1959, 1991; Merton 1972; Kroll et al. 1984). The traditional MV portfolio problem uses standard deviation as the measure of risk and assumes that the returns of the underlying assets are independent and identically distributed (i.i.d.).

S.F. Huang (✉)
Department of Applied Mathematics, National University of Kaohsiung, Kaohsiung, Taiwan
e-mail: huangsf@nuk.edu.tw

H.C. Lin · T.Y. Lin
Institute of Statistics, National University of Kaohsiung, Kaohsiung, Taiwan
e-mail: m1004408@mail.nuk.edu.tw

T.Y. Lin
e-mail: goodbaby781126@gmail.com

W.K. Härdle et al. (eds.), *Applied Quantitative Finance*, Statistics and Computing,
DOI 10.1007/978-3-662-54486-0_3

Recently many other risk measures are more commonly used by traders in reality, for example, the value-at-risk (VaR), the expected shortfall risk (ES) and a general class of coherent risk measures, called the spectral risk measure (SRM). Thus, the optimal portfolio selection problem with risk constraints rather than standard deviation attracts more attention for practical implementation (Acerbi and Simonetti 2002; Krokhmal et al. 2002; Chabaane et al. 2006; Huang and Lin 2017). Consequently, assessing the impact regarding the selection of different risk measures on portfolio allocation is of particular importance for asset managers.

When returns are Gaussian distributed, which is parameterized through the first two moments, one could therefore well rely upon the MV framework and the choice of a risk measure is purposeless (Härdle et al. 2014). The empirical study of Adam et al. (2008) based on the monthly returns of 16 hedge funds from January 1990 to July 2001 further showed the robustness of portfolio allocation with respect to the choice of risk measures even the samples are non-Gaussian distributed. Consequently, it seems that the risk managers do not need to worry about the choice of risk measures for portfolio allocation regardless of the Gaussian assumption if the asset returns are assumed to be i.i.d.. However, many empirical studies show that hedge fund returns often exhibit autocorrelation, and have significant negative skewness and excess kurtosis (Giamouridis and Vrontos 2007; Harris and Mazibas 2010, 2013). This motivates us to consider portfolio selection problem without the i.i.d. assumption for asset returns. Furthermore, we investigate the impacts of trader's risk attitude on the performance of optimal portfolios under assuming the asset returns following a multivariate time series model.

To model the autocorrelation and conditional heteroscedasticity of each underlying asset, we consider the following model:

$$\begin{cases} X_{i,t} &= f_{i,t}(\mathbf{X}_{i,t-1}, a_{i,t}), \\ a_{i,t} &= \sigma_{i,t}\varepsilon_{i,t}, \\ \sigma_{i,t} &= h_{i,t-1}(\sigma_{i,s}, \varepsilon_{i,s}; s = 0, \ldots, t-1), \end{cases} \tag{3.1}$$

where $X_{i,t}$ is the log return of the ith asset at time t, $f_{i,t}$ is a function of $\mathbf{X}_{i,t-1} = (X_{i,0}, X_{i,1}, \ldots, X_{i,t-1})$ and $a_{i,t}$ for $i = 1, \ldots, p$, $h_{i,t-1}$ is an $\mathcal{F}_{t-1}$ measurable function with $\mathcal{F}_{t-1}$ being the set of information from time 0 up to time $t-1$ and $\varepsilon_{i,t}$, $t = 0, 1, \ldots$, are i.i.d. innovations with zero mean and unit variance for the ith asset at time t. In addition, assets on the financial markets usually exhibit dependence. For example, the stock prices of two companies which have a complementary relationship may both increase or decrease simultaneously by public good or bad news (Zhang et al. 2015). Recent studies indicate that pair-copula decomposed models represent a more flexible way to construct multivariate distributions than standard multivariate copulae. Therefore, we model the joint distribution of $\varepsilon_{i,t}, i = 1, \ldots, p$, by a vine copula function. Vine copulae are able to model complex dependency patterns by using a cascade of bivariate copulae (see Aas et al. 2009; Brechmann and Schepsmeier 2013 and the references therein).

Assume that $X_{i,t}$, for $i = 1, \ldots, p$ and $t = 0, 1, \ldots$, follow model (3.1) and consider the following portfolio optimization problem:

$$\max_{\mathbf{c}_t} \pi(\mathbf{c}_t) = c_{1,t}\mathsf{E}_t(X_{1,t+1}) + c_{2,t}\mathsf{E}_t(X_{2,t+1}) + \cdots + c_{p,t}\mathsf{E}_t(X_{p,t+1}),$$

$$\text{subject to } \mathbf{c}_t \geq 0, \ \sum_{i=1}^{p} c_{i,t} \leq 1 \text{ and } \rho_t(\nu) \leq L, \tag{3.2}$$

where $\mathbf{c}_t = (c_{1,t}, \ldots, c_{p,t})^\top$ with $c_{i,t}$ being the holding position of $X_{i,t}$, $\mathbf{c}_t \geq 0$ is the no short-selling constraint, $\sum_{i=1}^{p} c_{i,t} \leq 1$ is the budget constraint, $\mathsf{E}_t(\cdot)$ denotes the conditional expectation given $\mathcal{F}_t$, $\rho_t(\nu)$ is the value of the time-t SRM with level ν, which reflects the degrees of risk aversion, and L is a pre-specified upper bound of risk. The main reason that we employ the SRM as the risk measure in this chapter is its link to investor's risk preference. The SRM is not only a general family of coherent risk measures (for example, the ES is a special case of the SRM), but also can reflect the degrees of risk aversion of investors since the generator of the SRM can be obtained by a trader's personal utility function. More details of the definition and properties of the SRM are introduced in Sect. 3.2.

Although model (3.1) is capable of depicting the dynamics of the underlying returns better than the traditional i.i.d. assumption, the corresponding computation of determining the optimal asset allocations in (3.2) becomes complicated. Harris and Mazibas (2013) considered a portfolio selection problem with the ES being the risk measure and employed an AR(1)-EGARCH(1,1) model to depict the marginal dynamics of the return process for each underlying asset. Moreover, they used copulae to model the dependence between the underlying assets. Since the linearization of the optimal portfolio selection problem under this realistic but complex model is difficult and not available yet in the literature, the method based on Monte Carlo simulation is proposed to obtain the optimal asset allocations. However, the simulation based method could be time consuming and the simulation biases could lead to wrong decision, especially when the optimal solution occurs on the boundary.

In the literature, linear programming (LP) is widely used in portfolio selection under the i.i.d. assumption. LP is a fast algorithm to obtain accurate estimates of the optimal asset allocations, especially when the optimal solution occurs on the boundary. Due to the principal that potential return rises with an increase in risk, the optimal solution of the portfolio selection problem usually occurs on the boundary and thus LP is a suitable technique for solving it. For example, Markowitz (1952) used LP to solve the MV portfolio selection problem. Rockafellar and Uryasev (2000) considered portfolio selection problem with ES and proposed a linearization to select the optimal portfolio by LP. Recently, Huang and Lin (2017) proposed a linearization scheme to approximate the original portfolio selection problem and then obtain the optimal asset allocations by LP when the SRM is used as the risk measure.

In the simulation study, we conduct several scenarios to investigate the accuracy of the proposed LP for obtaining the optimal allocations, the effects of the magnitude of tail dependence and the degrees of risk aversion on the performance of the optimal portfolio. We also conduct empirical studies by using the underlying stock prices included in FTSE TWSE Taiwan 100 Index. Our empirical results indicate that the optimal portfolios have different reactions to different economic situations.

The remainder of this chapter is organized as follows. Section 3.2 reviews some backgrounds including coherent measures of risk, utility functions, SRM and vine copulae. The LP of Huang and Lin (2017) for solving (3.2) with model (3.1) is introduced in Sect. 3.3. Simulation studies are presented in Sect. 3.4. Section 3.5 demonstrates empirical results by using the stock prices included in the FTSE TWSE Taiwan 100 Index. Concluding remarks are given in Sect. 3.6. Computational details are presented in the Appendix.

3.2 Backgrounds

3.2.1 *Coherent Measures of Risk*

Let $\mathcal{G}$ be the set of random portfolio returns, ρ be a risk measure, which is a mapping from $\mathcal{G}$ into $\mathbb{R}$, and X denote the return of an asset.

(A1) Translation invariance: If A is a deterministic portfolio with guaranteed return α, then for all $X \in \mathcal{G}$ we have $\rho(X + A) = \rho(X) - \alpha$.
(A2) Subadditivity: For all X and $Y \in \mathcal{G}$, $\rho(X + Y) \leqslant \rho(X) + \rho(Y)$.
(A3) Positive homogeneity: For all $\lambda \geq 0$ and all $X \in \mathcal{G}$, $\rho(\lambda X) = \lambda\rho(X)$
(A4) Monotonicity: For all X and $Y \in \mathcal{G}$ with $X \leq Y$, we have $\rho(Y) \leq \rho(X)$.
(A5) Law invariance: For any portfolio returns X and Y with distribution function F_X and F_Y, respectively, if $F_X = F_Y$, then $\rho(X) = \rho(Y)$.
(A6) Comonotonic additivity: For any comonotonic random variables X and Y, $\rho(X + Y) = \rho(X) + \rho(Y)$.

A risk measure satisfying (A1)–(A4) is called coherent (Artzner et al. 1999). Unfortunately, the popular risk measure, VaR, is not coherent since VaR fails to comply with the subadditivity property and thus does not provide good incentives with respect to portfolio diversification. In addition, it is not in general continuous with respect to the confidence level α. Consequently VaR is sensitive to small changes in α when it is applied to discontinuous distributions (Acerbi and Tasche 2002). On the other hand, Dhaene et al. (2004) showed that the ES is a coherent, law invariant (A5) and comonotonic additive (A6) risk measure. Thus, the ES can be treated as a coherent extension of the VaR.

3.2.2 *Utility Function*

When a consumer or an investor exposed to uncertainty, a risk-averse investor might choose to accept with a low but guaranteed payment, rather than choosing an investment with high expected returns but also with high risk of losing money. Let $U(x)$ be the utility function of a risk-averse investor, where x denotes the wealth. The aversion to risk implied by a utility function $U(\cdot)$ is to be assumed as a form of concavity (Pratt 1964). The more the curvature of a concave function $U(x)$, the more the risk aversion is there. Hence, a more risk-averse investor prefers a more conservative investment. In the following, three popular utility functions are briefly introduced through the absolute risk-aversion, denoted by $A(x) = -U''(x)/U'(x)$, and the relative risk-aversion, abbreviated as $R(x) = -xU''(x)/U'(x)$, (Leroy and Werner 2001):

1. Constant Absolute Risk-Aversion (CARA): If $A(x)$ is a positive constant which is independent of wealth x, then we call the corresponding utility function being CARA. For example, the negative exponential utility function defined by $U(x) = -e^{-\nu x}$ is a CARA utility.
2. Constant Relative Risk-Aversion (CRRA): If $R(x)$ is a positive constant R which is independent of wealth x, then we call the corresponding utility function being CRRA. If $R = 1$, then the utility function of CRRA can be written as $U(x) = \ln x$, for $x > 0$, which is called log utility. If $R \neq 1$, then $U(x) = \frac{x^{1-R}}{1-R}$, for $x > 0$, which is called power utility.
3. Hyperbolic Absolute Risk-Aversion (HARA): If a utility function satisfies $A(x) = -U''(x)/U'(x) = 1/(ax+b)$, which is a hyperbolic function of x, then it is called HARA. In particular, the HARA encompasses the CARA and CRRA cases since it reduces to the CARA if $a = 0$ and reduces to the CRRA if $b = 0$. In general, if $ab \neq 0$, the utility function of the HARA can be written as

$$U(x) = \begin{cases} \log(x - x_s), & \text{if } a = 1, \\ \dfrac{(x - x_s)^{1-R^*}}{1 - R^*}, & \text{otherwise,} \end{cases}$$

for $x > x_s$, and $U(x) = -\infty$, for $x \leq x_s$, where $R^* = 1/a$ and $x_s = -b/a$.

3.2.3 *Spectral Measures of Risk*

A general class of coherent risk measures, called spectral risk measure (SRM), is defined by

$$M_\phi(X) = -\int_0^1 \phi(p) F_X^{\leftarrow}(p) dp, \tag{3.3}$$

where $F_X^{\leftarrow}(p) = \inf\{x | F_X(x) \geq p\}$ and $\phi \in \mathscr{L}^1([0, 1])$ is called the risk aversion function of the risk measure $M_\phi(X)$. In addition, ϕ is said to be an "admissible" risk spectrum if it is non-negative, non-increasing and $\int_0^1 \phi(p)dp = 1$. SRM is a coherent measure of risk if ϕ is an admissible risk spectrum (Acerbi 2002). In the realm of spectral measures, an investor can optimize a portfolio in a more articulated way by expressing her subjective risk aversion via the function ϕ (Acerbi and Simonetti 2002).

Acerbi (2002) further mapped any rational investor's subjective risk aversion (or utility preference) onto a SRM. For example, if we consider the exponential utility function defined over random outcomes x by $U(x) = -e^{-\nu x}$, where $\nu > 0$, then the risk aversion function $\phi(\cdot)$ is defined by setting $\phi(p) \propto e^{-\nu p}$. To satisfy the constraint $\int_0^1 \phi(p)dp = 1$, we have $\phi(p) = \frac{\nu e^{-\nu p}}{1-e^{-\nu}}$, where $0 < p < 1$.
Additionally, since the ES can be expressed as

$$\mathrm{ES}(X) = -\frac{1}{\alpha}\int_0^{\alpha} F_X^{\leftarrow}(p)dp = -\int_0^1 \phi_{ES_\alpha}(p)F_X^{\leftarrow}(p)dp, \text{ for } 0 \le \alpha \le 1,$$

where $\phi_{ES_\alpha}(p) = \frac{1}{\alpha}\mathrm{I}_{\{p\le\alpha\}}$ with $\mathrm{I}_{\{\cdot\}}$ being an indicator function, thus the SRM defined in (3.3) can be expressed as a weighted average of expected shortfalls (Acerbi 2004).

3.2.4 Vine Copulae: C- and D-Vines

Traditionally, traders evaluate the performance and risk of a portfolio under the multivariate Gaussian assumption. However, many empirical studies found that this assumption is not adequate for financial data (Danielsson et al. 2006; Morton et al. 2006; Giamouridis and Vrontos 2007). Copulae help to release the Gaussian assumption and offer a general class of joint distributions. It uses a copula function to link the marginal distributions of individual asset returns to depict the dependence structure.

Copula has recently become increasingly popular in many fields of applications for constructing multivariate distributions (Choros et al. 2013, 2014). It establishes the link between the univariate margins and the multivariate distribution functions. The main concern in practical implementation is how to identify an adequate family of copulae. A rich variety of bivariate copula families is well-investigated in the literature (Joe 1997; Nelsen 2006). However, the choice of adequate families for higher dimensions is more challenging. Standard multivariate copulae such as the multivariate Gaussian, Student-t and Archimedean copulae lack the flexibility of accurately modeling the dependence among larger numbers of variables. In stead of generalizing the standard multivariate copulae by increasing the complexity of their structures, vine copulae propose to model multivariate dependency by using and benefiting from the rich variety of bivariate copulae as building blocks (Joe 1996; Bedford and Cooke 2001, 2002; Kurowicka and Cooke 2006).

Vine copulae are flexible graphical models for describing multivariate distributions by decomposing a multivariate density into a series of bivariate copulae, or called pair-copulae, where each pair-copula can be chosen independently from each others (Aas et al. 2009; Brechmann and Schepsmeier 2013). This decomposition allows for an enormous flexibility in modeling asymmetries and tail dependence of a large number of variables. Aas et al. (2009) proposed a method for statistical inference of pair-copula decomposed models. Brechmann and Schepsmeier (2013) established an R package, called CDVine, which provides functions and tools for statistical inference of canonical vine (C-vine) and D-vine copulae, where the C- and D-vines are two successful and popular vine copula families in many applications (see Brechmann and Schepsmeier 2013, and the references therein). In the following, we employ the multivariate distribution with 4 variables as an example to briefly illustrate the 4-dimensional C- and D-vines.

There are 12 different 4-dimensional C-vine forms and 12 different 4-dimensional D-vine forms, and none of them are the same. The 4-dimensional C-vine structure is generally represented as

$$\begin{aligned} f_{1234}(\mathbf{x}) = & f_1(x_1) \cdot f_2(x_2) \cdot f_3(x_3) \cdot f_4(x_4) \cdot \\ & c_{12}\{F_1(x_1), F_2(x_2)\} c_{13}\{F_1(x_1), F_3(x_3)\} c_{14}\{F_1(x_1), F_4(x_4)\} \cdot \\ & c_{23|1}\{F(x_2 \mid x_1), F(x_3 \mid x_1)\} c_{24|1}\{F(x_2 \mid x_1), F(x_4 \mid x_1)\} \cdot \\ & c_{34|12}\{F(x_3 \mid x_1, x_2), F(x_4 \mid x_1, x_2)\} \end{aligned} \tag{3.4}$$

and the 4-dimensional D-vine structure is represented as

$$\begin{aligned} f_{1234}(\mathbf{x}) = & f_1(x_1) \cdot f_2(x_2) \cdot f_3(x_3) \cdot f_4(x_4) \cdot \\ & c_{12}\{F_1(x_1), F_2(x_2)\} c_{23}\{F_2(x_2), F_3(x_3)\} c_{34}\{F_3(x_3), F_4(x_4)\} \cdot \\ & c_{13|2}\{F(x_1 \mid x_2), F(x_3 \mid x_2)\} c_{24|3}\{F(x_2 \mid x_3), F(x_4 \mid x_3)\} \cdot \\ & c_{14|23}\{F(x_1 \mid x_2, x_3), F(x_4 \mid x_2, x_3)\}, \end{aligned} \tag{3.5}$$

where $\mathbf{x} = (x_1, x_2, x_3, x_4)$, $f_{1234}(\mathbf{x})$ is the joint density of (X_1, X_2, X_3, X_4), $f_i(x_i)$ is the marginal density of X_i, $F_i(x_i)$ is the distribution function of X_i for $i = 1, 2, 3, 4$, $F(x_2 \mid x_1)$ is the conditional distribution function of X_2 given X_1, $c_{12}\{F_1(x_1), F_2(x_2)\}$ is a pair copula density of X_1 and X_2, $c_{23|1}\{F(x_2 \mid x_1), F(x_3 \mid x_1)\}$ is the conditional pair copula density of X_2 and X_3 given X_1 and so on. The details of the deviation of (3.4) and (3.5) are given in the Appendix.

The C- and D-vine trees help us to easily memorize the decompositions of (3.4) and (3.5). For example, the corresponding structure of a 4-dimensional C-vine including 3 trees is shown in Fig. 3.1a. In the first tree, the dependencies of the first and second variables, of the first and third, of the first and fourth, and so on, are modeled by pair copulae. That is, if we assign the orders 1,...,4 to the four random variables, then the pairs of $(1, 2)$, $(1, 3)$, $(1, 4)$, ... are modeled by bivariate copulae. In the second tree, $(2, j \mid 1)$ denotes the conditional dependence of the second and the jth variables given the first variable, for $j = 3, 4$, and a bivariate copula is employed to model each conditional distribution. In the third tree, we denote the conditional dependence of $(2, 3 \mid 1)$ and $(2, 4 \mid 1)$ by $(3, 4 \mid 1, 2)$ and again model the conditional joint distribution of $(3, 4 \mid 1, 2)$ by a bivariate copula. By comparing the C-vine trees with the decomposition given in (3.4), the pairs shown in the C-vine trees are exactly the same with the components of the pair copulae in (3.4). Similarly, Fig. 3.1b presents the corresponding 4-dimensional D-vine trees to (3.5).

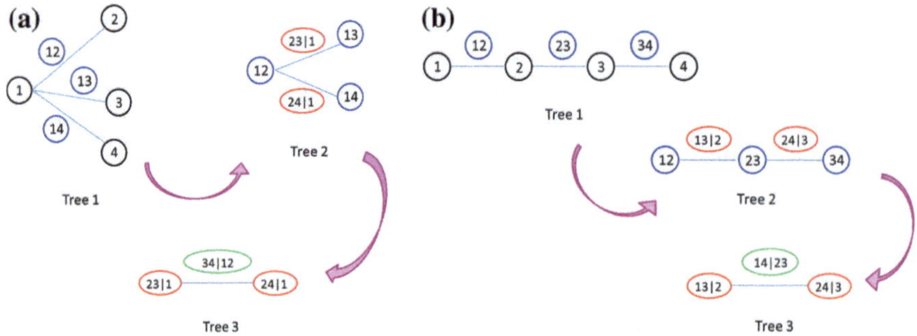

Fig. 3.1 **a** A 4-dimensional C-vine tree. **b** A 4-dimensional D-vine tree

3.3 Methodology

Rockafellar andUryasev (2000; 2002) proposed a scheme of linearization of the optimization problem (3.2) with the ES under the assumption of i.i.d. returns. In the following, we present their technique with the ES in our notation. First, rewrite the ES:

$$ES_{\alpha,t} = -\mathsf{E}_t(Y_{t+1} \mid -Y_{t+1} > \xi_{\alpha,t}) = \xi_{\alpha,t} + \frac{1}{\alpha}\mathsf{E}_t(-Y_{t+1} - \xi_{\alpha,t})^+,$$

where $Y_{t+1} = \sum_{m=1}^{p} c_{m,t} X_{m,t+1}$ is the portfolio return at time $t+1$ and $\xi_{\alpha,t}$ is the corresponding VaR of Y_{t+1} with respect to α level at time $t+1$ conditional on $\mathcal{F}_t$. Then, the optimization problem (3.2) with the ES can be rewritten as

$$\max_{\mathbf{c}_t, \xi_{\alpha,t}, z_1, \ldots, z_t} \quad \mathsf{E}_t(Y_{t+1}) \text{ subject to } \mathbf{c_t} \geq 0, \; \sum_{m=1}^{p} c_{m,t} \leq 1, \text{ and}$$

$$\begin{cases} \xi_{\alpha,t} + \frac{1}{t\alpha}\sum_{i=1}^{t} z_i & \leq L, \\ z_i & \geq 0, \\ z_i + \xi_{\alpha,t} & \geq -Y_i, \text{ for } i = 1, \ldots, t, \end{cases} \tag{3.6}$$

by incorporating z_i's to extend the set of unknown parameters. In (3.6), the objective function and the constraints are now linear functions of the unknown parameters $\{\mathbf{c}_t, \xi_{\alpha,t}, z_1, \ldots, z_t\}$ and thereby a LP technique can be used to obtain $\mathbf{c}_t$.

However, many empirical studies show that the return processes of the underlying assets in financial markets usually exhibit autocorrelation, negative skewness, kurtosis, conditional heteroscedasticity and tail dependence (Giamouridis and Vrontos 2007; Choros et al. 2013, 2014). It is of particular importance for asset managers to incorporate these features of the financial time series data when creating an investment or hedging portfolio. In order to model the autocorrelation and conditional heteroscedasticity, we assume that the mth underlying return process $X_{m,t}$, $m = 1, \ldots, p$, follows (3.1) and the joint distribution of $(\varepsilon_{1,t}, \ldots, \varepsilon_{p,t})$ is modeled by a C- or D-vine for depicting the multidimensional dependence among the underlying assets. Model (3.1) includes various financial time series models which are widely used in the market. For example, the ARMA-GARCH and ARMA-EGARCH models are two particular cases being commonly discussed in the economic, statistical, and financial literatures (see Bollerslev 1986; Nelson 1990; Duan 1995; Brandt and Jones 2006; Harveya and Sucarrat 2014).

Huang and Lin (2017) extended the i.i.d. scenario of Rockafellar and Uryasev (2000; 2002) to a more realistic situation as illustrated in (3.1) and linearize the nonlinear optimization problem in (3.2) with SRM. In particular, if we employ the ES, which is a special case of the SRM, as the risk measure, then the optimization problem (3.2) can be rewritten as

$$\max_{\mathbf{c}_t, \xi^*_{\alpha,t}, z_1, \ldots, z_t} \quad \mathsf{E}_t(Y_{t+1}), \text{ subject to } \mathbf{c_t} \geq 0, \ \sum_{m=1}^{p} c_{m,t} \leq 1, \text{ and}$$

$$\begin{cases} L \geq -\sum_{m=1}^{p} c_{m,t}\mu_{m,t} + \xi^*_{\alpha,t} + \frac{1}{t\alpha}\sum_{i=1}^{t} z_i, \\ z_i \geq 0 \\ z_i \geq -\xi^*_{\alpha,t} - \kappa_i, \ \text{for } i = 1, \ldots, t, \end{cases} \tag{3.7}$$

where $\mu_{m,t} = \mathsf{E}_t(X_{m,t+1})$, $\kappa_i = \sum_{m=1}^{p} c_{m,t}\sigma_{m,t+1}\varepsilon_{m,i}$, $\xi^*_{\alpha,t}$ is the corresponding VaR of κ_{t+1} with respect to α level at time $t+1$ conditional on $\mathcal{F}_t$. From comparing the expressions of (3.6) and (3.7), one can find the following three major changes:

1. The 1st term on the right-hand-side of the 1st inequality in (3.7) stands for the autocorrelated part.
2. On the right-hand-side of the 3rd inequality in (3.7) since the m-th summand of κ_i includes the conditional volatility $\sigma_{m,t+1}$, thus κ_i reflects the effect of conditional heteroscedasticity.
3. The role of the i.i.d. returns $X_{m,i}$ in (3.6) for each fixed m is replaced by the i.i.d. innovations $\varepsilon_{m,i}$ contained in κ_i in (3.7).

3.4 Simulation Study

In this section, we conduct several simulation scenarios to investigate the accuracy of the LP, the effects of the magnitude of tail dependence and the degrees of risk aversion on the performance of the optimal portfolio.

3.4.1 A 2-Dimensional Case

First, for the purpose of demonstration we concentrate on $p = 2$.

1. Generate observations of the mth underlying return process from the following AR(1)-EGARCH(1, 1) model, $m = 1, 2$,

$$\begin{cases} X_{m,t} &= \phi_{m,0} + \phi_{m,1} X_{m,t-1} + a_{m,t}, \\ a_{m,t} &= \sigma_{m,t} \varepsilon_{m,t}, \\ \log \sigma^2_{m,t} &= k_m + G_m \log \sigma^2_{m,t-1} + A_m[|\varepsilon_{m,t-1}| - \mathsf{E}(|\varepsilon_{m,t-1}|)] + L_m \varepsilon_{m,t-1}, \end{cases} \tag{3.8}$$

   where $(\varepsilon_{1,t}, \varepsilon_{2,t})$ are i.i.d. samples from a bivariate t distribution with zero means, unit variances, correlation ρ, and $\nu_1 = \nu_2 = \nu$. In particular,

$$\mathsf{E}(|\varepsilon_{m,t-1}|) = \frac{2\sqrt{\nu - 2}\Gamma[(\nu + 1)/2]}{(\nu - 1)\Gamma(\nu/2)\sqrt{\pi}}.$$

2. Solve the optimization problem defined in (3.1) and (3.2) for ES and SRM cases, where $\alpha = 0.05$ for the ES and the generating function $\phi(\cdot)$ for the SRM is set to be $\phi(p) = 10e^{-10p}/(1 - e^{-10})$ for $0 \leq p \leq 1$.

The expected returns (on the upper panel) and the values of risks (on the lower panel) of portfolios with different holding weights, c_1, of the 1st underlying asset under the model (3.8) are presented in Figs. 3.2 and 3.3 with ES and SRM, respectively.

The parameters in (3.8) are set to be $\rho = 0.5$, $\nu = 10$, $\phi_{1,0} = 0.01$, $\phi_{2,0} = 0.0105$, $\phi_{1,1} = 0.02$, $\phi_{2,1} = 0.0199$, $k_1 = k_2 = -0.3$, $A_1 = A_2 = 0.1776$, $G_1 = G_2 = 0.95$ and $L_1 = L_2 = -0.05$, and the upper bound L of the ES (or SRM) is set up to be the value of the ES (or SRM) of the portfolio with $c_1 = c_2 = 0.5$. Figure 3.2 plots the results of the ES case and Fig. 3.3 presents the results of the SRM with $T = 250$ on the left panel and $T = 500$ on the right panel. The red dashed lines in the lower panel denote the predetermined upper bound of the risk. If the value of the risk of a specified c_1 is below the red dashed line, then we plot the corresponding point in green, otherwise we mark the point in blue. The red circles on the upper panel denote the optimal solution calculated from the LP, which are close to the optimal selection of c_1 shown in Figs. 3.2 and 3.3, especially we increase the number of observations T to 500. This phenomenon confirms the accuracy of the proposed method in this 2-dimensional case.

3.4.2 The Impacts of Tail-Dependence

In this section, we investigate the impacts of tail-dependence under bear or bull markets. Consider the case of 10 assets, where assets 1–5 are independent and assets 6–10 have nonlinear tail dependency. We employ a 5-dimensional D-vine to model

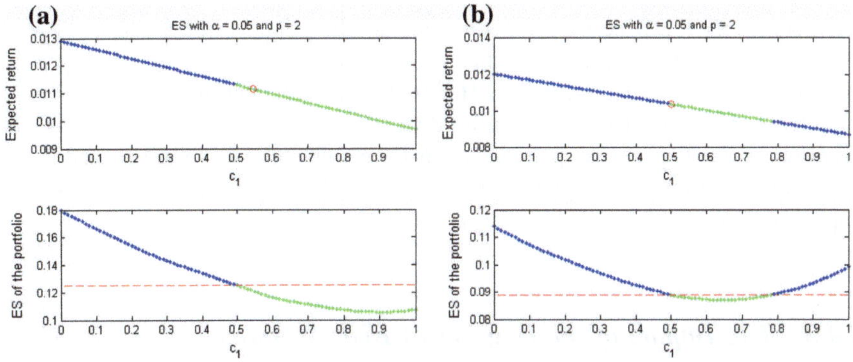

Fig. 3.2 The expected returns and the values of the ES of portfolios with different holding weights c_1 of the 1st underlying asset under model (3.8), where the numbers of observations are **a** $T = 250$ and **b** $T = 500$. XFGexp_rtn_ES_2d

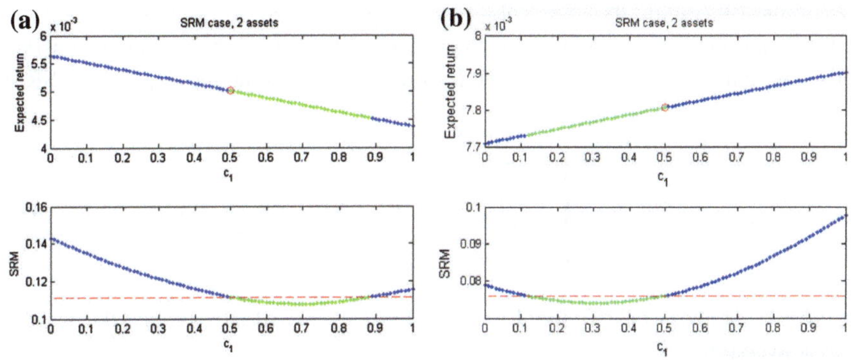

Fig. 3.3 The expected returns and the values of the SRM of portfolios with different holding weights c_1 of the 1st underlying asset under model (3.8), where the numbers of observations are **a** $T = 250$ and **b** $T = 500$. XFGexp_rtn_SRM_2d

the joint distribution of the dependent assets 6–10. In particular, we employ bivariate Clayton and Gumbel copulae to describe the nonlinear tail dependency between assets 6–10 in the first tree of the D-vine for bear and bull markets, respectively, where the copula parameters are randomly chosen from a $U(3,5)$ random variable. By using the same settings as in Sect. 3.4.1, except for setting $\phi_{i,0} = 0.1$, for $i = 1, \ldots, 5$, and $(\phi_{i,0}, \phi_{i,1}, k_i) = (0.11, 0.02, -0.28)$, for $i = 6, \ldots, 10$, to enlarge the expected returns of the assets in the bull market case, the optimal allocations are solved by the proposed LP method with ES under the bear and bull markets, separately.

We compute the sums of the weights of the assets 6–10 under bear and bull markets separately. The average of the holding proportions of assets 6–10 in the optimal portfolio based on 100 random replications is around 37% for the bear market and is around 90% for the bull market. These values reveal interesting and reasonable phenomenon. In a bear market, since the lower tail dependence of the assets 6–10 are modeled by a D-vine with Clayton copulae, the prices of the assets 6–10 tend to decrease simultaneously. In practice, diversification strategies are employed by investors in tough economic times. Hence, the independent assets 1–5 are more attractive to investors than the lower tail-dependent assets 6–10 in bear markets. On the contrary, the upper tail dependent assets have higher chance to be selected in the optimal portfolio than the independent assets in bull markets since the assets with upper tail dependencies tend to increase simultaneously.

3.4.3 *The Impact of the Degrees of Risk Aversion*

In this section, we investigate the performance of the optimal portfolios with different degrees of risk aversion, where each asset return process is assumed to follow an AR(1)-EGARCH(1,1) process. Consider that an investor plans to construct a portfolio by solving (3.2) with 30 assets subject to his personal risk attitude with a HARA utility function $U(x) = \log(x + b)$, where $b \in (-1, 0)$. Let ε be a positive constant satisfying $\max(0, b) < \varepsilon < 1 + b$ and set the generating function $\phi(p)$ of the SRM to be

$$\phi(p) = \begin{cases} \dfrac{-\log\varepsilon}{\eta}, & 0 \le p < \varepsilon - b, \\ \dfrac{-\log(p+b)}{\eta}, & \varepsilon - b \le p \le 1, \end{cases} \tag{3.9}$$

where $\eta = b\log\varepsilon - (1+b)\log(1+b) + (1+b) - \varepsilon > 0$ and b reflects the degrees of risk aversion of an investor.

Figure 3.4a presents boxplots of the optimal expected returns obtained by solving (3.2) with a generating function of the SRM defined in (3.9), where $b = -0.2, -0.3, -0.5$, $\varepsilon = 10^{-4}$ and the number of replications is 100. Figure 3.4b presents the corresponding utility functions, where other parameters in model (3.8) are set to be the same as in Sect. 3.4.1. In Fig. 3.4b, the solid lines are the tangents at $x = 0.7$ for the 3 utility functions. Since the slope of the tangent line in the case of $b = -0.5$ is larger than the others, thus investors having the utility function with $b = -0.5$ are more aggressive than those with $b = -0.2$ and -0.3. Figure 3.4a indicates that less risk-averse or more aggressive investors have larger expected returns than conservative investors.

3.5 Empirical Studies

We carry out our empirical investigation by using underlying assets stock price data included in the FTSE TWSE Taiwan 100 Index. We selected 79 stocks from 100 underlying assets included in the Taiwan 100 Index, where the daily returns from 1,

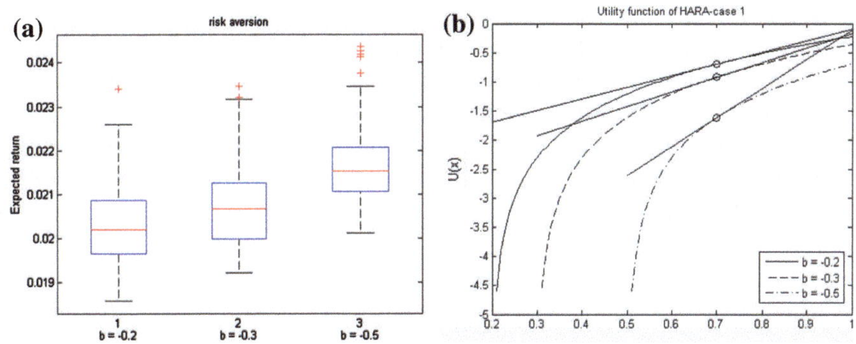

Fig. 3.4 **a** Boxplots of the optimal expected returns obtained by solving (3.2) with a generating function of the SRM defined in (3.9), where $b = -0.2$, –0.3 and –0.5. **b** The corresponding utility functions for $b = -0.2$, -0.3 and -0.5. XFGexp_rtn_SRM

December 2004 through 3, July 2014 (2365 observations) are used for investigation. This period includes a number of financial crises, for example, the subprime lending, stagflation, the Lehman crisis, the Greek government-debt crisis as well as the U.S. monetary policy-QE2. These events caused financial markets to have large volatility variation. In the following, we divide the time period into three sub-periods for the investigation: December 2004 to November 2007 (denoted by P1), representing relatively favorable market conditions (737 observations), December 2007 to December 2010 (denoted by P2), representing more extreme market conditions (764 observations) and January 2011 to 3, July 2014 (denoted by P3), representing improved market conditions (864 observations). We construct a self-financing trading strategy by using the proposed LP method to daily rebalance the portfolio with the 79 stocks for each of the three sub-periods. In particular, the FTSE TWSE Taiwan 100 Index is used as our benchmark for comparison. In the following, we use P1 as an example to illustrate the details of the investigation:

1. Let $P_{m,t}$ and $FTSE_t$ be the price of the mth asset and FTSE TWSE Taiwan 100 Index at time t, where $t = 0$ stands for the date of 1, December 2004.
2. Let V_t denote the value of our portfolio at time t and V_{250} be the same with the value of FTSE TWSE Taiwan 100 Index on 5, December 2005. That is,

$$V_{250} = FTSE_{250} = b^{(250)} \sum_{m=1}^{p} c_{m,250} P_{m,250} + Cash_{250},$$

where $Cash_{250} = FTSE_{250}(1 - \sum_{m=1}^{p} c_{m,250})$ is the amount invested in the bank, $b^{(250)} = FTSE_{250} \sum_{m=1}^{p} c_{m,250} / \sum_{m=1}^{p} c_{m,250} P_{m,250}$ is a scalar such that $V_{250} = FTSE_{250}$, $c_{m,250}$ are obtained by solving (3.2) with ES of level $\alpha = 0.05$ by the proposed LP method, and each underlying return process is modeled by an AR(1)-EGARCH(1,1) based on $X_{m,t} = \ln P_{m,t} - \ln P_{m,t-1}$ for $t = 1, \ldots, 250$ and $m = 1, \ldots, 79$.

3. At time $t = 251$, the value of our portfolio is $V_{251^-} = b^{(250)} \sum_{m=1}^{p} c_{m,250} P_{m,251} + e^{r_{\text{day}}} Cash_{250}$ prior to adjusting the allocations, where r_{day} is the daily riskfree interest rate and is set up as 0.01/250 in our investigation. By using the data $P_{m,t}$, $t = 1, \ldots, 251$, we reestimate the dynamic models of each return process and compute the updated optimal allocations, which are proportional to $c_{m,251}$ obtained from solving (3.2) by LP, where the value of the updated portfolio, denoted by V_{251^+}, is the same as V_{251^-}. That is,

$$V_{251^+} = b^{(251)} \sum_{m=1}^{p} c_{m,251} P_{m,251} + Cash_{251}, \tag{3.10}$$

where $b^{(251)} = V_{251^-} \sum_{m=1}^{p} c_{m,251} / \sum_{m=1}^{p} c_{m,251} P_{m,251}$ is a scalar such that $V_{251^-} = V_{251^+}$ for satisfying self-financing, and $Cash_{251} = V_{251^-} (1 - \sum_{m=1}^{p} c_{m,251})$ is the amount invested in the bank after the reallocation.
4. Repeat Step 3 until the end of P1.

Figure 3.5a–c plot the values of our trading strategy and the FTSE TWSE Taiwan 100 Index for P1, P2 and P3, respectively, where the black line is the values of the Taiwan 100 Index and the upper bound L of the risk is set to be 0.02, 0.03 or 0.05. In Fig. 3.5a–c, the values of the self-financing portfolio with $L = 0.05$ (green line) fluctuate more than those of $L = 0.02$ (red lines) and 0.03 (blue lines) no matter which economic situation is since a more aggressive trading strategy (with larger L) could gain more profits by taking more risks. In particular, the optimal portfolio tends to be more aggressive (with larger L) in bull markets and be more conservative (with smaller L) in bear markets. For example, during the financial crisis from December 2007 to June 2009 in Fig. 3.5b, the optimal portfolios with smaller L perform better than those with larger L.

In practice, investors would not use a fixed L for selecting their optimal portfolio, but rely on constructing the efficient frontier with various L instead. The discussion of how to construct the optimal portfolio through the efficient frontier framework is beyond the scope of this chapter. The objective of this chapter is to demonstrate that the proposed LP is useful to obtain the optimal allocations under conditional heteroscedastic models with more general risk measures than standard deviation. The

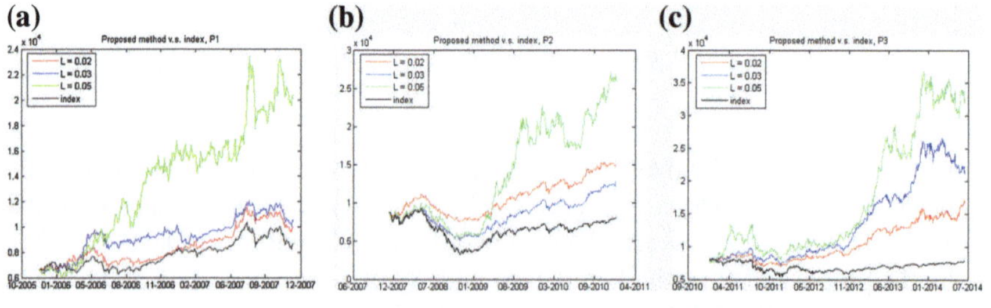

Fig. 3.5 The values of the self-financing trading strategy and the FTSE TWSE Taiwan 100 Index for **a** P1 **b** P2 **c** P3 with different fixed upper bounds of risk. XFGTWSE100_strategy_fixedESlevel

empirical study is designed to investigate whether the optimal portfolio would react to different economic situations if we consider a more complex but more realistic model. Please note though that we did not consider transaction costs in the daily reallocation and also allow to hold fractional numbers of shares of assets. What we have done is to provide an accurate and fast computational method for the investors who use model (3.1) to depict the dynamics of the underlying assets and obtain their optimal allocations of the assets by solving (3.2).

3.6 Concluding Remarks

In this chapter, we considered a portfolio optimization problem with the SRM, where the dynamics of the underlying return processes are depicted by autoregressive and conditional heteroscedastic models. The tail-dependence of the underlying assets is modeled by a CD-vine copula. A linearization of the optimal portfolio selection problem is used to compute the optimal asset allocations accurately and quickly. Simulation studies are conducted to investigate several interesting economic phenomena. First, we demonstrate the accuracy of the LP method for solving the optimal portfolio problem by using the case of two underlying assets. Second, we reveal that the optimal portfolio tends to diversify the investing risk by selecting the independent assets in bear market. Third, the less risk-averse investors achieve larger expected returns than conservative investors. The empirical study indicates that the optimal portfolio tends to be aggressive in bull markets and be conservative in bear markets.

Appendix

Derivation of (3.4) **and** (3.5)
To show the 4-dimensional C-vine, first note that

$$f_{1234}(\mathbf{x}) = f_1(x_1) f(x_2 \mid x_1) f(x_3 \mid x_1, x_2) f(x_4 \mid x_1, x_2, x_3), \tag{3.11}$$

where $\mathbf{x} = (x_1, x_2, x_3, x_4)$, $f_{1234}(\mathbf{x})$ is the joint density of (X_1, X_2, X_3, X_4), $f_i(x_i)$ is the marginal density of X_i for $i = 1, 2, 3, 4$, $f(x_2 \mid x_1)$ is the conditional density of X_2 given X_1 and so on. In addition, we have the following identities

$$\begin{aligned}
f(x_2 \mid x_1) &= c_{12}\{F_1(x_1), F_2(x_2)\} f_2(x_2), \\
f(x_3 \mid x_1, x_2) &= \frac{f(x_2, x_3 \mid x_1)}{f(x_2 \mid x_1)} \\
&= c_{23|1}\{F(x_2 \mid x_1), F(x_3 \mid x_1)\} f(x_3 \mid x_1) \\
&= c_{23|1}\{F(x_2 \mid x_1), F(x_3 \mid x_1)\} c_{13}\{F_1(x_1), F_3(x_3)\} f_3(x_3),
\end{aligned}$$

and

$$
\begin{aligned}
f(x_4 \mid x_1, x_2, x_3) &= \frac{f(x_3, x_4 \mid x_1, x_2)}{f(x_3 \mid x_1, x_2)} \\
&= c_{34|12}\{F(x_3 \mid x_1, x_2), F(x_4 \mid x_1, x_2)\} f(x_4 \mid x_1, x_2) \\
&= c_{34|12}\{F(x_3 \mid x_1, x_2), F(x_4 \mid x_1, x_2)\} \frac{f(x_2, x_4 \mid x_1)}{f(x_2 \mid x_1)} \\
&= c_{34|12}\{F(x_3 \mid x_1, x_2), F(x_4 \mid x_1, x_2)\} c_{24|1}\{F(x_2 \mid x_1), F(x_4 \mid x_1)\} f(x_4 \mid x_1) \\
&= c_{34|12}\{F(x_3 \mid x_1, x_2), F(x_4 \mid x_1, x_2)\} c_{24|1}\{F(x_2 \mid x_1), F(x_4 \mid x_1)\} \\
&\quad c_{14}\{F_1(x_1), F_4(x_4)\} f_4(x_4).
\end{aligned}
$$

By substituting the above identities into (3.11), we have

$$
\begin{aligned}
f_{1234}(\mathbf{x}) = &\ f_1(x_1) f_2(x_2) f_3(x_3) f_4(x_4) \\
&\ c_{12}\{F_1(x_1), F_2(x_2)\} c_{13}\{F_1(x_1), F_3(x_3)\} c_{14}\{F_1(x_1), F_4(x_4)\} \\
&\ c_{23|1}\{F(x_2 \mid x_1), F(x_3 \mid x_1)\} c_{24|1}\{F(x_2 \mid x_1), F(x_4 \mid x_1)\} \\
&\ c_{34|12}\{F(x_3 \mid x_1, x_2), F(x_4 \mid x_1, x_2)\}.
\end{aligned}
$$

Therefore, (3.4) holds.

On the other hand, the 4-dimensional D-vine is obtained through the following representation:

$$
f_{1234}(\mathbf{x}) = f_2(x_2) f(x_3 \mid x_2) f(x_1 \mid x_2, x_3) f(x_4 \mid x_1, x_2, x_3). \tag{3.12}
$$

By using a similar argument to the derivation of the C-vine, we have the following identities:

$$
\begin{aligned}
f(x_3 \mid x_2) &= c_{23}\{F_2(x_2), F_3(x_3)\} f(x_3), \\
f(x_1 \mid x_2, x_3) &= c_{13|2}\{F(x_1 \mid x_2), F(x_3 \mid x_2)\} f(x_1 \mid x_2) c_{12}\{F_1(x_1), F_2(x_2)\} f_1(x_1), \\
f(x_4 \mid x_1, x_2, x_3) &= c_{14|23}\{F(x_1 \mid x_2, x_3), F(x_4 \mid x_2, x_3)\} c_{24|3}\{F(x_2 \mid x_3), F(x_4 \mid x_3)\}, \\
&\quad c_{34}\{F_3(x_3), F_4(x_4)\} f_4(x_4).
\end{aligned}
$$

Therefore, (3.12) can be rewritten as

$$
\begin{aligned}
f_{1234}(\mathbf{x}) = &\ f_1(x_1) f_2(x_2) f_3(x_3) f_4(x_4) \\
&\ c_{12}\{F_1(x_1), F_2(x_2)\} c_{23}\{F_2(x_2), F_3(x_3)\} c_{34}\{F_3(x_3), F_4(x_4)\} \\
&\ c_{13|2}\{F(x_1 \mid x_2), F(x_3 \mid x_2)\} c_{24|3}\{F(x_2 \mid x_3), F(x_4 \mid x_3)\} \\
&\ c_{14|23}\{F(x_1 \mid x_2, x_3), F(x_4 \mid x_2, x_3)\}
\end{aligned}
$$

and (3.5) holds.

References

Aas, K., Czado, C., Frigessi, A., & Bakken, H. (2009). Pair-copula constructions of multiple dependence. *Insurance: Mathematics and Economics*, *44*, 182–198.
Acerbi, C. (2002). Spectral measures of risk: A coherent representation of subjective risk aversion. *Journal of Banking and Finance*, *26*, 1505–1518.
Acerbi, C. (2004). Coherent representation of subjective risk aversion. In G. Szego (Ed.), *Risk measures for the 21st Centry*. New York: Wiley.
Acerbi, C., & Simonetti, P. (2002). Portfolio optimization with spectral measures of risk, Working paper, Abaxbank, Italy.
Acerbi, C., & Tasche, D. (2002). On the coherence of expected shortfall. *Journal of Banking and Finance*, *26*, 1487–1503.
Adam, A., Houkari, M., & Laurent, J.-P. (2008). Spectral risk measures and portfolio selection. *Journal of Banking and Finance*, *32*, 1870–1882.
Artzner, P., Delbaen, F., Eber, J.-M., & Heath, D. (1999). Coherent measures of risk. *Mathematical Finance*, *9*, 203–208.
Bedford, T., & Cooke, R. M. (2001). Probability density decomposition for conditionally dependent random variables modeled by vines. *Annals of Mathematics and Artificial Intelligence*, *32*, 245–268.
Bedford, T., & Cooke, R. M. (2002). Vines-a new graphical model for dependent random variables. *Annals of Statistics*, *30*, 1031–1068.
Brandt, M. W., & Jones, C. S. (2006). Volatility forecasting with range-based EGARCH models, *Journal of Business and Economic Statistics, 24*, 470–486.
Brechmann, E. C., & Schepsmeier, U. (2013). Modeling dependence with C- and D-vine copulas: The R package CDVine. *Journal of Statistical Software*, *52*, 1–27.
Bollerslev, T. (1986). Generalized autoregressive conditional heteroskedasticity. *Journal of Econometrics*, *31*, 307–327.
Chabaane, A., Laurent, J.-P., Malevergne, Y., & Turpin, F. (2006). Alternative risk measures of alternative investments. *Journal of Risk*, *8*, 1–32.
Choros, B., Härdle, W., & Okhrin, O. (2013). Valuation of collaterablized debt obligations with hierarchical Archimedean copulae. *Empirical Finance*, *24*, 42–62. doi:10.1016/j.jempfin.2013.08.001.
Choros, B., Härdle, W., & Overbeck, L. (2014). Copula dynamics in CDOs. *Quantitative Finance*, *14*, 1573–1585. doi:10.1080/14697688.2013.847280.
Danielsson, J., Jorgensen, B. N., Sarma, M., & De Vries, C. G. (2006). Comparing downside risk measures for heavy tailed distributions. *Economics Letters, 92*, 202–208.
Dhaene, J., Goovaerts, M. J., Kaas, R., Vanduffel, S., & Vyncke, D. (2004). Capital requirements, risk measures and comonotonicity. *Belgian Actuarial Bulletin*, *4*, 53–61.
Duan, J. C. (1995). The GARCH option pricing model. *Mathematical Finance, 5*, 13–32.
Giamouridis, D., & Vrontos, I. D. (2007). Hedge fund portfolio construction: A comparison of static and dynamic approaches. *Journal of Banking and Finance*, *31*, 199–217.
Härdle, W., Nasekin, S., Lee, D., & Phoon, K. F. (2014). TEDAS - Tail event driven asset allocation. *Journal of Financial Econometrics*, submitted, SFB 649 DP 2014-32.
Harris, R. D. F., & Mazibas, M. (2010). Dynamic hedge fund portfolio construction. *International Review of Financial Analysis*, *19*, 351–357.
Harris, R. D. F., & Mazibas, M. (2013). Dynamic hedge fund portfolio construction: A semi-parametric approach. *Journal of Banking and Finance*, *37*, 139–149.
Harveya, A., & Sucarrat, G. (2014). EGARCH models with fat tails, skewness and leverage. *Computational Statistics & Data Analysis, 76*, 320–338.
Huang, S. F., & Lin, T. Y. (2017). A linearization of portfolio optimization problem with general risk measures under multivariate conditional heteroscedastic models. *Journal of Forecasting*, submitted.

Joe, H. (1996). Families of m-variate distributions with given margins and $m(m-1)/2$ bivariate dependence parameters. In L. Rüschendorf, B. Schweizer & M.D. Taylor (Eds.), *Distributions with fixed marginals and related topics*. California: Institute of Mathematical Statistic.
Joe, H. (1997). *Multivariate models and dependence concepts*. London: Chapman & Hall.
Krokhmal, P., Palmquist, J., & Uryasev, S. (2002). Portfolio optimization with conditional value-at-risk objective and constraints. *Journal of Risk, 4*.
Kroll, Y., Levy, H., & Markowitz, H. M. (1984). Mean-variance versus direct utility maximization. *Journal of Finance, 39*, 47–61.
Kurowicka, D., & Cooke, R. M. (2006). *Uncertainty analysis with high dimensional dependence modelling*. New York: Wiley.
Leroy, S. F., & Werner, J. (2001). *Principles of financial economics*. New York: Cambridge University Press.
Markowitz, H. M. (1952). Portfolio selection. *Journal of Finance, 7*, 77–91.
Markowitz, H. M. (1959). *Portfolio selection*. New York: Wiley.
Markowitz, H. M. (1991). *Portfolio selection: Efficient diversification of investment*. Cambridge: Blackwell.
Merton, R. C. (1972). An analytic derivation of the efficient portfolio frontier. *Journal of Financial and Quantitative Analysis, 7*, 1851–1872.
Morton, D., Popova, E., & Popova, I. (2006). Efficient fund of hedge funds construction under downside risk measures. *Journal of Banking & Finance, 30*, 503–518.
Nelson, D. B. (1990). Stationarity and persistence in the GARCH(l, l) model. *Econometric Theory, 6*, 318–334.
Nelsen, R. B. (2006). *An introduction to copulas* (2nd ed.). Berlin: Springer.
Pratt, J. W. (1964). Risk aversion in the small and in the large. *Econometrica, 32*, 122–136.
Rockafellar, R. T., & Uryasev, S. (2000). Optimization of conditional value-at-risk. *Journal of Risk, 2*, 21–41.
Rockafellar, R. T., & Uryasev, S. (2002). Conditional value-at-risk for general loss distributions. *Journal of Banking and Finance, 26*, 1443–1471.
Zhang, J. Z., Härdle, W., Chen, Y. C., & Bommes, E. (2015). Distillation of news flow into analysis of stock reactions. *Journal of Business & Economic Statistics, 34*, 547–563.

Chapter 4
Implementation of Local Stochastic Volatility Model in FX Derivatives

J. Zheng and X. Yuan

Abstract In this paper, we present our implementations of the Local Stochastic Volatility (LSV) Model in pricing exotic options in FX Market. Firstly, we briefly discuss the limitations of the Black-Scholes model, the Local Volatility (LV) Model and the Stochastic Volatility (SV) Model. To overcome the drawbacks of the above three models, a more generalized LSV model has been proposed to describe the dynamics of implied volatilities. Secondly, we present the details of LSV Model calibration in terms of the Forward Kolgomorov equation. Thirdly, we introduce the numerical methods of option pricing using the LSV model, including both the Backward Partial Differential Equation (PDE) method and Forward Monte Carlo method. Finally, based on our implementations, we compare the calibration and pricing results of the LSV model with the LV model and the SV model, lower calibration errors and relatively accurate pricing results are achieved, which demonstrates the effectiveness of the methods presented in the paper.

4.1 Introduction

Traditional Black-Scholes model (Black and Scholes 1973) is broadly used in European vanilla option pricing for both FX and equity markets. In the Black-Scholes model for FX market, the FX spot rate S_t is assumed to follow the Stochastic Differential Equation (SDE) as below

$$dS_t = \left(r_d - r_f\right) S_t dt + \sigma S_t dW_t \tag{4.1}$$

J. Zheng (✉) · X. Yuan
Global Market Department, Industrial and Commercial Bank of China,
Peking, People's Republic of China
e-mail: zhjune@gmail.com, jun.zheng@icbc.com.cn

X. Yuan
e-mail: yuanxun@ustc.edu

W.K. Härdle et al. (eds.), *Applied Quantitative Finance*, Statistics and Computing,
DOI 10.1007/978-3-662-54486-0_4

where r_d and r_f denote the domestic interest rate and the foreign interest rate respectively, and volatility σ is assumed to be constant.

However, in real market (e.g. the FX market), the volatility is not constant across different strikes and maturity dates, which is quite important for pricing barrier options. To tackle the problem of volatility smile and to describe the dynamics of implied volatilities, several models have been developed to generalize Black-Scholes model.

The Local Volatility (LV) model was firstly proposed by Dupire (1994). In the LV model, the diffusion coefficient is a deterministic function of time and the FX spot rate, $\sigma_{LV}(S_t, t)$, the corresponding SDE is as below

$$dS_t = \left(r_d - r_f\right) S_t dt + \sigma_{LV}(S_t, t) S_t dW_t \tag{4.2}$$

Theoretically, the LV model is able to provide a perfect fit to the quoted market implied volatilities. However, it still has several drawbacks. Firstly, it has been pointed out that the delta of an option computed from the LV model is far away from precise, because of an improper implied volatility dynamics (Hagan et al. 2002). Secondly, the forward implied volatility smile generated by the LV model is almost flat (Fengler 2005), but the smile persists over time in the reality. Thirdly, the LV model generates the volatility smile using a deterministic function $\sigma_{LV}(S_t, t)$, which depends on the spot level S_t. Therefore, the LV model is sticky-strike, which seldom happened in the FX market (Clark 2011).

Based on an empirical observation of FX market, it is more reasonable to model the instaneous volatility via a stochastic process, which leads to the Stochastic Volatility (SV) model. In a SV model, the diffusion coefficient is a function of a stochastic process v_t, $a(v_t)$, the corresponding SDE is as following

$$dS_t = \left(r_d - r_f\right) S_t dt + a(v_t) S_t dW_t \tag{4.3a}$$
$$dv_t = b(v_t) dt + c(v_t) dZ_t \tag{4.3b}$$
$$dW_t dZ_t = \rho dt$$

where ρ represents the correlation between the Brownian motions W_t and Z_t. In most cases, the stochastic variance v_t is assumed to be mean-reverting, continuous, and positive. For example, in the well-known Heston model (Heston 1993), the Cox-Ingersoll-Ross (CIR) process is used to model the variance process v_t:

$$dS_t = \left(r_d - r_f\right) S_t dt + \sqrt{v_t} S_t dW_t \tag{4.4a}$$
$$dv_t = \kappa (m - v_t) dt + \alpha \sqrt{v_t} dZ_t \tag{4.4b}$$
$$dW_t dZ_t = \rho dt$$

where κ is the mean-reverting speed, m is the mean-reverting level, and α corresponds to the volatility of variance. Compared with the LV model, the SV model is able to imply a more realistic forward implied volatility smile. However, it still has several drawbacks. Firstly, the SV model is not able to fit the implied volatility surface perfectly as the LV does. Secondly, the SV model generates the same smile irrespective of initial level of the spot, and is therefore "sticky-delta", which is not the reality in FX market either (Clark 2011).

To overcome the drawbacks of the LV model and the SV model, a more generalized model, named Local Stochastic Volatility (LSV) model was introduced. In the LSV model, the diffusion coefficient is the multiplication of a deterministic local volatility component $\sigma_{LSV}(S_t, t)$ and a stochastic volatility component v_t. For example, the SDE for a Heston-type LSV model is as below.

$$dS_t = \left(r_d - r_f\right) S_t dt + \sigma_{LSV}(S_t, t)\sqrt{v_t} S_t dW_t \tag{4.5a}$$
$$dv_t = \kappa(m - v_t)\,dt + \alpha\sqrt{v_t} dZ_t \tag{4.5b}$$
$$dW_t dZ_t = \rho dt$$

In the LSV model, part of the volatility smile is generated by the deterministic local volatility term $\sigma_{LSV}(S_t, t)$, while the rest part of the smile is generated by the stochastic volatility term v_t. Therefore, the LSV model is the model between"sticky-delta" and "sticky-strike", which is actually useful in the FX market. Moreover, it fits the implied volatility surface quite well as the LV model does, and meanwhile implies a more realistic forward implied volatility smile assumed by the SV model.

The rest of this paper is organized as following. In Sect. 4.2, we detail the LSV model calibration process through solving a Fokker–Planck Equation (FPE) iteratively. In Sect. 4.3, two different numerical methods for pricing exotic options using the LSV model are introduced, Backward PDE, and Forward Monte Carlo. Numerical results for model calibration and barrier option pricing are presented in Sect. 4.4, followed by the conclusion remarks and future works in Sect. 4.5.

4.2 Model Calibration

As mentioned in Sect. 4.1, by choosing different stochastic processes for v_t, we can get different types of the LSV model. For simplicity, we limit our discussions to Heston-type LSV model. The calibration of other types of LSV model can be performed similarly.

Generally speaking, the calibration of the LSV model consists of two main steps. In step 1, the parameters of the SV part are calibrated to fit a certain proportion of volatility smile. The proportion is controlled by a mixing fraction parameter, which is between 0 and 1. In step 2, the parameters of LV part are added to calibrate the LSV model to the whole volatility smile.

Step 1: Calibrate the parameters for the SV part, this step is performed infrequently. Specify a mixing weight η, which controls the proportion of volatility smile generated by the SV part and the proportion generated by the LV part. The mixing weight is used to mark down the implied volatility smile and skew, which can be done in two ways. One way is to multiply the market quotes of Butterfly and Risk Reversal by the factor

η. The Butterfly quotes correspond to the volatility smile, while the Risk Reversal quotes correspond to the volatility skew. Since the multiplication will reduce the volatility smile and skew, we calibrate a purely SV model to the market quotes with a reduced smile and skew. The other way is to calibrate a purely SV model to the true market quotes firstly, and then multiply the volatility of variance α and correlation ρ by the factor η, because the volatility of variance parameter corresponds to the volatility smile, and the correlation parameter corresponds to the volatility skew.

Step 2: Calibrate the leverage function $\sigma_{LSV}(S_t, t)$ so that the LSV model can fit the market quotes of vanilla options. This step is usually performed more frequently than step 1. We will detail the implementations of this step in the later part of this Section.

In our experiments, we set the mixing fraction empirically as described in Clark (2011). However, please note that the mixing fraction can also be calibrated using the quoted prices of liquid barrier options, as described in Tian (1993).

The calculation of the leverage function $\sigma_{LSV}(S_t, t)$ is based on the following important result: there exists only one LSV surface $\sigma_{LSV}(S_t, t)$ so that the LSV model can mimic the LV model, and $\sigma_{LSV}(S_t, t)$ must follow

$$\sigma_{LV}(s,t)^2 = E\left[\sigma_{LSV}(s,t)^2 v_t \mid S_t = s\right] = \sigma_{LSV}(s,t)^2 E\left[v_t \mid S_t = s\right] \tag{4.6}$$

For the proof the above important result, please refer to Ren et al. (2007), Tachet (2011). Based on the result, we can compute $\sigma_{LSV}(S_t, t)$ as the ratio between local volatility and conditional expectation of stochastic volatility:

$$\sigma_{LSV}(s,t) = \frac{\sigma_{LV}(s,t)}{\sqrt{E\left[v_t \mid S_t = s\right]}} = \sigma_{LV}(s,t)\sqrt{\frac{\int_v p(s,v,t)\,dv}{\int_v v\cdot p(s,v,t)\,dv}} \tag{4.7}$$

where $\sigma_{LV}(S_t, t)$ can be acquired from the LV model. Therefore, the key of calculating $\sigma_{LSV}(S_t, t)$ is to compute the joint probability distribution $p(s, v, t)$. Ren, Madan, and Qian (2007) firstly proposed to calculate $p(s, v, t)$ by solving the Fokker–Planck Equation (FPE) of the LSV model through a Finite Difference Method. After their pioneering work, Tachet (2011), Tian (1993), and Clark (2011) also solved the FPE with the Finite Difference Method, while Engelmann (2012) used the finite volume method, and Cozzi (2012) used the finite element method.

Let $X_t = \ln(S_t)$, the FPE for Heston-type LSV is as following

$$\begin{aligned}\frac{\partial p}{\partial t} &= \frac{1}{2}\frac{\partial^2\left[v\sigma_{LSV}^2(X,t)\,p\right]}{\partial X^2} + \rho\alpha\frac{\partial^2\left[v\sigma_{LSV}(X,t)\,p\right]}{\partial X\partial v} + \frac{1}{2}\alpha^2\frac{\partial^2[vp]}{\partial v^2}\\ &+ \frac{\partial}{\partial X}\left[\left(\frac{1}{2}v\sigma_{LSV}^2(X,t) - \left(r_d - r_f\right)\right)p\right] + \kappa\frac{\partial[(v-m)\,p]}{\partial v}\end{aligned} \tag{4.8}$$

where, for simplicity, $\sigma_{LSV}(S_t, t) = \sigma_{LSV}(X_t, t)$ refers to the leverage function of LSV model either in logspot or spot coordinates.

To solve the FPE (4.8), an Alternating-Direction-Implicit (ADI) method is used. Tataru and Fisher (2010) suggest to use a modified Douglas scheme, which was used by Hout and Foulson (2010) to solve the Backward pricing PDE for Heston model. The modified Douglas scheme is as below.

$$\begin{aligned} Y_0 &= p_{n-1} + \Delta t \left[F_0 (p_{n-1}, t_{n-1}) + F_1 (p_{n-1}, t_{n-1}) + F_2 (p_{n-1}, t_{n-1}) \right] \\ Y_1 - \theta \Delta t F_1 (Y_1, t_n) &= Y_0 - \theta \Delta t F_1 (p_{n-1}, t_{n-1}) \\ Y_2 - \theta \Delta t F_2 (Y_2, t_n) &= Y_1 - \theta \Delta t F_2 (p_{n-1}, t_{n-1}) \\ p_n &= Y_2 \end{aligned} \tag{4.9}$$

where p_n denotes the transition probability $p(s, v, t_n)$ at time t_n. The parameter θ affects the stability and accuracy of the ADI method, which lies in the range [0, 1]. F_0, F_1, and F_2 refer to derivative terms in mixed derivative, v-direction, and X-direction respectively.

$$\begin{aligned} F_0 (p, t) &= \rho \alpha \frac{\partial^2 \left[v \sigma_{LSV} (X, t) p \right]}{\partial X \partial v} \\ F_1 (p, t) &= \frac{1}{2} \alpha^2 \frac{\partial^2 [vp]}{\partial v^2} + \kappa \frac{\partial [(v - m) p]}{\partial v} \\ F_2 (p, t) &= \frac{1}{2} \frac{\partial^2 \left[v \sigma^2_{LSV} (X, t) p \right]}{\partial X^2} + \frac{\partial}{\partial X} \left[\left(\frac{1}{2} v \sigma^2_{LSV} (X, t) - \left(r_d - r_f \right) \right) p \right] \end{aligned} \tag{4.10}$$

The initial value for the FPE is $p_0 = p(X, v, 0) = \delta(X - X_0) \delta(v - v_0)$, where the $\delta()$ is the Dirac Delta function. According to Eq. (4.7), the leverage function at time zero is $\sigma_{LSV}(X_0, 0) = \frac{\sigma_{LV}(X_0, 0)}{\sqrt{v_0}}$. At time t_n, we have p_n and $\sigma_{LSV}(X, t_n)$, then we can solve FPE (4.8) forward one step to get p_{n+1}, and then use Eq. (4.7) to get the leverage function $\sigma_{LSV}(X, t_{n+1})$ at time t_{n+1}. This process is repeated through time, and we can solve p_n and $\sigma_{LSV}(X, t_n)$ for all time points:

The solved $\sigma_{LSV}(X, t)$ can be used to price derivative products, either by a backward PDE or a forward Monte Carlo approach. We will detail the two pricing methods for the LSV model in the next section (Fig. 4.1).

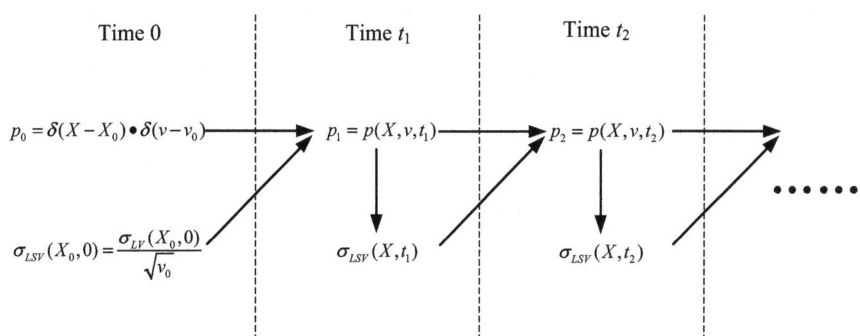

Fig. 4.1 Solve the FPE Iteratively to get the Leverage Function $\sigma_{LSV}(X, t)$

4.3 Pricing (Backward PDE and Forward Monte Carlo)

Let $V(X, v, t)$ denotes the option value as a function of time to expiry t, log-Spot level X, and the instaneous variance v. The backward pricing PDE for Heston-type LSV model are as following.

$$\begin{aligned}\frac{\partial V}{\partial t} &= \frac{1}{2}v\sigma_{LSV}^2(X,t)\frac{\partial^2 V}{\partial X^2} + \frac{1}{2}\alpha^2 v\frac{\partial^2 V}{\partial v^2} + \rho\alpha v\sigma_{LSV}(X,t)\frac{\partial^2 V}{\partial X\partial v}\\ &+ \left(r_d - r_f - \frac{1}{2}v\sigma_{LSV}^2(X,t)\right)\frac{\partial V}{\partial X} + \kappa(m-v)\frac{\partial V}{\partial v} - r_d V \end{aligned} \tag{4.11}$$

Note that $t = 0$ corresponds to the option expiry, and $t = T$ corresponds to today. This is different from the FPE in Sect. 4.2, where we use $t = 0$ for today, and $t = T$ for option expiry.

We can also solve the backward pricing PDE using the modified Douglas scheme as shown in Eq. (4.9). Instead, the mixed derivative operator F_0, v-direction derivative operator F_1, and X-direction derivative operator F_2 for the pricing PDE are as follow.

$$\begin{aligned} F_0(V,t) &= \rho\alpha v\sigma_{LSV}(X,t)\frac{\partial^2 V}{\partial X\partial v}\\ F_1(V,t) &= \frac{1}{2}\alpha^2 v\frac{\partial^2 V}{\partial v^2} + \kappa(m-v)\frac{\partial V}{\partial v} - \frac{1}{2}r_d V\\ F_2(V,t) &= \frac{1}{2}v\sigma_{LSV}^2(X,t)\frac{\partial^2 V}{\partial X^2} + \left(r_d - r_f - \frac{1}{2}v\sigma_{LSV}^2(X,t)\right)\frac{\partial V}{\partial X} - \frac{1}{2}r_d V \end{aligned} \tag{4.12}$$

In the backward pricing PDE, we start from the terminal condition, i.e., the payoff at expiry ($t = 0$). Based on the pricing PDE and some boundary conditions, we can propagate $V(X, v, t)$ backward to today ($t = T$), where we get the option value $V(X_T, v_T, T)$ by interpolation. The terminal condition and other boundary conditions are all determined by the option characteristics.

Besides the backward pricing PDE method, Monte Carlo method (Glasserman 2003) can also be utilized to price the options based on the LSV model. One key problem of Monte Carlo method for the LSV model is the discretization scheme of the SDE (4.5) for LSV model. A tradeoff between the computation complexity and accuracy should be found in the discretization scheme. Let $X_t = \ln(S_t)$, Eq. (4.5) can be rewritten as follow.

$$dX_t = \left[r_d - r_f - \frac{1}{2}V_t\sigma_{LSV}^2(X_t,t)\right]dt + \sigma_{LSV}(S_t,t)\sqrt{V_t}\left(\rho dW_v(t) + \sqrt{1-\rho^2}dW_x(t)\right) \tag{4.13a}$$

$$dV_t = \kappa(m - V_t)dt + \alpha\sqrt{V_t}dW_v(t) \tag{4.13b}$$

where $dW_x(t)$ and $dW_v(t)$ denote independent Brownian motions. When Feller condition $2\kappa m \geq \alpha^2$ is not satisfied, the variance process can become negative with non-zero probability in the Euler discretization. Therefore, we adopt the QE (Quadratic Exponential) (Andersen 2008) scheme for the discretization of the variance process. For the discretization of the log-spot process, we adopt the local-freezing of $\sigma_{LSV(X,t)}$, introduced by Van etc. (2014). More specifically, the discretization scheme for log-spot process is as follow.

$$x_{i+1} = x_i + r\Delta - \frac{1}{2}\sigma_{LSV}^2 (x_i, t_i) v_i \Delta + \frac{\rho}{\alpha}\sigma_{LSV}^2 (x_i, t_i) (v_{i+1} - \kappa m \Delta + v_i c_1) + Z_x \cdot \sqrt{1-\rho^2} \cdot \sqrt{\sigma_{LSV}^2 (x_i, t_i) v_i \Delta} \tag{4.14}$$

where $Z_x \sim N(0, 1)$, $c_1 = \kappa\Delta - 1$.

4.4 Empirical Results

For the implementations of LSV model, one strives to solve the FPE accurately with low calibration errors w.r.t the market prices of vanilla options. In our empirical results, the low calibration errors for LSV model are achieved, which demonstrate the effectiveness of the methods presented in this paper. Moreover, we also compare the pricing results of reverse knock-out barrier options using the LV, the SV, and the LSV respectively. Among the three models, the price derived from the LSV model is the closest one to the market prices.

As a representative example, we calibrate the LV model (Dupire model), the SV model (Heston model), and the LSV model (as described above, its SV part is Heston-type) from market data in June 22, 2016 (data source: Bloomberg Terminal). Both the calibrated model parameters and calibration errors for the three models are discussed as following.

The implied volatility market data is shown in Table 4.1, while in Table 4.2 we present the calibrated implied volatilities of LV model with corresponding errors in the bracket. One can see that the calibration errors are very small, suggesting that the LV model is able to provide a perfect fit to the quoted market implied volatilities, as stated in Sect. 4.1. Theoretically the errors can be zero, however in practice there are usually some small errors remained when numerical methods are used. The model parameter, i.e. leverage surface $\sigma_{LV}(S_t, t)$ in the LV model is shown in Fig. 4.2.

The calibrated implied volatilities of the Heston model, with corresponding errors in the bracket, is shown in Table 4.3. Comparing Tables 4.2 and 4.3, we can find that the calibration errors for the Heston model are larger than LV model, which demonstrates that the Heston model is not able to fit the implied volatility surface perfectly as the LV does, as stated in Sect. 4.1. The corresponding model parameters for Heston model is shown in Table 4.4.

In Table 4.5, we present the calibrated implied volatilities of the LSV model with corresponding errors in the bracket. Comparing Tables 4.2, 4.3 and 4.5, one can find that the LSV model and the LV model can achieve much lower calibration errors than

Table 4.1 EUR/USD market implied volatility (in%)

Maturity	10-Delta put	25-Delta put	ATM	25-Delta call	10-Delta call
1W	22.554	19.756	17.333	15.944	15.531
2W	17.814	15.585	13.505	12.42	12.111
3W	16.466	14.176	12.217	11.304	11.334
1M	15.135	13.334	11.555	10.676	10.375
6W	14.463	12.744	11.049	10.231	10.023
2M	13.304	11.725	10.175	9.465	9.416
3M	12.894	11.298	9.855	9.302	9.416
4M	12.897	11.272	9.841	9.315	9.475
5M	12.901	11.243	9.825	9.33	9.542
6M	12.905	11.215	9.81	9.345	9.61
9M	12.79	11.088	9.733	9.32	9.662
1Y	12.666	10.951	9.65	9.294	9.719
18M	12.58	10.971	9.793	9.519	9.94
2Y	12.478	10.99	9.885	9.67	10.083

Table 4.2 Calibrated implied volatility of the LV model for EUR/USD (in%)

Maturity	10-Delta put	25-Delta put	ATM	25-Delta call	10-Delta call
1W	22.534[−0.020]	19.799[0.043]	17.255[−0.078]	15.919[−0.025]	15.525[−0.006]
2W	18.259[0.445]	16.007[0.422]	13.813[0.308]	12.726[0.306]	12.413[0.302]
3W	16.765[0.299]	14.481[0.305]	12.427[0.210]	11.513[0.209]	11.513[0.179]
1M	15.406[0.271]	13.579[0.245]	11.666[0.111]	10.799[0.123]	10.533[0.158]
6W	14.303[−0.160]	12.630[−0.114]	10.889[−0.160]	10.081[−0.150]	9.884[−0.139]
2M	13.550[0.246]	11.963[0.238]	10.355[0.180]	9.636[0.171]	9.575[0.159]
3M	12.960[0.066]	11.371[0.073]	9.866[0.011]	9.320[0.018]	9.434[0.018]
4M	12.867[−0.030]	11.288[0.016]	9.818[−0.023]	9.326[0.011]	9.475[0.000]
5M	12.915[0.014]	11.286[0.043]	9.858[0.033]	9.379[0.049]	9.585[0.043]
6M	12.944[0.039]	11.256[0.041]	9.799[−0.011]	9.347[0.002]	9.623[0.013]
9M	12.572[−0.218]	10.926[−0.162]	9.593[−0.140]	9.216[−0.104]	9.545[−0.117]
1Y	12.687[0.021]	10.975[0.024]	9.641[−0.009]	9.296[0.002]	9.723[0.004]
18M	12.736[0.156]	11.120[0.149]	9.883[0.090]	9.615[0.096]	10.042[0.102]
2Y	12.491[0.013]	11.003[0.013]	9.870[−0.015]	9.665[−0.005]	10.085[0.002]

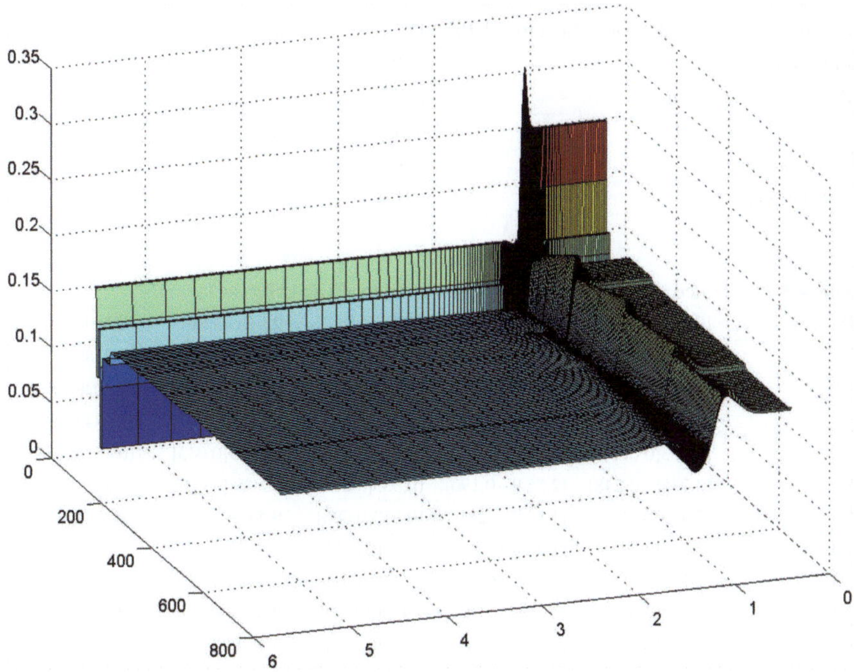

Fig. 4.2 Leverage surface in LV Model for EUR/USD

Table 4.3 Calibrated implied volatility of the heston model for EUR/USD (in%)

Maturity	10-Delta put	25-Delta put	ATM	25-Delta call	10-Delta call
1W	14.797[−7.757]	13.681[−6.075]	12.772[−4.561]	12.223[−3.721]	12.003[−3.528]
2W	14.819[−2.995]	13.558[−2.027]	12.515[−0.990]	11.926[−0.494]	11.752[−0.359]
3W	14.906[−1.560]	13.455[−0.721]	12.270[0.053]	11.648[0.344]	11.544[0.210]
1M	14.874[−0.261]	13.332[−0.002]	12.008[0.453]	11.369[0.693]	11.341[0.966]
6W	14.904[0.441]	13.146[0.402]	11.628[0.579]	10.989[0.758]	11.095[1.072]
2M	14.711[1.407]	12.797[1.072]	11.172[0.997]	10.566[1.101]	10.792[1.376]
3M	14.538[1.644]	12.384[1.086]	10.611[0.756]	10.063[0.761]	10.488[1.072]
4M	14.416[1.519]	12.107[0.835]	10.241[0.400]	9.742[0.427]	10.291[0.816]
5M	14.247[1.346]	11.828[0.585]	9.908[0.083]	9.455[0.125]	10.101[0.559]
6M	14.104[1.199]	11.629[0.414]	9.692[−0.118]	9.269[−0.076]	9.970[0.360]
9M	13.666[0.876]	11.171[0.083]	9.266[−0.467]	8.885[−0.435]	9.641[−0.021]
1Y	13.316[0.650]	10.878[−0.073]	9.036[−0.614]	8.664[−0.630]	9.418[−0.301]
18M	12.853[0.273]	10.565[−0.406]	8.831[−0.962]	8.460[−1.059]	9.167[−0.773]
2Y	12.535[0.057]	10.399[−0.591]	8.763[−1.122]	8.378[−1.292]	9.017[−1.066]

Table 4.4 Heston model parameters for EUR/USD

Initial variance	0.017
Mean-reverting speed κ	2.486
Mean-reverting level m	0.00953
Vol of variance α	0.57
Correlation ρ	−0.4

Table 4.5 Calibrated implied volatility of LSV model for EUR/USD (in%)

Maturity	10-Delta put	25-Delta put	ATM	25-Delta call	10-Delta call
1W	22.097[−0.457]	19.482[−0.274]	16.998[−0.335]	15.711[−0.233]	15.338[−0.193]
2W	17.868[0.054]	15.653[0.068]	13.520[0.015]	12.481[0.061]	12.203[0.092]
3W	16.502[0.036]	14.248[0.072]	12.240[0.023]	11.357[0.053]	11.384[0.050]
1M	15.233[0.098]	13.434[0.100]	11.555[0.000]	10.719[0.043]	10.463[0.088]
6W	14.215[−0.248]	12.554[−0.190]	10.833[−0.216]	10.056[−0.175]	9.888[−0.135]
2M	13.419[0.115]	11.853[0.128]	10.270[0.095]	9.575[0.110]	9.531[0.115]
3M	12.997[0.103]	11.425[0.127]	9.923[0.068]	9.387[0.085]	9.505[0.089]
4M	12.915[0.018]	11.352[0.080]	9.882[0.041]	9.391[0.076]	9.543[0.068]
5M	12.954[0.053]	11.355[0.112]	9.930[0.105]	9.449[0.119]	9.645[0.103]
6M	12.946[0.041]	11.279[0.064]	9.815[0.005]	9.359[0.014]	9.628[0.018]
9M	12.779[−0.011]	11.150[0.062]	9.788[0.055]	9.374[0.054]	9.679[0.017]
1Y	12.645[−0.021]	10.936[−0.015]	9.566[−0.084]	9.191[−0.103]	9.617[−0.102]
18M	12.500[−0.080]	10.879[−0.092]	9.606[−0.187]	9.313[−0.206]	9.741[−0.199]
2Y	12.414[−0.064]	10.916[−0.074]	9.735[−0.150]	9.492[−0.178]	9.899[−0.184]

the Heston model does. Theoretically, the LV model and the LSV model are more likely to achieve zero calibration errors, whereas the Heston model can't. In practice, there are still some small errors remained for the LV model and the LSV model due to numerical methods. Usually these numerical errors of the LSV model are larger than the LV model, because the LSV model involves more complex numerical methods than the LV model. In Table 4.5, the calibration errors for LSV model are very low, which demonstrate the effectiveness of the numerical methods presented in Sects. 4.2 and 4.3.

The model parameters for the SV part of the LSV model are acquired from the calibrated Heston model, except that the volatility of variance is multiplied by the mixing fraction parameter, which is set to 0.4 here. The model parameter, i.e. leverage surface $\sigma_{LSV}(S_t, t)$ in the LSV model is shown in Fig. 4.3.

As stated above, the key problem of the LSV model implementations is to solve the FPE accurately to get low calibration errors. The FPE (4.8) is about the transition probability p. To show the numerical stability, we export the time evolution of the transition probability p in Eq. (4.8) to Fig. 4.4. From Fig. 4.4, we can see that the evolution of transition probability is stable. It is noted that for numerical stability,

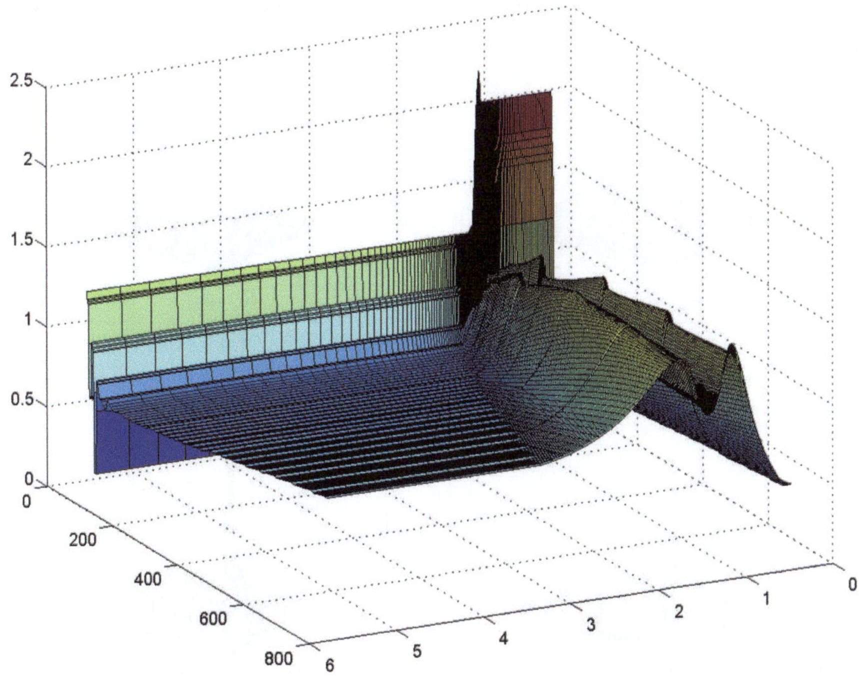

Fig. 4.3 Leverage surface in LSV Model for EUR/USD

we make the following transformation to spot s and variance v when calculating Eq. (4.8) numerically: $Y_t = \ln(S_t/S_0)$, $Z_t = \ln(V_t/V_0)$.

We also compare the pricing results of reverse knock-out barrier options, which are up-and-out single-barrier call options, and traded quite frequently in the market. The pricing method is the backward PDE introduced in Sect. 4.3. The prices of the three different models, as well as the market prices, are summarized in Table 4.6. The market prices are collected from Bloomberg. We can see that the LSV model provides the prices which are closest to the market prices.

4.5 Conclusion and Future Works

In this paper, we detail our implementations of a Heston-type LSV model. The model calibration is based on solving a Fokker–Planck Equation iteratively. For derivatives pricing, both the backward PDE method and Forward Monte Carlo method are introduced. In numerical results, the low calibration errors and relatively accurate pricing results demonstrate the effectiveness of the methods presented in this paper. For future works, the most important task is to improve the calibration stability. In our implementations, we face the similar problem described in Ait (2013): the calibration

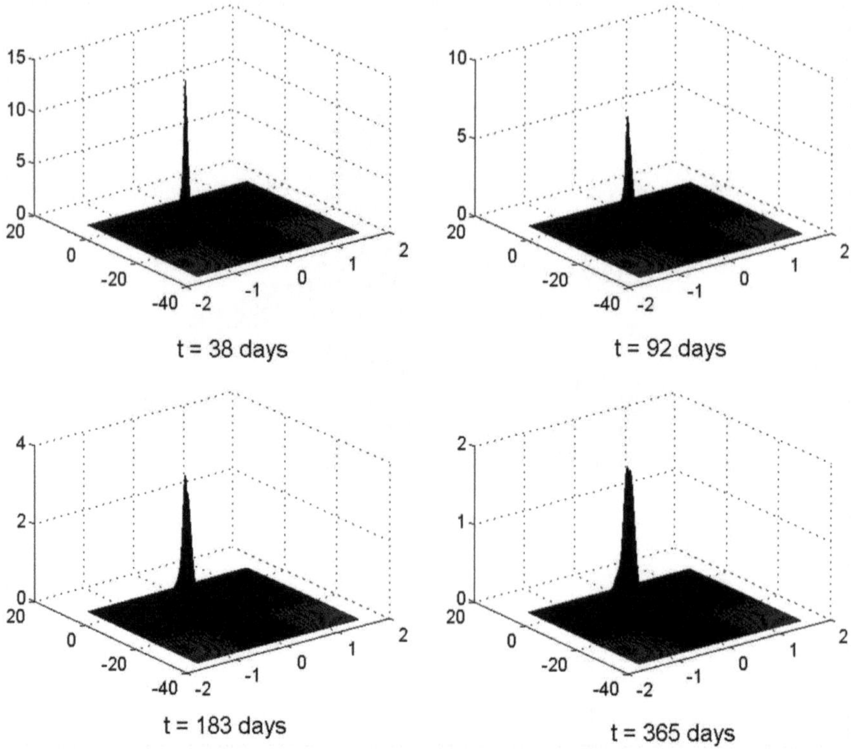

Fig. 4.4 Time evolution of the transition probability

Table 4.6 Pricing results of reverse knock-out barrier options

Tenor	Strike	Barrier	Heston	LV	LSV	Market price
1M	1.1269	1.15	1430	962	1054	1035
1M	1.1269	1.17	6001	5920	6072	5997
3M	1.1293	1.16	2262	922	1173	1087
3M	1.1293	1.2	10855	9287	9812	9901
6M	1.1331	1.19	6925	3328	4323	3883
6M	1.1331	1.24	16907	14170	15322	15102
1Y	1.1417	1.22	9787	4477	6103	5212
1Y	1.1417	1.3	25057	20601	22530	21936

becomes instable for large volatility-of-variance and longer maturity. Two ways are supposed to improve the calibration stability: one is to add a zero-flux boundary condition when solving the FPE (Lucic 2013; Gottker and Spanderen 2014); the other is to perform forward induction of backward PDE (Andreasen and Huge 2010).

References

Ait, Z. (2013). Alternate direction implicit method for a stochastic local volatility model: High performance computing platforms comparison.

Andersen, L. (2008). Simple and efficient simulation of the heston stochastic volatility model. *Journal of Computational Finance*, *11*, 1–42.

Andreasen, J., & Huge, B. N. (2010). Finite difference based calibration and simulation.

Black, F., & Scholes, M. (1973). The pricing of options and corporate liabilities. *Journal of Political Economy*, *81*, 637–654.

Clark, I. (2011). Foreign exchange option pricing: A practitioners guide. West Sussex: Wiley.

Cozzi, D. (2012). Local stochastic volatility models. Solving the smile problem with a nonlinear partial integro-differential equation.

Dupire, B. (1994). Pricing with a smile. *Risk Magazine*, *7*, 18–20.

Engelmann, B., Koster, F., & Oeltz, D. (2012). Calibration of the heston stochastic local volatility model: A finite volume scheme. Available online, Ecole Centrale Paris.

Fengler, M. (2005). Semi-parametric modeling of implied volatility. Lecture notes. Berlin: Springer.

Glasserman, P. (2003). *Monte Carlo methods in financial engineering*. New York: Springer.

Göttker, S. J., & Spanderen, K. (2014). Stochastic local volatility in quantlib.

Hagan, B., Dumar, D., Lesniewski, A., & Woodward, D. (2002). Managing smile risk. *Wilmott Magazine*, *7*, 84–108.

Heston, S. (1993). A closed-form solution for options with stochastic volatility with applications to bond and currency options. *Review of Financial Studies*, *6*, 327–343.

Hout, K. J., & Foulon, S. (2010). ADI finite difference schemes for option pricing in the heston model with correlation. *International Journal of Numerical Analysis and Modeling*, *7*, 303–320.

Lucic, Z. (2013). Alternate direction implicit method for a stochastic local volatility model: High performance computing platforms comparison.

Ren, Y., Dilip, M., & Qian, M. (2007). Calibrating and pricing with embedded local volatility models. *Risk Magazine*, *20*, 138.

Tachet, R. (2011). Non-parametric model calibration in finance. Ph.D. thesis, Ecole Centrale Paris.

Tataru, G., & Fisher, T. (2010). *Stochastic local volatility*. Bloomberg.

Tian, Y. (1993). Hybrid stochastic-local volatility model with applications in pricing FX options. Ph.D. thesis, Monash University.

Van, A., Gryelak, L., & Oosterlee, C. (2014). The heston stochastic-local volatility model: efficient monte carlo simulation. *International Journal of Theoretical and Applied Finance*.

Part II
Credit Risk

Chapter 5
Estimating Distance-to-Default with a Sector-Specific Liability Adjustment via Sequential Monte Carlo

J.-C. Duan and W.-T. Wang

Abstract Distance-to-Default (DTD), a widely adopted corporate default predictor, arises from the classical structural credit risk model of Merton (1974). The modern way of estimating DTD applies the model on an observed time series of equity values along with the default point definition made popular by the commercial KMV model. It is meant to be a default trigger level one year from the evaluation time, and is assumed to be the short-term debt plus 50% of the long-term debt. This default point assumption, however, leaves out other corporate liabilities, which can be substantial and particularly so for financial firms. Duan et al. (2012) rectified it by adding other liabilities after applying an unknown but estimable haircut. Typical DTD estimation uses a one-year long daily time series. With at most four quarterly balance sheets, the estimated haircut is bound to be highly unstable. Post-estimation averaging of the haircuts being applied to a sector of firms is thus sensible for practical applications. Instead of relying on post-estimation averaging, we assume a common haircut for all firms in a sector and devise a novel density-tempered expanding-data sequential Monte Carlo method to jointly estimate this common and other firm-specific parameters. Joint estimation is challenging due to a large number of parameters, but the benefits are manifold, for example, rigorous statistical inference on the common parameter becomes possible and estimates for asset correlations are a by-product. Four industry groups of US firms in 2009 and 2014 are used to demonstrate this estimation method. Our results suggest that this haircut is materially important, and

J.-C. Duan (✉)
Business School, Risk Management Institute, and Department of Economics,
National University of Singapore, Singapore, Singapore
e-mail: bizdjc@nus.edu.sg

W.-T. Wang
Risk Management Institute, National University of Singapore,
Singapore, Singapore
e-mail: rmiww@nus.edu.sg

W.K. Härdle et al. (eds.), *Applied Quantitative Finance*, Statistics and Computing,
DOI 10.1007/978-3-662-54486-0_5

varies over time and across industries; for example, the estimates are 78.97% in 2009 and 66.4% in 2014 for 40 randomly selected insurance firms, and 0.76% for all 31 engineering and construction and 83.92% for 40 randomly selected banks in 2014.

5.1 Introduction

Corporate credit risk is a common concern for all financial institutions due to their natural exposures to firms through lending activities. From the perspective of banks, the Basel Capital Accord and its compliance adds further importance to modeling credit risks. The investment community cares about corporate credit risk too due to potential losses to their portfolios. Policy makers/regulators also pay a great deal of attention to corporate credit risk because of the destabilizing effect on the economy/markets when massive corporate defaults occur. Since the seminal credit risk model of Merton (1974), viewing corporate capital structure as an option-like arrangement has gained a wide acceptance in assessing corporate default probabilities. Typically, fundamental information from the balance sheet and equity prices from the stock market are utilized in estimating the model. A particularly important risk measure out of Merton's model is distance-to-default (DTD), whose practical usage has been made popular by the commercial KMV model.

DTD is a widely adopted corporate default predictor. Its empirical estimate is typically obtained by using an observed time series of equity values along with some capital structure attributes. For practical applications, a typically complex capital structure must be simplified. This is usually done through the default point definition made popular by the KMV model. The default point is meant to be the default trigger level one year from the evaluation time, and the KMV default point, according to (Crosbie and Bohn, 2003), equals short-term debt plus 50% of the long-term debt. This default point definition, however, leaves out a firm's other liabilities, which can be substantial and particularly so for financial firms. Duan et al. (2012) proposed to add to the default point all remaining liabilities subject to a haircut, and estimated this haircut by applying the transformed-data maximum likelihood method of Duan (1994, 2000). In typical applications involving one-year long daily time series, only four quarterly balance sheets are available, which offer limited information in identifying the haircut. Thus, averaging the estimates for firms in the same corporate sector and then applying the same haircut to all firms in a two-stage estimation seems to be a sensible and practical solution. The two-stage approach has in fact been adopted by the Credit Research Initiative's live corporate default prediction system at the Risk Management Institute, National University of Singapore.

We propose a density-tempered expanding-data sequential Monte Carlo (SMC) method to estimate the haircut without relying on ad hoc averaging. This haircut is estimated jointly along with all other parameters for individual firms in the same corporate sector. This estimation task is technically challenging because of its high dimensionality (easily over one hundred parameters). Our method progressively adds a block of firms to the sample, and each time the likelihood function due to the additional data is density-tempered in a way that a somewhat arbitrary initial SMC sample of the parameters for these additional firms can be brought through a sequence of steps (reweighting, resampling and support boosting) to eventually arrive at a sample of parameters representing the distribution implied by the target likelihood.

Our method combines the two recently emerged SMC techniques: (1) density-tempered SMC by Del Moral et al. (2006) and Duan and Fulop (2015), and (2) expanding-data SMC by Chopin et al. (2013) and Fulop and Li (2013). Our method is not a simple combination of the two SMC techniques, however. Expanding data in our context is to increase the number of firms as opposed to increasing the number of observations on the same set of firms, and thus it is accompanied by an increase in the number of parameters. The second key difference is our frequentist interpretation of the estimation problem as in Chernozhukov and Hong (2003), and for which we in effect assume an improper prior, meaning that all parameters are treated equally likely before seeing the data. On the methodological front, our innovation is to do away with the need for a prior distribution in the sequential technique, which is accomplished by introducing a somewhat arbitrary but sensible initialization sampler with an analytical density function; for example, multivariate normal or truncated normal when some parameters are subject to domain restrictions. The density associated with this initialization sampler is then absorbed into the importance weight.

Joint estimation with this density-tempered expanding data SMC method is demonstrated with four sectors of US firms (insurance, banks, airlines, and engineering and construction) in 2009 and 2014, respectively. Our results suggest that this haircut is materially important, for example, the estimate is 52.19% for all 37 Engineering and Constructions in 2009 and 83.92% for 40 randomly selected banks in 2014. Joint estimation also yields estimates materially different from those obtained with the two-stage estimation method; for example, 92% for banks in 2009 under the former versus 72.61% under the latter, and the difference is way outside the 95% confidence interval obtained with the SMC method.

In addition to its methodological rigor, joint estimation has another advantage of generating asset correlations among members of a corporate sector. For example, banks and insurers show a significantly heightened level of asset correlations in 2009 as compared to 2014, which is consistent with 2009 being in the midst of a global financial crisis. For the airlines and engineering and construction sectors, a similar pattern exists but the magnitude of the difference in asset correlations are far less pronounced.

The DTD estimates generated by the two-stage method are sometimes comparable to those by our joint estimation method, for example, the engineering and construction industry in both years. For banks, however, the DTDs from the two methods are quite different. The magnitude aside, the correlations (Kendall or Pearson) between the estimates of the DTDs from the two methods exceed 80% except for banks which exhibit substantial but lower correlations as compared to other sectors.

5.2 DTD Subject to a Sector-Specific Liability Adjustment

Typical DTD estimation using a time series of equity values is performed on a firm-by-firm basis. If a corporate sector is to share a common parameter, estimation will require one to stack together all equity time series in that sector in order to reflect asset correlations among firms. To address asset correlations, we modify the Merton (1974) model by incorporating a latent common risk factor for the sector. This modification will, however, retain the Merton model's original results on a firm-by-firm basis.

5.2.1 The Structural Credit Risk Model with a Common Liability Adjustment

Let $V_{i,t}$ be the unobserved asset value of firm i at time t. Per usual, it follows a geometric Brownian motion, but we assume a common factor to allow for asset correlations:

$$\frac{dV_{i,t}}{V_{i,t}} = \mu_i dt + \beta_i dB_t^c + \nu_i dB_{i,t} \tag{5.1}$$

where B_t^c and $B_{i,t}$ are two independent standard Brownian motions, β_i is the firm specific coefficient used to capture how firm i responds to the common risk factor, B_t^c, and ν_i is a volatility coefficient to reflect the idiosyncratic risk of firm i. The total variance naturally becomes $\sigma_i^2 = \beta_i^2 + \nu_i^2$. Let F_{it} denote the default point at time T below which firm i will default, and $F_{i,t}$ is known at time t. The Merton (1974) model gives rise to the following equity value of firm i:

$$E_{i,t} = V_{i,t}\Psi(d_{i,t}) - F_{i,t}e^{-r(T-t)}\Psi(d_{i,t} - \sigma_i\sqrt{T-t}) \tag{5.2}$$

where $\Psi(.)$ is the standard normal cumulative distribution function and

$$d_{i,t} = \frac{\ln\left(\frac{V_{i,t}}{F_{i,t}}\right) + \left(r + \frac{\sigma_i^2}{2}\right)(T-t)}{\sigma_i\sqrt{T-t}}. \tag{5.3}$$

The time-t probability of default equals $\Psi(-DTD_{i,t})$, where

$$DTD_{i,t} = \frac{\ln\left(\frac{V_{i,t}}{F_{i,t}}\right) + \left(\mu_i - \frac{\sigma_i^2}{2}\right)(T-t)}{\sigma_i\sqrt{T-t}}. \tag{5.4}$$

The above DTD formula is, however, rarely used in practice because parameter μ is well known to be subject to huge sampling errors when daily time series is used in estimation. A modified DTD formula avoiding μ is typically used in practice; for example, Crosbie and Bohn (2003) and Duan and Wang (2012). This modified formula has also been adopted by the live corporate default prediction system of the Credit Research Initiative at the Risk Management Institute, National University of Singapore (NUS-RMI 2015). Specifically, this modified formula, denoted by DTD* is:

$$DTD_t^* := \frac{\ln\left(\frac{V_{i,t}}{F_{i,t}}\right)}{\sigma_i\sqrt{T-t}} \tag{5.5}$$

Following Duan et al. (2012), the default point is assumed to be sector-specific; that is, $F_{i,t} = SD_{i,t} + 0.5LD_{i,t} + \delta OL_{i,t}$ where the the short term debt ($SD_{i,t}$) is taken as total, the long term debt ($LD_{i,t}$) is halved, and other liabilities ($OL_{i,t}$) is subject to a unknown haircut common to all firms in the industry sector. This default point formula reduces to the KMV model's default point definition when $\delta = 0$. The ideal behind the KMV default point is a recognition that the debts of a firm typically cover a wide range of maturities, and a simple way of adapting the reality to the single-maturity set-up of the Merton model is to apply a 50% haircut to the longer-term debts. As noted in Duan et al. (2012) and further elaborated in Duan and Wang (2012), financial firms tend to have an extremely large amount of other liabilities vis-a-vis short-term and long-term debts (e.g., deposits for banks and policy obligations for insurers can amount to about 80% of their total liabilities). Thus, leaving other liabilities out of the default point will significantly distort the DTD estimate. However, the appropriate haircut is unknown and has to be estimated.

Estimating a firm-specific δ is not a sensible approach, because corporate balance sheets are available at best quarterly. The typical application of using one-year time series of daily equity values only offers three change points in liabilities, leading to a highly noisy estimate of δ. Common δ for a corporate sector is obviously a sensible compromise, but the joint estimation becomes too numerically challenging. Thus, Duan et al. (2012) employed a two-stage approach, which first estimates δ along with other model parameters for each firm in a sector, then averages all δ estimates in the sector, and finally fixing at the average δ, re-estimates other parameters for each firm in the sector. As mentioned earlier, this two-stage approach has also been adopted for the live corporate default prediction system maintained by the CRI team at the Risk Management Institute, National University of Singapore. We show later that joint estimation with all firms in a sector, instead of the two-stage approach, is actually feasible by adapting the modern density-tempered SMC technique to this specific estimation problem.

5.2.2 *The Transformed-Data Likelihood*

Duan (1994, 2000) proposed the transformed-data maximum likelihood estimation method for estimating parameters using derivative contract while the asset values are not directly observable. We apply the method to our joint estimation problem. Let $\hat{V}_{i,t}(\sigma_i, \delta)$ denote the implied asset value computed at (σ_i, δ) using the observed equity value, $E_{i,t}$, where the inverse exists and is unique, because Eq. (5.2) is monotonically increasing in $V_{i,t}$. By the process in Eq. (5.1), the N-firm one-period joint distribution at time $t-1$ is of multivariate normality with mean vector $\boldsymbol{\mu}_{1:N}$ and covariance matrix $\boldsymbol{\Sigma}_{1:N,1:N}$:

$$\begin{bmatrix} \ln(\frac{\hat{V}_{1,t}(\sigma_1,\delta)}{\hat{V}_{1,t-1}(\sigma_1,\delta)}) \\ \ln(\frac{\hat{V}_{2,t}(\sigma_2,\delta)}{\hat{V}_{2,t-1}(\sigma_2,\delta)}) \\ \vdots \\ \ln(\frac{\hat{V}_{N,t}(\sigma_N,\delta)}{\hat{V}_{N,t-1}(\sigma_N,\delta)}) \end{bmatrix} \sim \Psi(\boldsymbol{\mu}_{1:N}, \boldsymbol{\Sigma}_{1:N,1:N}). \tag{5.6}$$

where

$$\boldsymbol{\mu}_{1:N} = \begin{bmatrix} \mu_1 - \frac{1}{2}\sigma_1^2 \\ \mu_2 - \frac{1}{2}\sigma_2^2 \\ \vdots \\ \mu_N - \frac{1}{2}\sigma_N^2 \end{bmatrix} \quad \text{and} \quad \boldsymbol{\Sigma}_{1:N,1:N} = \begin{bmatrix} \beta_1^2+\nu_1^2 & \beta_1\beta_2 & \cdots & \beta_1\beta_N \\ \beta_2\beta_1 & \beta_2^2+\nu_2^2 & \cdots & \beta_2\beta_N \\ \vdots & & \ddots & \vdots \\ \beta_N\beta_1 & \beta_N\beta_{12} & \cdots & \beta_N^2+\nu_N^2 \end{bmatrix}$$

Also evident from the above, changing the sign of $\{\beta_i, i = 1, 2\cdots, N\}$ all at once will not change the density function of the above system. For identification, therefore, one can impose a positive sign on any one of them, say, β_1, as long as it is not equal to zero.

As argued in Duan et al. (2012), a firm's asset value may change dramatically due to major investment and financing activities. Hence, the asset value implied from the observed equity value is better standardized using the corresponding book value of assets. This adjustment is to remove the scale effect so as to better capture the dynamics for the assets in place instead of reacting to jumps caused by capital structure changes. Let $\hat{\boldsymbol{W}}_{t,1:N} = [\ln(\hat{V}_{1,t}(\sigma_1,\delta)/A_{1,t}), \ln(\hat{V}_{2,t}(\sigma_2,\delta)/A_{2,t}), \cdots, \ln(\hat{V}_{N,t}(\sigma_N,\delta)/A_{N,t})]'$, where $A_{i,t}$ is book asset value of firm i at time t. The transformed-data log-likelihood function can be derived by taking into account the Jacobian of the transformation from equity value to asset value. We introduce $\boldsymbol{\theta}_{i:j} = \{(\mu_k, \beta_k, \nu_k), k = i, \cdots, j\}$ to stand for the set of the firm-specific parameters from Firm i to j inclusive. Note again that σ_k is a deduced parameter where $\sigma_k^2 = \beta_k^2 + \nu_k^2$.

For a time series sample of equity values on N firms over $t = 1, 2, \cdots, T$, denoted by $\boldsymbol{E}_{1:N}$, the log-likelihood function is

$$\begin{aligned} &\ln \mathcal{L}(\delta, \boldsymbol{\theta}_{1:N}; \boldsymbol{E}_{1:N}) \\ &= -\frac{N(T-1)}{2}\ln(2\pi) - \frac{T-1}{2}\ln(\det(\boldsymbol{\Sigma}_{1:N})) \\ &-\frac{1}{2}\sum_{t=2}^{T}\left(\Delta\hat{\boldsymbol{W}}_{t,1:N} - \boldsymbol{\mu}_{1:N}\right)' \boldsymbol{\Sigma}^{-1}{}_{1:N,1:N}\left(\Delta\hat{\boldsymbol{W}}_{t,1:N} - \boldsymbol{\mu}_{1:N}\right) \end{aligned}$$

$$-\sum_{t=2}^{T} \mathbf{1}'_N \hat{\boldsymbol{W}}_{t,1:N} - \sum_{t=2}^{T}\sum_{i=1}^{N} \ln \Psi \left(d_{i,t}(\hat{V}_{i,t}(\sigma_i, \delta), F_{i,t}(\delta), \sigma_i) \right) \tag{5.7}$$

In the above, we have made explicit some elements of the $d_{i,t}$ function defined in Eq. (5.3) so that it is understood as a function of those model parameters. Note that $\Delta\hat{\boldsymbol{W}}_{t,1:N} = \hat{\boldsymbol{W}}_{t,1:N} - \hat{\boldsymbol{W}}_{t-1,1:N}$ and $\mathbf{1}_N$ is an N-dimensional column vector of 1.

Note that directly inverting $\boldsymbol{\Sigma}_{1:N,1:N}$ would create a heavy computational burden when n is relatively large. Under our model specification, this matrix is easily invertible with the Sherman–Morrison formula. Specifically, $\boldsymbol{\Sigma}_{1:N,1:N}$ can be decomposed into the sum of a diagonal matrix $A = diag(\nu_1^2, \ldots, \nu_N^2)$ and the outer product of a column vector $\upsilon = [\beta_1, \ldots, \beta_N]'$ with itself. If $1 + \upsilon' A^{-1} \upsilon \neq 0$, then

$$\boldsymbol{\Sigma}_{1:N,1:N}^{-1} = \left(A + \upsilon\upsilon'\right)^{-1} = A^{-1} - \frac{A^{-1}\upsilon\upsilon' A^{-1}}{1 + \upsilon' A^{-1} \upsilon}. \tag{5.8}$$

Missing data invariably occur in the real-life data sample. In our case, missing data can occur due to some required items in the balance sheet are occasionally absent or stock prices are not available for some firms at some time points. The likelihood function in Eq. (5.7) can be modified to allow for missing data. Specifically, one adjusts the number of firms, i.e., N, according to data availability at time t; for example, there are s firms with missing data at time t. Once the remaining $N - s$ implied asset values are computed according to Eq. (5.2), the implied asset returns of these firms again follows a multivariate normal distribution with an $(N - s)$ sub-vector of $\boldsymbol{\mu}_{1:N}$ and an $(N - s)$ sub-matrix of $\boldsymbol{\Sigma}_{1:N,1:N}$. Since missing data may occur differently over time, the adjustment to the likelihood function in Eq. (5.7) will have to be time-dependent. To make the computer code run efficiently, it will be useful to first sequence those firms without missing data and follow by those with missing data. Particularly, firms with similar missing data patterns are better grouped together so that the likelihood function of multiple firms can be evaluated in a larger time block.

5.3 Parameter Estimation by the Density-Tempered Expanding-Data Sequential Monte Carlo

The number of parameters in the likelihood function can be quite large; for example, there were 327 banks in the US in December 2009 giving rise to 982 parameters. Even for the relatively small airlines industry, there were 12 firms in December 2014 totaling 37 parameters to be jointly estimated. The density-tempered expanding-data SMC seems to be the only practical way for estimating such large systems.

Our density-tempered expanding-data SMC method combines the two recently emerged SMC techniques: (1) density-tempered SMC by Del Moral et al. (2006) and Duan and Fulop (2015), and (2) expanding-data SMC by Chopin et al. (2013) and Fulop and Li (2013). The common thread in these methods is to find a bridge linking the prior to the posterior distribution in the Bayesian context of parameter estimation. In the case of density-tempering, the likelihood is raised to a power between 0 (corresponding to the prior) and 1 (corresponding to the posterior) so that by applying a simple self-adapted control, one can sure-footedly migrate from a set of parameter particles representing the prior to the final set of particles for the posterior. The expanding-data SMC in the language of Duan and Fulop (2015), on the other hand, creates a bridge by gradually adding data so that the sequence of intermediate posteriors, represented by different sets of parameter particles and corresponding to various intermediate likelihoods, eventually goes to the final posterior distribution. As argued and demonstrated in Duan and Fulop (2015), density-tempering is a far more stable SMC scheme than the expanding-data approach. In our case, expanding data gradually is because handling a large number of firms all at once is not necessary and in fact not ideal in the earlier stage of estimation due to the extra computational load involved. By sequentially expanding the data set, one in effect only approximately density-temper the incremental likelihood to ensure proper distribution migrations along the way.

Our method is not a simple combination of the two SMC techniques. Expanding data in our context is to increase the number of firms as opposed to increasing the number of observations on the same set of firms, and thus it is accompanied by an increase in the number of parameters. The second key difference is our frequentest interpretation of the estimation problem, and for which we in effect assume an improper prior, meaning that all parameters are treated equally likely before seeing the data. Our methodological innovation is to do away with the prior distribution, and is done by introducing a somewhat arbitrary but sensible initial sampler with an analytical density function; for example, multivariate normal or truncated normal when some parameters are subject to domain restrictions. The corresponding methodological change needed is to replace the likelihood function, used in density-tempering or expanding-data, with the ratio of the likelihood over the initialization density.

We first define the log-likelihood function for the new firms conditional on the firms already being added ($N_s < N_q$); that is,

$$\begin{aligned}
&\ln \mathcal{L}(\delta, \boldsymbol{\theta}_{1:N_q}; \hat{\mathbf{W}}_{t,N_s+1:N_q}, t = 1, \cdots, T \mid \boldsymbol{E}_{1:N_s}) \\
&= -\frac{(N_q - N_s)(T-1)}{2} \ln(2\pi) - \frac{T-1}{2} \ln\left(\det(\boldsymbol{\Sigma}_{N_s+1:N_q|N_s})\right) \\
&\quad -\frac{1}{2} \sum_{t=2}^{T} \left(\Delta \hat{\mathbf{W}}_{t,N_s+1:N_q} - \boldsymbol{\mu}_{t,N_s+1:N_q|N_s}\right)' \boldsymbol{\Sigma}^{-1}_{N_s+1:N_q|N_s} \\
&\qquad \left(\Delta \hat{\mathbf{W}}_{t,N_s+1:N_q} - \boldsymbol{\mu}_{t,N_s+1:N_q|N_s}\right) \\
&\quad -\sum_{t=2}^{T} \mathbf{1}'_{N_q-N_s} \hat{\mathbf{W}}_{t,N_s+1:N_q} - \sum_{t=2}^{T} \sum_{i=N_s+1}^{N_q} \ln \Psi\left(d_{i,t}(\hat{V}_{i,t}(\sigma_i, \delta), F_{i,t}(\delta), \sigma_i)\right)
\end{aligned} \tag{5.9}$$

where

$$\boldsymbol{\mu}_{t,N_s+1:N_q|N_s} = \boldsymbol{\mu}_{t,N_s+1:N_q} + \boldsymbol{\Sigma}_{N_s+1:N_q,1:N_s}\boldsymbol{\Sigma}^{-1}_{1:N_s,1:N_s}\left(\Delta\hat{\boldsymbol{W}}_{t,1:N_s} - \boldsymbol{\mu}_{t,1:N_s}\right) \tag{5.10}$$

$$\boldsymbol{\Sigma}_{N_s+1:N_q|N_s} = \boldsymbol{\Sigma}_{N_s+1:N_q,N_s+1:N_q} - \boldsymbol{\Sigma}_{N_s+1:N_q,1:N_s}\boldsymbol{\Sigma}^{-1}_{1:N_s,1:N_s}\boldsymbol{\Sigma}'_{N_s+1:N_q,1:N_s} \tag{5.11}$$

The above two items are respectively the covariance matrix and mean vector for the $(N_q - N_s)$-dimensional asset returns corresponding to the new firm block, conditional on the asset returns of the existing N_s firms.

Our model's parameters can be divided into to two groups – common (i.e., δ) and firm-specific (i.e., $\boldsymbol{\theta}_{1:N}$). We are interested in the recursive exploration of the sequence of intermediate distributions with the recursion associated with data expansion and density-tempering. The initialization sampler's density for the firm-specific parameters from Firm i to j is denoted by $I_0(\boldsymbol{\theta}_{i:j})$, whereas the one for the common parameter is denoted by $I_0(\delta)$. For the first block of N_1 firms, its initialization sampler is independent of $I_0(\delta)$ so that the joint sampling density is $I_0(\delta)I_0(\boldsymbol{\theta}_{1:N_1})$. The intermediate distribution (up to a proportional constant) while tempering the likelihood with γ is defined as

$$f_{N_1,\gamma}(\delta, \boldsymbol{\theta}_{1:N_1}; \boldsymbol{E}_{1:N_1}) \propto \left(\frac{\mathcal{L}(\delta, \boldsymbol{\theta}_{1:N_1}; \boldsymbol{E}_{1:N_1})}{I_0(\delta)I_0(\boldsymbol{\theta}_{1:N_1})}\right)^{\gamma} \times I_0(\delta)I_0(\boldsymbol{\theta}_{1:N_1}). \tag{5.12}$$

The term raised to power γ on the right-hand side of (5.12) is nothing but the importance weight in a sampling sense. Different SMC schemes depart in how the importance weight is controlled so as to obtain a quality sample to represent the final target distribution. Evidently, when $\gamma = 0$, $f_{N_1,0}(\delta, \boldsymbol{\theta}_{1:N_1}; \boldsymbol{E}_{1:N_1})$ is the initialization density. When $\gamma = 1$, $f_{N_1,1}(\delta, \boldsymbol{\theta}_{1:N_1}; \boldsymbol{E}_{1:N_1}) = \mathcal{L}(\delta, \boldsymbol{\theta}_{1:N_1}; \boldsymbol{E}_{1:N_1})$, which is the likelihood function, up to a proportional constant, for the data sample up to N_1 firms.

When a new block of firms is added (taking from N_s to N_{s+1} firms), it will be more efficient to take advantage of the knowledge about the common parameter already implied by the first N_s firms and the firm-specific parameters of these N_s firms conditional on the common parameter. In real applications, the common parameter, if it were implied solely by the newly added firms, might be quite different from the common parameter suggested by the first N_s firms. When new firms are added, an ideal re-initialization sampler for the common parameter and the firm-specific parameters of the first N_s firms will be a mixture distribution combining the updated distribution revealed by these firms and the original initialization distribution. Specifically, we use the mixture distribution: $I_s^{(m)}(\delta, \boldsymbol{\theta}_{1:N_s}) = [\lambda I_s(\delta) + (1-\lambda)I_0(\delta)]I_s(\boldsymbol{\theta}_{1:N_s} \mid \delta)$ where $I_s(\delta)$ and $I_s(\boldsymbol{\theta}_{1:N_s} \mid \delta)$ denotes the distribution of the common parameter and the firm-specific parameters conditional on the common parameter derived from the SMC sample of the first N_s firms. A natural way of sampling with the conditional distribution, $I_s(\boldsymbol{\theta}_{1:N_s} \mid \delta)$, is to run regressions of $\boldsymbol{\theta}_{1:N_s}$ on δ using the SMC sample already obtained.

The tempered distribution (up to a proportional constant) when reaching N_s firms (for $s \geq 2$) is defined as

$$f_{N_s,\gamma}(\delta, \boldsymbol{\theta}_{1:N_s}; \boldsymbol{E}_{1:N_s})$$
$$\propto \left(\frac{\mathcal{L}(\delta, \boldsymbol{\theta}_{1:N_{s-1}}; \boldsymbol{E}_{1:N_{s-1}})\mathcal{L}(\delta, \boldsymbol{\theta}_{N_{s-1}+1:N_s}; \hat{\boldsymbol{W}}_{t,N_{s-1}+1:N_s}, t = 1, \cdots, T \mid \boldsymbol{E}_{1:N_{s-1}})}{I_{s-1}^{(m)}(\delta, \boldsymbol{\theta}_{1:N_{s-1}})I_0(\boldsymbol{\theta}_{N_{s-1}+1:N_s})} \right)^{\gamma}$$
$$\times I_{s-1}^{(m)}(\delta, \boldsymbol{\theta}_{1:N_{s-1}})I_0(\boldsymbol{\theta}_{N_{s-1}+1:N_s}). \tag{5.13}$$

The initialization sampler for the firm-specific parameters associated with the newly added firms naturally uses the original initialization sampler.

The terms raised to power γ on the right-hand side of (5.13) is again the importance weight in a sampling sense, controlling sample migration from an initial distribution to the target distribution. If one can obtain a simulated sample of parameter values properly representing $\mathcal{L}(\delta, \boldsymbol{\theta}_{1:N_s}; \boldsymbol{E}_{1:N_s})$, this Bayesian posterior with an improper prior, i.e., the likelihood function, shall converge to the asymptotic distribution. Hence, their sample means become the parameter estimates, and the confidence intervals can be straightforwardly obtained. Alternatively, one can use the result of Chernozhukov and Hong (2003) to justify the use of the SMC sample means and covariances in inference because the information equality holds when the correctly specified likelihood function is the target.

Advancing the tempered density will experience two cases. For the initial set of firms (i.e., N_1), moving γ to 1 can be accomplished by applying the following incremental important weight:

$$\frac{f_{N_1,\gamma^{(2)}}(\delta, \boldsymbol{\theta}_{1:N_1}; \boldsymbol{E}_{1:N_1})}{f_{N_1,\gamma^{(1)}}(\delta, \boldsymbol{\theta}_{1:N_1}; \boldsymbol{E}_{1:N_1})} \propto \left(\frac{\mathcal{L}(\delta, \boldsymbol{\theta}_{1:N_1}; \boldsymbol{E}_{1:N_1})}{I_0(\delta)I_0(\boldsymbol{\theta}_{1:N_1})} \right)^{\gamma^{(2)}-\gamma^{(1)}} \tag{5.14}$$

Advancing from N_{s-1} to N_s firms ($s \geq 2$) can be executed with the following incremental importance weight:

$$\frac{f_{N_s,\gamma^{(2)}}(\delta, \boldsymbol{\theta}_{1:N_s}; \boldsymbol{E}_{1:N_s})}{f_{N_s,\gamma^{(1)}}(\delta, \boldsymbol{\theta}_{1:N_s}; \boldsymbol{E}_{1:N_s})} \propto$$
$$\left(\frac{\mathcal{L}(\delta, \boldsymbol{\theta}_{1:N_{s-1}}; \boldsymbol{E}_{1:N_{s-1}})\mathcal{L}(\delta, \boldsymbol{\theta}_{N_{s-1}+1:N_s}; \hat{\boldsymbol{W}}_{t,N_{s-1}+1:N_s}, t = 1, \cdots, T \mid \boldsymbol{E}_{1:N_{s-1}})}{I_{s-1}^{(m)}(\delta, \boldsymbol{\theta}_{1:N_{s-1}})I_0(\boldsymbol{\theta}_{N_{s-1}+1:N_s})} \right)^{\gamma^{(2)}-\gamma^{(1)}} \tag{5.15}$$

While maintaining a minimum effective sample size by a self-adaptive control on γ, one must resample the parameters to even the important weights, and then follow up with several Metroplis-Hastings (MH) moves to boost the empirical support that has been reduced due to resampling. At any stage of (N_s, γ), the MH move targets $f_{N_s,\gamma}(\delta, \boldsymbol{\theta}_{1:N_s}; \boldsymbol{E}_{1:N_s})$ and replaces, if accepted, a subset of $(\delta, \boldsymbol{\theta}_{1:N_s})$. In fact, we need to run block MH moves, because proposing a good-quality parameter vector of a very high dimension without dividing them into blocks would be difficult. We first replace the common parameter, δ, and then proceed to replace firm-specific parameters sequentially according to how blocks of firms are added. Suppose that we have 23 firms and 5 firms are added at a time. The MH moves will comprise first proposing δ for replacement, then 15 parameters associated with the first 5 firms, then another 15 parameters for next 5 firms, and finally the last block of 9 parameters for 3 firms.

We compute the realized acceptance rates for the common parameter and each block of the firm-specific parameters after completing the MH move for the M parameter particles. The MH move will be repeated for the common parameter and blocks of firm-specific parameters, but a particular element (i.e., the common parameter or a block of firm-specific parameters) will be skipped when its cumulative realized acceptance rate has reached a target level, say, 100%. This is to ensure that the empirical support has been properly boosted but without running excessive MH moves.

A suitable proposal sampler for the common parameter or any block of firm-specific parameters is fairly easy to come by, and is typically of high quality. This is because a sample of size, say, M representing $f_{N_s,\gamma}(\delta, \boldsymbol{\theta}_{1:N_s}; \boldsymbol{E}_{1:N_s})$ is already available. The proposal density for the common parameter, δ, is defined as a linear regression model with normally distributed errors on a subset of m parameters, denoted by $\{\theta_1, \theta_2, \cdots, \theta_m\}$, randomly selected from the firm-specific parameters, $\boldsymbol{\theta}_{1:N_s}$; that is, δ^* is sampled based on the following regression model estimated to the parameter sample of size M:

$$\delta = a_0 + \sum_{j=1}^{m} a_j \theta_j + \epsilon, \quad \text{where} \quad \epsilon \sim N(0, \omega). \tag{5.16}$$

Naturally, a sampled δ should be discarded if it is outside of the [0, 1] interval.

For the firm-specific parameters, the proposal sampler is based on a set of regression models. Consider replacing the firm-specific parameters of a block of firms from $N_a + 1$ to N_b when the estimation has already been advanced to N_s firms. For each k between $N_a + 1$ and N_b, we use δ as regressor and estimate the following set of regressions:

$$\begin{aligned} \mu_k &= b_{k,0} + b_{k,1}\delta + \epsilon_{1,k} \\ \beta_k &= c_{k,0} + c_{k,1}\delta + \epsilon_{2,k} \\ \nu_k &= d_{k,0} + d_{k,1}\delta + \epsilon_{3,k} \end{aligned} \tag{5.17}$$

where $\epsilon_{1,k}$, $\epsilon_{2,k}$ and $\epsilon_{3,k}$ are normally distributed with mean zeros and their covariance matrix is computed from the regression residuals. Over different k's, $(\epsilon_{1,k}, \epsilon_{2,k}, \epsilon_{3,k})$ are treated as independent. In short, the proposal sampler takes the three firm-specific parameters as correlated for a firm but independent across different firms in a replacement block.

The regression parameters in effect define the proposal sampler, and these regression parameters are a function of the parameter sample of size M. So, we will use

$\mathcal{M}_{\delta,\boldsymbol{\theta}_{1:N_s}}$ to stand for these sufficient statistics. The proposed new parameters are denoted by $(\delta^*, \boldsymbol{\theta}^*_{1:N_s})$. Since we only propose a subset each time, $(\delta^*, \boldsymbol{\theta}^*_{1:N_s})$ is same as $(\delta, \boldsymbol{\theta}_{1:N_s})$ except for a particular subset being proposed for replacement.

$$\begin{aligned}&\alpha_{N_s,\gamma}\{(\delta, \boldsymbol{\theta}_{1:N_1}) \Rightarrow (\delta^*, \boldsymbol{\theta}^*_{1:N_s})\}\\ &= \min\left(1, \frac{f_{N_s,\gamma}(\delta^*, \boldsymbol{\theta}^*_{1:N_s}; \boldsymbol{E}_{1:N_s})}{f_{N_s,\gamma}(\delta, \boldsymbol{\theta}_{1:N_s}; \boldsymbol{E}_{1:N_s})} \frac{h(\delta, \boldsymbol{\theta}_{1:N_s} | \mathcal{M}_{\delta,\boldsymbol{\theta}_{1:N_s}})}{h(\delta^*, \boldsymbol{\theta}^*_{1:N_s} | \mathcal{M}_{\delta,\boldsymbol{\theta}_{1:N_s}})}\right)\end{aligned} \tag{5.18}$$

By the standard argument, the target intermediate distribution in (5.13) is the stationary solution to the Markov kernel defined by the above acceptance probability. Note that we are using independent proposal, because $\mathcal{M}_{\delta,\boldsymbol{\theta}_{1:N_s}}$ reflects the whole sample of M parameter values as opposed to an individual element, $(\delta, \boldsymbol{\theta}_{1:N_s})$.

Operationally speaking, the MH acceptance probability falls into one of two cases, and each can be simplified differently.

Case (1): $s = 1$ when the operation is still on the first block of firms (i.e., $1 : N_1$)

The first ratio in (5.18) can be expressed as

$$\frac{f_{N_1,\gamma}(\delta^*, \boldsymbol{\theta}^*_{1:N_1}; \boldsymbol{E}_{1:N_1})}{f_{N_1,\gamma_1}(\delta, \boldsymbol{\theta}_{1:N_1}; \boldsymbol{E}_{1:N_1})} = \left(\frac{\mathcal{L}(\delta^*, \boldsymbol{\theta}^*_{1:N_1}; \boldsymbol{E}_{1:N_1})}{\mathcal{L}(\delta, \boldsymbol{\theta}_{1:N_1}; \boldsymbol{E}_{1:N_1})}\right)^{\gamma} \left(\frac{I_0(\delta^*)I_0(\boldsymbol{\theta}^*_{1:N_1})}{I_0(\delta)I_0(\boldsymbol{\theta}_{1:N_1})}\right)^{1-\gamma} \tag{5.19}$$

Case (2): $s \geq 2$ when one adds another block of k firms $(N_s = N_{s-1} + k)$

The first ratio in (5.18) can be expressed as

$$\begin{aligned}&\frac{f_{N_s,\gamma}(\delta^*, \boldsymbol{\theta}^*_{1:N_s}; \boldsymbol{E}_{1:N_s})}{f_{N_s,\gamma}(\delta, \boldsymbol{\theta}_{1:N_s}; \boldsymbol{E}_{1:N_s})}\\ &= \left(\frac{\mathcal{L}(\delta^*, \boldsymbol{\theta}^*_{1:N_s}; \boldsymbol{E}_{1:N_s})}{\mathcal{L}(\delta, \boldsymbol{\theta}_{1:N_s}; \boldsymbol{E}_{1:N_s})}\right)^{\gamma} \left(\frac{I^{(m)}_{s-1}(\delta^*, \boldsymbol{\theta}^*_{1:N_{s-1}})I_0(\boldsymbol{\theta}^*_{N_{s-1}+1:N_s})}{I^{(m)}_{s-1}(\delta, \boldsymbol{\theta}_{1:N_{s-1}})I_0(\boldsymbol{\theta}_{N_{s-1}+1:N_s})}\right)^{1-\gamma}\end{aligned} \tag{5.20}$$

Some of the above ratios may be further simplified to speed up calculation by utilizing the fact that $\boldsymbol{\theta}^*_{1:N_s}$ typically shares the same value with $\boldsymbol{\theta}_{1:N_s}$ over some initial segment of variable length. Assume that the firm-specific parameters to be replaced corresponds to the block of firms from $N_a + 1$ to N_b. Note that $\delta^* = \delta, \boldsymbol{\theta}^*_{1:N_a-1} = \boldsymbol{\theta}_{1:N_a-1}$, and $\boldsymbol{\theta}^*_{N_b+1:N_s} = \boldsymbol{\theta}_{N_b+1:N_s}$. Hence,

$$\frac{\mathcal{L}(\delta^*, \boldsymbol{\theta}^*_{1:N_s}; \boldsymbol{E}_{1:N_s})}{\mathcal{L}(\delta, \boldsymbol{\theta}_{1:N_s}; \boldsymbol{E}_{1:N_s})} = \frac{\mathcal{L}(\delta^*, \boldsymbol{\theta}^*_{1:N_s}; \hat{\boldsymbol{W}}^*_{t,N_a+1:N_s}, t = 1, \cdots, T \mid \boldsymbol{E}_{1:N_a})}{\mathcal{L}(\delta, \boldsymbol{\theta}_{1:N_s}; \hat{\boldsymbol{W}}_{t,N_a+1:N_s}, t = 1, \cdots, T \mid \boldsymbol{E}_{1:N_a})}.$$

Finally, the second ratio in (5.18) in connection with the proposal density can naturally be simplified because sampling is only for the firm-specific parameters pertaining to a specific block of firms and the densities for the parameters outside the block are never invoked.

To summarize, the whole density-tempered expanding-data SMC algorithm along with our specific implementation parameters goes as follows:

- **Step 1: Initialization**
Sample $(\delta, i = 1, 2, \cdots, M)$ according to the initialization density, $I_0(\delta)$, which is taken as a normal distribution with mean 0.5 and standard deviation 0.3, truncated to [0,1]. Similarly, sample $(\boldsymbol{\theta}^{(i)}_{1:N_1}, i = 1, 2, \cdots, M)$ for the first N_1 firms based on $I_0(\boldsymbol{\theta}_{1:N_1})$. We set the initial block size to 5 firms, i.e., $N_1 = 5$. $I_0(\boldsymbol{\theta}_{1:N_1})$ is a product of normal densities, and they are taken as *i.i.d.* across firms and over the three firm-specific parameters of a firm. For μ_i, the mean and standard deviation are set to 0.2 and 0.2, respectively. In the case of β_i, the mean and standard deviation are 0.15 and 0.05. β_1 is restricted to be positive because of the identification issue discussed earlier in Sect. 2.2, and its sampling is carried out with a truncated normal distribution. Finally for $\ln \nu_i$, the mean and standard deviation are set to $\ln(0.1)$ and 0.05, respectively. The initial sample is of course equally weighted, i.e., $1/M$, and M is set to 1,024.
- **Step 2: Reweighting and resampling**
Set $\gamma^{(0)} = 0$. Start from $j = 0$ and compute the tempered incremental importance weight:

$$w_{\gamma,\gamma^{(j)}}(\delta^{(i)}, \boldsymbol{\theta}^{(i)}_{1:N_1}) = \left(\frac{\mathcal{L}(\delta^{(i)}, \boldsymbol{\theta}^{(i)}_{1:N_1}; \boldsymbol{E}_{1:N_1})}{I_0(\delta^{(i)}) I_0(\boldsymbol{\theta}^{(i)}_{1:N_1})} \right)^{\gamma-\gamma^{(j)}}$$

and find γ^* such that the Effective Sample Size (ESS) is no less than B where B is set to $M/2 = 512$. This can be done with a simple grid search to find γ^* to meet the condition, which need not be exact. Note that ESS $= \frac{\left(\sum_{i=1}^M w_{\gamma,\gamma^{(j)}}(\delta^{(i)}, \boldsymbol{\theta}^{(i)}_{1:N_1})\right)^2}{\sum_{i=1}^M w^2_{\gamma,\gamma^{(j)}}(\delta^{(i)}, \boldsymbol{\theta}^{(i)}_{1:N_1})}$.
Resample with the incremental weights to obtain an equally weighted sample of size M.
- **Step 3: Support boosting**
If ESS $\geq 0.9M$, this support boosting step will be skipped. Otherwise, apply the Metropolis-Hastings (MH) move to remove duplicates so as to boost the empirical support (i.e., increase the ESS). Block MH moves are run per the earlier discussion. First, δ is replaced, and then firm-specific parameters $\boldsymbol{\theta}_{1:N_1}$ are replaced in blocks with k firms at a time, and k is set to 5. Compute the realized acceptance rates (over M) for the common parameter and different blocks of firm-specific parameters. The MH move will be repeated for the common parameter and blocks of firm-specific parameters, but a particular element (i.e., the common parameter or a block of firm-specific parameters) will be skipped when its cumulative realized acceptance rate has reached a target level of 100%.
- **Step 4: Advance γ to 1**
Set $\gamma^{(j+1)} = \gamma^*$. With the support-boosted sample in place, one computes the tempered incremental important weight and finds γ^* again as in Step 2. Reweight,

resample, and follow with support boosting according to the acceptance probability in (5.18). Repeat the operations until reaching $\gamma = 1$.

- **Step 5: Add more firms**
Add more firms to take from N_{s-1} to N_s, where $N_s = N_{s-1} + k$ and k is set to 5 unless less than 5 firms are left. Perform re-initialization by sampling δ using $I_{s-1}^{(m)}(\delta) = \lambda I_{s-1}(\delta) + (1-\lambda)I_0(\delta)$ and $\boldsymbol{\theta}_{1:N_{s-1}}$ from $I_{s-1}(\boldsymbol{\theta}_{1:N_{s-1}} \mid \delta)$, where λ is set at 0.8 and $I_{s-1}(\delta)$ is similar to the truncated normal sampler used in the initialization, i.e., $I_0(\delta)$, except for using the sample mean and variance of δ in the SMC sample up to N_{s-1} firms. Sampling $\boldsymbol{\theta}_{1:N_{s-1}}$ conditional on $\delta^{(i)}$ relies on the following three-dimensional multivariate regression:

$$\boldsymbol{\theta}_j = \boldsymbol{\eta}_{j,0} + \boldsymbol{\eta}_{j,1}\delta^{(i)} + \epsilon_j, \qquad \text{where } \epsilon \sim N(\mathbf{0}, \boldsymbol{\Lambda}_j) \text{ and } j = 1, 2, \cdots, N_{s-1}.$$

Independence across firms is assumed for this sampler, which means $I_{s-1}(\boldsymbol{\theta}_{1:N_{s-1}} \mid \delta)$ is a product of N_{s-1} three-dimensional multivariate normal densities. Again, β_1 must be restricted to be positive for the identification purpose. Thus, $\boldsymbol{\theta}_{1:1}$ is treated differently where its three elements are sampled only using their sample means and covariances obtained from the previous stage so as to avoid the complication arising from the point-specific truncation probability.
Finally, sample the additional parameters, $(\boldsymbol{\theta}_{N_{s-1}+1:N_s}, i = 1, 2, \cdots, M)$, using the initialization sampler $I_0(\boldsymbol{\theta}_{N_{s-1}+1:N_s})$, which are normally distributed independent across firms and over different parameters for a firm. Append it to $\boldsymbol{\theta}_{1:N_{s-1}}$ to become $\boldsymbol{\theta}_{1:N_s}$. Set $\gamma^{(0)} = 0$. Start from $j = 0$, and compute the incremental important weight as in Eq. (5.15):

$$v_{\gamma,\gamma^{(j)}}(\delta^{(i)}, \boldsymbol{\theta}_{1:N_s}^{(i)}) = \left(\frac{\mathcal{L}(\delta^{(i)}, \boldsymbol{\theta}_{1:N_{s-1}}^{(i)}; \boldsymbol{E}_{1:N_{s-1}})\mathcal{L}(\delta^{(i)}, \boldsymbol{\theta}_{1:N_s}^{(i)}; \hat{\boldsymbol{W}}_{t,N_{s-1}+1:N_s}, t = 1, \cdots, T \mid \boldsymbol{E}_{1:N_{s-1}})}{I_{s-1}^{(m)}(\delta^{(i)}, \boldsymbol{\theta}_{1:N_{s-1}}^{(i)}) I_0(\boldsymbol{\theta}_{N_{s-1}+1:N_s}^{(i)})} \right)^{\gamma - \gamma^{(j)}}$$

Find γ^* such that the ESS is no less than B, and follow with reweighting, resampling and support boosting again. Repeat until reaching $\gamma = 1$.

- **Step 6: Repeat adding more firms**
Repeat Step 5 to take N_s to N_{s+1} until finally reaching N firms.

5.4 Empirical Implementation

5.4.1 Data

We obtain the data from the RMI-CRI database (National University of Singapore, Risk Management Institute, CRI database. Available at: http://rmicri.org [Accessed August 2015]). The data include (1) the daily market capitalization based on closing share price and number of shares outstanding on a subset of US firms in four sectors,

(2) the 3-month US Treasury interest rate series, and (3) the book values of assets and liabilities (short-term, long-term and the remainder) from quarterly balance sheets for these US firms. Share prices and interest rates are available daily, but balance sheets are released quarterly. For a given day, the relevant items are taken from the most recently available quarterly balance sheet. The firms are classified into 76 industry groups by Bloomberg Industry Classification System (BICS). To demonstrate our estimation method for the common liability adjustment factor (i.e., δ), we select four industry groups: Insurance (BICS 10008-20055), Banks (BICS 10008-20051), Airlines (BICS 10004-20018), and Engineering and Construction (BICS 10011-20082) and focus on two years: 2009 and 2014. Our sample size is 250 daily observations for each firm up to the end of the year. According to the δ estimates produced by the RMI-CRI system in its first stage of the two-stage estimation, these four industry sectors show a range of δ's that helps in gaining a better understanding of our proposed method.

Table 5.1 presents the capital structures of these four industry sectors in 2009 and 2014. The firms considered must have consecutive data for at least 22 days in a year. The smallest number of firm is 12 for the airlines industry in 2014 whereas the largest sector is banks with 327 firms in 2009. Evidently from this table, other liabilities being left out of the KMV default point formula can be quite substantial, measured as a fraction of total liabilities. This is particularly so for financial firms such as insurers and banks with other liabilities being around 80% of the total liabilities. If the haircut, i.e., δ, is not negligible, DTD of financial firms will be seriously distorted.

As Table 5.1 shows, there are many banks and insurers in their respective sectors. In the following estimation, we randomly select 40 firms common to 2009 and 2014, and do so for each of these two sectors. In these cases, we in effect jointly

Table 5.1 Capital structure of four industry sectors of US firms

	Airlines		Engineering & Construction		Banks		Insurance	
	2009	2014	2009	2014	2009	2014	2009	2014
# of firms	18	12	37	31	327	312	132	120
Average value								
Market capitalization	1979.77	13428.39	1247.59	1950.87	3311.06	5826.30	4538.22	9657.12
Short-term debt (SD)	2317.60	4608.96	629.86	783.56	8657.02	9056.90	2433.34	3074.60
Long-term debt (LD)	3299.49	3863.44	152.52	502.10	6854.90	5607.53	2711.79	2110.10
Other liabilities (OL)	2673.93	3767.06	136.28	173.28	21408.08	29162.43	24998.16	34612.81
Total liabilities (TL)	8667.21	15495.66	1597.43	2300.46	40789.69	49114.01	35191.69	47934.07
Total assets (TA)	8291.01	12239.46	918.66	1458.94	36920.00	43826.85	30143.29	39797.51
OL/TL	24.10%	23.34%	11.32%	11.72%	83.84%	89.06%	79.16%	78.80%

estimate 121 parameters (1 common plus 40 sets of 3 firm-specific parameters for each firm). Going all the way to jointly estimate using, say, 327 banks in 2009 (close to 1,000 parameters) would be methodologically feasible, but would require a GPU parallel computing implementation to complete the estimation task within a reasonable amount of time.

5.4.2 *Results*

Table 5.2 presents the results of comparing the estimated haircuts from the density-tempered expanding-data SMC method with those from the two-stage approach. The number of firms refers to the firms used in the joint estimation, not the total number of firms in that sector; for example, banks and insurers are capped at 40. The data missing rate is computed as the ratio of the number of missing day-firm observations over the maximum number of day-firms in a particular year. Missing data causes some algorithmic complications. One missing equity value, for example, results in two consecutive missing returns. Missing returns can be easily handled when a single firm is involved. Jointly estimating all firms in a sector as in this paper requires making adjustments to the conditioning set along the time dimension in order to evaluate the conditional likelihood function in Eq. (5.9). To improve computational efficiency, one needs to arrange firms with similar missing data patterns into the same group, and then leaves groups with more missing data to later processing in the sequential optimization scheme.

For the two-stage estimation, the average δ of a sector is computed over the firms in a sector (or 40 firms in the banking or insurance sector) with the haircut values generated by RMI-CRI in its first stage of the two-stage estimation. Also reported and labelled as "Used by CRI" are the haircut actually employed by the RMI-CRI live system, which are averages over a very broad division into financial and non-

Table 5.2 The haircut parameter, δ, for four industry sectors in 2009 and 2014

	Airlines		Engineering & Construction		Banks		Insurance	
	2009	2014	2009	2014	2009	2014	2009	2014
# of firms used in estimation	18	12	37	31	40	40	40	40
Missing data rate	3.64%	1.93%	1.48%	1.26%	7.53%	6.69%	0.96%	0.18%
Two-stage estimation								
Average over firms	0.3493	0.3666	0.5990	0.5009	0.7261	0.6667	0.6262	0.3136
Used by RMI-CRI	0.5671	0.3537	0.5671	0.3537	0.6898	0.5417	0.6898	0.5417
Joint estimation by SMC								
Mean	0.1693	0.0074	0.5219	0.0076	0.9200	0.8392	0.7897	0.6640
$Q_{2.5}$	0.0826	0.0002	0.2847	0.0002	0.8532	0.8170	0.7479	0.6257
$Q_{97.5}$	0.2527	0.0251	0.7450	0.0277	0.9856	0.8627	0.8356	0.6985

financial sectors as opposed to more specific sub-sectors used in this study. The joint estimation results reported in the same table provide the point estimates for the haircut for different sectors in 2009 and 2014. Also presented in the table are upper and lower values of the 95% confidence interval. These confidence intervals suggest that only engineering and construction sector has their estimated haircuts in 2009 from the two-stage method to be statistically indistinguishable from their corresponding haircuts obtained under the joint estimation method.

Table 5.3 is used to highlight the difference in the firm-specific parameters. For the two-stage estimation method, there are only two parameters (μ and σ), and their sector average values in 2009 and 2014 are reported. In contrast, the joint estimation method yields β and ν estimates in addition to μ. Note that β and ν can be combined

Table 5.3 Firm-specific parameters for four industry sectors in 2009 and 2014

	Airlines		Engineering & Construction		Banks		Insurance	
	2009	2014	2009	2014	2009	2014	2009	2014
Two-stage estimation								
μ	0.0972	0.1527	0.1271	−0.0597	−0.0528	0.0084	−0.0273	−0.0226
σ	0.2722	0.2295	0.4631	0.2893	0.1258	0.0618	0.1963	0.1188
Joint estimation by SMC								
μ								
Mean	0.2548	0.1610	0.3405	0.0223	0.6243	0.0081	0.0795	0.0091
Median	0.2364	0.2384	0.1903	0.0368	0.2449	0.0117	0.0355	0.0101
Min	−0.0056	−0.3828	−0.2527	−0.3455	−0.1439	−0.0948	−0.0776	−0.1247
Max	0.5785	0.5093	2.8807	0.5819	3.0980	0.1101	0.7667	0.2155
β								
Mean	0.3719	0.1584	0.1989	0.1248	0.7237	0.0256	0.1624	0.0574
Median	0.3644	0.1710	0.2156	0.1343	0.4694	0.0271	0.1352	0.0455
Min	0.0164	0.0677	−0.1017	0.0227	−0.2294	−0.0048	0.0657	0.0123
Max	0.6770	0.2732	0.3928	0.2272	2.6869	0.0900	0.5246	0.1648
ν								
Mean	0.2480	0.1781	0.4041	0.2567	0.2361	0.0392	0.1508	0.0959
Median	0.2095	0.1392	0.3221	0.2100	0.1560	0.0315	0.0860	0.0650
Min	0.1181	0.1066	0.1305	0.0957	0.0523	0.0199	0.0404	0.0114
Max	0.5752	0.3404	1.3683	0.9328	1.0132	0.1143	0.4923	0.6636
Asset volatility: $\sigma = \sqrt{\beta^2 + \nu^2}$								
Mean	0.4717	0.2431	0.4809	0.2952	0.7821	0.0492	0.2308	0.1142
Median	0.4738	0.2246	0.3889	0.2639	0.4971	0.0445	0.1735	0.0803
Min	0.2314	0.1362	0.2158	0.1150	0.0679	0.0267	0.1007	0.0167
Max	0.7087	0.3988	1.3721	0.9362	2.8716	0.1455	0.7130	0.6838
Asset correlation								
Mean	0.6096	0.4310	0.2627	0.2375	0.7055	0.2476	0.5828	0.3557
Median	0.6870	0.4066	0.2613	0.2273	0.8217	0.2111	0.5986	0.3465
Min	0.0186	0.1815	−0.0641	0.0032	−0.8382	−0.0895	0.0517	0.0737
Max	0.9261	0.6897	0.7452	0.6323	0.9782	0.7290	0.8768	0.7298

to produce σ estimate and also asset correlations. For some sectors, two methods yield distinctively different σ estimates; for example, airlines and banks in 2009. In general, the σ estimates by the joint estimation method are higher than those obtained by the two-stage method. The summary statistics on asset correlations suggest that asset were much more correlated in 2009 as compared to 2014. This is in agreement with the common perception of increased correlations during the 2008–2009 global financial crisis period.

Table 5.4 summarizes the DTDs generated by two estimation methods for the four sectors in 2009 and 2014. The DTD estimates generated by the two-stage method are in some cases comparable to those by the joint estimation method; for example, the engineering and construction industry in both years. For banks, however, the DTD estimates from the two methods are quite different. Generally speaking, the two-stage method yields higher DTD estimates for all sectors in 2009, when markets were more volatile then. A higher DTD implies a higher solvency, and thus the two-stage method leads to a conclusion that firms were safer than they actually were. The magnitude aside, Kendall's τ or Pearson correlation of the two set of DTD estimates exceed 80% except for banks. The correlations for banks are much lower in magnitude but still substantial. Take together, we can conclude that the DTDs from two estimation methods are materially different. When used as a default predictor in a reduced-form model, different estimation methods likely yield different prediction performances. It is reasonable to conjecture that the joint estimation will generate a better default predictor, either judging intuitively from its characteristics over the financial crisis period or simply based on its methodological rigor.

Table 5.4 DTD comparison for four industry sectors in 2009 and 2014

	Airlines		Engineering & Construction		Banks		Insurance	
	2009	2014	2009	2014	2009	2014	2009	2014
Two-stage estimation (RMI-CRI values)								
Mean	1.5749	4.6966	2.7687	4.4324	0.8708	4.1698	2.2356	5.9887
Median	1.4447	4.2140	3.0445	3.9109	0.8303	4.1269	2.4128	5.5646
Min	−0.7147	2.8421	−0.0122	0.5205	−1.2278	0.9697	−0.4286	2.4827
Max	4.0182	7.4182	6.0303	12.8785	3.0753	7.8429	6.9532	11.3443
Joint estimation by SMC								
Mean	0.8738	4.7830	2.6743	4.1897	−0.3748	3.8005	1.7170	5.5910
Median	0.6886	4.4713	2.8580	3.6562	−0.5219	3.8501	1.9077	5.0447
Min	−0.7657	2.8093	−0.0433	0.4990	−1.4501	0.8166	−0.6529	2.3012
Max	3.2526	7.6093	5.8595	13.1316	2.0759	6.5182	6.3411	10.1488
Correlation of the two methods								
Kendall	0.8382	0.9091	0.9670	0.9901	0.6410	0.8063	0.9190	0.9568
Pearson	0.9635	0.9944	0.9992	0.9988	0.7841	0.9681	0.9889	0.9972

References

Chernozhukov, V., & Hong, H. (2003). An MCMC approach to classical estimation. *Journal of Econometrics*, *115*, 293–346.

Chopin, N., Jacob, P. E., & Papaspiliopoulos, O. (2013). SMC^2: a sequential Monte Carlo algorithm with particle Markov Chain Monte Carlo updates. *Journal of the Royal Statistical Society: Series B*, *75*, 397–426.

Crosbie, P., Bohn, J. 2003. Modeling Default Risk, Moodys KMV technical document.

Del Moral, P., Doucet, A., & Jasra, A. (2006). Sequential Monte Carlo samplers. *Journal of the Royal Statistical Society: Series B*, *68*(3), 411–436.

Duan, J.-C. (2000). Correction: maximum likelihood estimation using price data of the derivative contract. *Mathematical Finance*, *10*, 461–462.

Duan, J.-C. (1994). Maximum likelihood estimation using price data of the derivative contract. *Mathematical Finance*, *4*(2), 155–167.

Duan, J.-C., & Fulop, A. (2015). Density-tempered marginalized sequential Monte Carlo samplers. *Journal of Business and Economic Statistics*, *33*(2), 192–202.

Duan, J.-C., Sun, J., & Wang, T. (2012). Multiperiod corporate default prediction ? *A Forward Intensity Approach, Journal of Econometrics*, *170*(1), 191–209.

Duan, J.-C., & Wang, T. (2012). Measuring distance-to-default for financial and non-financial firms. *Global Credit Review*, *2*, 95–108.

Fulop, A., & Li, J. (2013). Efficient learning via simulation: a marginalized resample-move approach. *Journal of Econometrics*, *176*, 146–161.

Merton, R. (1974). On the pricing of corporate debt: the risk structure of interest rates. *Journal of Finance*, *29*, 449–470.

Staff, N. U. S.-R. M. I. (2015). NUS-RMI credit research initiative technical report version: 2015 update 1. *Global Credit Review*, *5*, 113–203.

Chapter 6
Risk Measurement with Spectral Capital Allocation

L. Overbeck and M. Sokolova

Abstract Spectral risk measures provide the framework to formulate the risk aversion of a firm specifically for each quantile of the loss distribution of a portfolio. More precisely the risk aversion is codified in a weight function, weighting each quantile. Since the basic coherent building blocks of spectral risk measures are expected shortfall measures, the most intuitive approach comes from combinations of those. For investment decisions the marginal risk or the capital allocation is the sensible approach. Since spectral risk measures are coherent there exists also a sensible capital allocation based on the notion of derivatives or more in the light of the coherency approach as an expectation under a generalized maximal scenario.

6.1 Introduction

Portfolio modeling has two main objectives: the quantification of portfolio risk, which is usually expressed as the economic capital of the portfolio, and its allocation to subportfolios and individual transactions. The standard approach in credit portfolio modeling is to define the economic capital in terms of a quantile of the portfolio loss distribution

$$q_\alpha(L) = F_L^{-1}(\alpha)$$

The capital charge of an individual transaction is traditionally based on a covariance technique and called volatility contribution. We refer to Bluhm et al. (2002) for a survey on credit portfolio modeling and capital allocation.

L. Overbeck (✉)
Institut für Mathematik, Universität Gießen, Giessen, Germany
e-mail: Ludger.Overbeck@math.uni-giessen.de

M. Sokolova
Royal Bank of Scotland and Imperial College London, London, UK
e-mail: Maria.Sokolova@rbs.com

W.K. Härdle et al. (eds.), *Applied Quantitative Finance*, Statistics and Computing,
DOI 10.1007/978-3-662-54486-0_6

Since the work by Artzner et al. (1997) coherent risk measures are discussed intensively in finance and risk management. More recent is the question of a more coherent capital allocation. Especially the use of expected shortfall allocation as an allocation rule is recommend in Overbeck (2000), Denault (2001), Bluhm et al. (2002), Kurth and Tasche (2003) and Kalkbrener et al. (2004).
Expected shortfall measures

$$ES_\alpha(L) = \frac{1}{1-\alpha}\int_\alpha^1 qu(L)du$$

are the building blocks of more general coherent risk measures, the spectral risk measure ρ. These are convex mixtures of expected shortfall measures. They can be represented by their spectral measure μ through

$$\rho = |\rho_\mu = \int_0^t ES_\alpha(1-\alpha)\mu(da) \tag{6.1}$$

or as a weighted sum of quantiles with $w(\alpha) = \mu([0,\alpha])$,

$$\rho = \rho_\mu = \rho_w = \int_0^1 q_\alpha(\cdot)w(\alpha)d\alpha. \tag{6.2}$$

In this paper we apply the allocation rules associated with a spectral risk measure to a credit portfolio and point out, which consequences to risk management the choice of the weight function w, the spectral measure μ or the measure

$$\tilde{\mu} \stackrel{\text{def}}{=} (1-\alpha)\mu(d\alpha),$$

which we call mixing measure and thought to be the most easily one to calibrate and implement. The theoretical basis of the approach can be found in the basic papers Kalkbrener (2002), Kalkbrener et al. (2004) and the explicit application to spectral capital allocation is provided by Overbeck (2004). We will first present the theoretical foundation of the proposed risk and allocation measures and then discuss general impact of the choice of the weight or mixing function and finally exhibits the differences on a concrete credit portfolio example.

6.2 Review of Coherent Risk Measures and Allocation

6.2.1 Coherent Risk Measures

It is well-known that the following four conditions define a coherent risk measure, Artzner et al. (1997, 1999), Delbaen (2000).

Formally, a risk measure is nothing else as a positive real valued function r defined on the set of random variable (potential losses) V. The number $r(X)$ denotes the risk in portfolio X. r is called coherent if it obeys the following 4 rules.

⊡ Subadditivity (Diversification)

$$r(X+Y) < r(X) + r(Y)$$

⊡ Positive homogenous (Scaling)

$$r(aX) = ar(X),\ a > 0$$

⊡ Monotone

$$r(X) < r(Y) \text{ if } X < Y \text{ (almost surely)}$$

⊡ Translation property

$$r(X+a) = r(X) - a$$

Convex analysis gives already that a sub-additive positive homogenous function r can be point wise written as the maximal value of all linear functions which are below r (Delbaen 2000; Kalkbrener 2002; Kalkbrener et al. 2004). For risk measures this means that the first two axioms above lead to the following representation

$$r(X) = \max\{l(X) | l < r,\ l \text{ linear function}\} \tag{6.3}$$

The risk measure evaluate at a loss variable X takes the same value as the largest value of all linear function which lies below r on V evaluated on X.

Conceptually, this is similar to the gradient of the function r evaluated at the point X or as the best linear approximation of r which coincides with r at the point X. We will later see that this intuition gives rise to a sensible capital allocation.

A typical linear function for random variable is the expectation operator. Hence the basic result by Artzner et al. (1997), Delbaen (2000)

$$r(X) = \sup\{\mathsf{E}_Q[X] | Q \in \mathcal{Q}\} \tag{6.4}$$

$\mathcal{Q}, = \mathcal{Q}_r$, a suitable set of probability measures of absolutely continuous probability measures $Q << P$ with density dQ/dP, is similar to the representation (6.3).

The set $\mathcal{Q}$ is called the generalized scenarios associated with r. If the supremum is actually taken at some probability measure, this probability measure or its density with respect to P is called the generalized scenario associated with r. These approach also fits into the intuitive feature of risk measurement, namely scenario or stress analysis. For the interpretation in terms of scenarios the formulation with probability measure is more natural, but for the axiomatic approach to capital allocation the representation (6.3) is very useful.

The currently most prominent example of a coherent risk measure is Expected Shortfall (sometimes called Conditional VaR /tail conditional expectation). It is denoted by ES_α and measures the average loss above the α-quantile of the loss distribution. The associated generalized scenarios can be explained as follows:

To each loss variable Y define the scenario as the "historical" calibrated objective scenario constraint on the condition that the loss variable exceeded its quantile. The expected shortfall coincides with the largest mean loss in these scenarios. Intuitively,

$$\mathsf{E}\{L|L > q_\alpha(L)\} = \max\{\mathsf{E}\{L|Y > q_\alpha(Y)\}| \text{ all } Y \in L_\infty\}$$

Even if generalized scenarios are defined as a supremum, in the case of Expected Shortfall we can identify the density of the maximal "scenario". For this we need the formally correct definition of Expected Shortfall at level α. The problem with the intuitive definition above is the possible positive mass at the quantile itself. The exact definition of the Expected Shortfall at level α is therefore Acerbi and Tasche (2002), Kalkbrener et al. (2004):

Definition 6.1

$$ES_\alpha(L) \stackrel{\text{def}}{=} (1-\alpha)^{-1}(\mathsf{E}[L\mathbf{1}\{L > \mathrm{q}_\alpha(L)\}] + \mathrm{q}_\alpha(L) \cdot [\mathrm{P}\{L \le \mathrm{q}_\alpha(L)\} - \alpha]).$$

Here we take the quantile defined by

$$q_u(L) = \inf\{x|P(L \le x) \ge u\}$$

the smallest u-quantile

Since $ES_\alpha(L) = \mathsf{E}\{L_{g\alpha}(L)\}$ with the function

$$g_\alpha(Y) \stackrel{\text{def}}{=} (1-\alpha)^{-1}[\mathbf{1}\{Y > \mathrm{q}_\alpha(Y)\} + \beta_Y \mathbf{1}\{Y = \mathrm{q}_\alpha(Y)\}], \tag{6.5}$$

where β_Y is a real number and

$$\beta_Y \stackrel{\text{def}}{=} \frac{\mathrm{P}\{Y \le \mathrm{q}_\alpha(Y)\} - \alpha}{\mathrm{P}\{Y = \mathrm{q}_\alpha(Y)\}} \text{ if } \mathrm{P}\{Y = \mathrm{q}_\alpha(Y)\} > 0.$$

the density of the associated maximal scenario turns out to be the function g_α. Note that $ES_\alpha(Y) = \mathsf{E}\{Y \cdot g(Y)\}$ and $ES_\alpha(X) \ge \mathsf{E}\{X \cdot g(Y)\}$ for every $X,\ Y \in V$.

6.2.2 Spectral Risk Measures

For the interpretation of this density function (6.5) in terms of risk aversion as outlined in Acerbi (2002), let us reformulate the expected shortfall as an integral over the quantile function, the inverse of the distribution of L. It is well-known that

$$ES_\alpha(L) = (1-\alpha)^{-1} \int_\alpha^1 qu(L)du.$$

The implicit risk aversion with expected shortfall is, that all quantiles below α or all losses below the α quantile have no weights, i.e. there is no risk aversion and all losses above the α-quantile have the same risk aversion. Therefore the risk aversion weight function associated with ES_α turns out to be

$$w_{ES_\alpha}(u) = (1-\alpha)^{-1} 1(u > \alpha). \tag{6.6}$$

From a risk management point of view there might be many other weights given to some confidence levels u. If the weight function is increasing, which is reasonable since higher losses should have larger risk aversion weight, then we arrive at spectral risk measures.

Definition 6.2 Let w be an increasing function from [0, 1] such that $\int_0^1 w(u)du = 1$, then the map r_w defined by

$$r_w(L) = \int_0^1 w(u)qu(L)du$$

is called a spectral risk measure with weight function w.

The name spectral risk measure comes from the representation

$$r_w(X) = \int_0^1 ES_\alpha(1-\alpha)\mu_u(da) \tag{6.7}$$

$$\text{with the spectral measure } \mu([0, b]) = w(b). \tag{6.8}$$

This representation is very useful when we want to find the scenario function representing a spectral risk measure r_w.

Proposition 6.1 *The density of the scenario associated with the risk measure equals*

$$L_w \stackrel{\text{def}}{=} gw(L) \stackrel{\text{def}}{=} \int_0^1 g_\alpha(L)(1-\alpha)\mu(d\alpha). \tag{6.9}$$

Here $g\alpha(L)$ is defined in formula (6.5). In particular

$$r_w(L) = E(LL_w) \tag{6.10}$$

Proof We have

$$
\begin{aligned}
r_w(L) &= \int_0^1 ES_\alpha(L)(1-\alpha)\mu(d\alpha) \\
&= \int_0^1 \mathsf{E}(LL_\alpha)(1-\alpha)\mu(d\alpha) \\
&= \int_0^1 \max[\mathsf{E}\{Lg_\alpha(Y)\}|Y \in L_\infty](1-\alpha)\mu(d\alpha) \\
&\geq \max\left[\int_0^1 \mathsf{E}\left\{L\int_0^1 g\alpha(Y)(1-\alpha)\mu(d\alpha)\right\}|Y \in L_\infty\right] \\
&= \max[\mathsf{E}\{Lg_w(Y)\}|\forall Y \in L_\infty] \\
&\geq \mathsf{E}\{Lg_w(L)\}
\end{aligned}
$$

Hence

$$
r_w(L) = \max[\mathsf{E}\{Lg_w(Y)\}|\forall Y \in L_\infty] = \mathsf{E}\{Lg_w(L)\}
$$

□

6.2.3 Coherent Allocation Measures

Starting with the representation (6.3) one can now find for each Y a linear function $h_Y = h_Y^r$ which satisfies

$$
r(Y) = h_Y(Y) \text{ and } h_Y(X) \leq r(X),\ \forall X. \tag{6.11}
$$

A "diversifying" capital allocation associated with r is given by

$$
\Lambda_r(X, Y) = h_Y(X). \tag{6.12}
$$

The function Λ_r is then *linear* in the first variable and *diversifying* in the sense that the capital allocated to a portfolio X is always bounded by the capital of X viewed as its own subportfolio

$$
\Lambda(X, Y) \leq \Lambda(X, X).
$$

$\Lambda(X, X)$ can be called the standalone capital or risk measure of X. In general we have the following two results: A linear and diversifying capital allocation Λ, which is continuous, i.e. $\lim_{\epsilon\to 0}\Lambda(X, Y+\epsilon X) = \Lambda(X, Y)\forall X$, at a portfolio Y, is uniquely determined by its associated risk measure, i.e. the diagonal values of Λ. More specifically, given the portfolio Y then the capital allocated to a subportfolio X of Y is the derivative of the associated risk measure ρ at Y in the direction of X.

Proposition 6.2 *Let Λ be a linear, diversifying capital allocation. If Λ is continuous at $Y \in V$ then for all $X \in V$*

$$\Lambda(X, Y) = \lim_{\epsilon \to 0} \frac{r(Y + \epsilon X) - \rho(Y)}{\epsilon}.$$

The following proposition states the equivalence between positively homogeneous, sub-additive risk measures and linear, diversifying capital allocations.

Proposition 6.3 *(a) If there exists a linear, diversifying capital allocation Λ with associated risk measure r, i.e. $r(X) = \Lambda(X, X)$, then r is positively homogeneous and sub-additive.*

(b) If r is positively homogeneous and sub-additive then Λ_r as defined in (6.12) is a linear, diversifying capital allocation with associated risk measure r.

6.2.4 Spectral Allocation Measures

Since in the case of spectral risk measures r_w the maximal linear functional in (6.11) can be identified as an integration with respect to the probability measure with density (6.9) from Proposition 6.1, we obtain $h_Y(X) = \mathsf{E}\{X g_w(Y)\}$ and therefore the following capital allocation

$$\Lambda_w(X, Y) = \mathsf{E}\{X g_w(Y)\} = \int_0^1 ESC_\alpha(X, Y)(1 - \alpha)\mu(d\alpha) \tag{6.13}$$

$$= \int_0^1 ESC_\alpha(X, Y)\tilde{\mu}(d\alpha) \tag{6.14}$$

$$\text{where} \quad ESC_\alpha(X, Y) = \mathsf{E}\{X g_\alpha(Y)\} \tag{6.15}$$

is the Expected Shortfall Contribution and $\tilde{\mu}$ is defined in (6.16). Intuitively, the capital allocated to transaction or subportfolio X in a portfolio Y equals its expectation under the generalized maximal scenario associated with w.

6.3 Weight Function and Mixing Measure

One might try to base the calibration or determination of the spectral risk measure based on the spectral measure μ or the weight function w. Since the weight function w is nothing else as the distribution function of μ, there is also a 1-1 correspondence to the more intuitive mixing measure

$$\tilde{\mu}(d\alpha) = (1 - \alpha)\mu(d\alpha). \tag{6.16}$$

If we define more generally for an arbitrary measure $\tilde{\mu}$ the functional

$$\tilde{\rho} = \int_0^1 ES_\alpha \tilde{\mu}(da) \tag{6.17}$$

then $\tilde{\rho}$ is coherent iff $\tilde{\mu}$ is a probability measure. Since

$$\begin{aligned} 1 = \tilde{\mu}([0,\ 1]) &= \int_0^1 (1-u)\mu(du) \\ &= \int_0^1 \int_0^1 1[u,1](v)dv\mu(du) = \int_0^1 \int_0^1 1[0,v](u)\mu(du)dv \\ &= \int_0^1 w(v)dv. \end{aligned}$$

If we have now a probability measure $\tilde{\mu}$ on [0, 1] the representing μ and w in (6.1), (6.2) can be obtained by

$$\frac{d\mu}{d\tilde{\mu}} = \frac{1}{1-\alpha} \tag{6.18}$$

$$w(b) = \mu([0,b]) = \int_0^b \frac{1}{1-\alpha}\tilde{\mu}(d\alpha). \tag{6.19}$$

6.4 Risk Aversion

If we assume a discrete measure

$$\tilde{\mu} = \sum_{i=1}^{n} p_i \delta_{\alpha i} \tag{6.20}$$

then the risk aversion function w is an increasing step function with step size of $p_i/(1-\alpha_i)$ at the points α_i

$$w(b) = \sum_{\alpha_i \le b} \frac{p_i}{1-\alpha_i}. \tag{6.21}$$

This has to be kept in mind. If we assume equal weights for the two expected shortfall at 99 and 90% then the increase in risk aversion at the first quantile 90% is $0.5/0.1 = 5$ and $0.5/0.01 = 50$. The risk aversion against losses above the 99% is therefore 11 times higher than against those between the 90 and 99% quantile. It is therefore sensible to assume quite small weights on ES_α with large αs.

6.5 Implementation

There are several ways to implement a spectral contribution in a portfolio model. According to Acerbi (2002) a Monte-Carlo-based implementation of the spectral risk measure would work as follows:

Let L^n be the n-th realization of the portfolio loss. If we have generated N loss distribution scenario, let us denote by n: N index of the n-th largest loss which itself is then denote by $L^{n:N}$, i.e. the indices $1 : N, 2 : N, \ldots, N : N \in N$ are defined by the property that

$$L^{1:N} < L^{2:N} < \ldots < L^{N:N}$$

The approximative spectral risk measure is then defined by

$$\sum_{n=1}^{N} L^{n:N} w(n/N) / \sum_{k=1}^{N} w(k/N)$$

Therefore a natural way to approximate the spectral contribution of another random variable L_i, which specifically might be a transaction in the portfolio represented by L or a subportfolio of L, is

$$\sum_{n=1}^{N} L_i^{n:N} \frac{w(n/N)}{\sum_{k=1}^{N} w(k/N)}, \tag{6.22}$$

where $L_i^{n,N}$ denotes the loss in transaction i in the scenario $n : N$, i.e. in the scenario where the portfolio loss was the n-th largest. It is then expected that

$$\mathsf{E}(L_i L_w) = \lim_{N \to \infty} \sum_{n=1}^{N} L_i^{n:N} \frac{w(n/N)}{\sum_{k=1}^{N} w(k/N)}.$$

As in most applications we assume that

$$L = \sum_i L_i$$

with the transaction loss variable L_i and in the example later we will actually calculate within a multi-factor Merton-type credit portfolio model.

6.5.1 Mixing Representation

Let us review the standard implementation of the expected shortfall contribution. In the setting of the previous setting we can see that for $w(u) = \frac{1}{1-\alpha} 1[\alpha, 1](u)$ the weights for all scenarios with $\frac{n}{N} < \alpha$ is 0 and for all others it is

$$\frac{\frac{1}{1-\alpha}}{\sum_{k=\{(\alpha)N\}} \frac{1}{1-\alpha}} \cong \frac{1}{(1-\alpha)N}$$

(Here $[\cdot]$ denote the Gauss brackets.) Therefore the expected shortfall contribution equals

$$\frac{1}{\{(1-\alpha)N\}} \sum_{n=(\alpha n)}^{N} L_i^{n:N} \tag{6.23}$$

or more intuitively the average of the counterparty i losses in all scenarios where the portfolio losses was higher or equal than the $[\alpha N]$ largest portfolio loss.

Due to the fact that we have chosen a finite convex combination of Expected Shortfall, i.e. the mixing measure

$$\tilde{\mu}(du) = \sum_{k=1}^{K} p_i \delta_{\alpha_i}$$

and formulae (6.23) and (6.17) we will take for a transaction Li the approximation

$$\mathrm{SCA}(L_i,\ L)_{\mathrm{vec}p,\,\mathrm{vec}\alpha,\,N} = \sum_{k=1}^{K} p_i \left[\frac{1}{\{(1-\alpha_i)N\}} \sum_{n=[\alpha_i N]}^{N} L_i^{n:N} \right] \tag{6.24}$$

as the Spectral Capital Allocation with discrete mmixing measure represented by the vectors $\mathrm{vec}p = (p_1, \ldots, p_K)$, $\mathrm{vec}\alpha = (\alpha_1, \ldots, \alpha_K)$ for a Monte- Carlo-Sample of length N.

6.5.2 Density Representation

Another possibility is to rely on the approximation of the Expected Shortfall Contribution as in Kalkbrener et al. (2004) and to integrate over the spectral measure μ:

$$\mathsf{E}(L_i L_w) = \lim_{N\to\infty} \int_0^1 \left\{ \sum_{n=1}^{N} L_i^{n:N} \frac{w_\alpha(i/N)}{\sum_{k=1}^{N} w_\alpha(k/N)} (1-\alpha) \right\} \mu(da) \tag{6.25}$$

If L has a continuous distribution than we have that

$$\begin{aligned} \mathsf{E}(L_i L_w) &= \mathsf{E}\{L_i \int_0^1 L_\alpha \mu(d\alpha)\} \\ &= \int_0^1 \mathsf{E}[L_i 1\{L > q_\alpha(L)\}](1-\alpha)^{-1} \mu(d\alpha) \\ &= \lim_{N\to\infty} N^{-1} \sum_{n=1}^{N} L_i^n \int_0^1 1\{L^n > q_\alpha(L)\}(1-\alpha)^{-1} \mu(d\alpha) \end{aligned} \tag{6.26}$$

If L has not a continuous distribution we have to use the density function (6.9) and might approximate the spectral contribution by

$$\mathsf{E}(L_i L_w) \sim N^{-1} \sum_{n=1}^{N} L_i^n g_w(L^n). \tag{6.27}$$

The actual calculation of the density g_w in (6.27) might be quite involved. On the other hand the integration with respect to μ in (6.25) and (6.26) is also not easy. If w is a step function as in the example 1 above, then μ is a sum of weighted Dirac-measure and the implementation of spectral risk measure as in (6.22) is straightforward.

6.6 Credit Portfolio Model

In the examples below we apply the presented concepts to a standard default only type model with a normal copula based on an industry and region factor model, with 27 factors mainly based on MSCI equity indices. We assume fixed recovery and exposure-at-default. For a specification of such a model, we could refer to Bluhm et al. (2002) or other text books on credit risk modeling.

6.7 Examples

6.7.1 Weighting Scheme

Lets take 5 quantile 50, 90, 95, 99, 99.9% and the 99.98% quantile. We like now to find weighting scheme for Expected Shortfall, which still gives a nice risk aversion function. Or inversely we start with a sensible risk aversion as in (6.28) and then solve for the suitable convex combination of expected shortfall measures.

As a first step in the application of spectral risk measures one might think to give to different loss probability levels different weight. This is a straightforward extension of expected shortfall. One might view Expected Shortfall at the 99%-level view as a risk aversion which ignores losses below the 99%-quantile and all losses above the 99%-quantile have the same influence. From an investors point of view this means that only senior debts are cushioned by risk capital. One might on the other hand also be aware of losses which occur more frequently, but of course with a lower aversion than those appearing rarely.

As a concrete example one might set that losses up to the 50% confidence level should have zero weights, losses between 50 and 99% should have a weight w_0 and losses above the 99%-quantile should have a weight of $k_1 w_0$ and above the 99.9% quantile it should have a weight of $k_2 w_0$. The first tranch from 50 to 99% correspond to an investor in junior debt, and the tranch from 99 to 99.9% to a senior investor and above the 99.9% a super senior investor or the regulators are concerned. This gives a step function for w:

$$\begin{aligned} w(u) = {} & w_0 1(0.99 > u > 0.5) + k_1 w_0 1(0.999 > u > 0.99) \\ & + k_2 w_0 1(1 > u > 0.999) \end{aligned} \tag{6.28}$$

The parameter w_0 should be chosen such that the integral over w is still 1.

6.7.2 *Concrete Example*

The portfolio consists of 279 assets with total notional EUR 13.7bn and the following industry and regions breakdown:

The portfolio correlation structure is obtained from the R^2 and the correlation structure of the industry and regional factors. The R^2 is the R^2 of the one-dimensional regression of the asset returns with respect to its composite factor, modeled as the sum of industry and country factor. The underlying factor model is based on 24 MSCI Industries and 7 MSCI Regions (Fig. 6.1). The weighted average R^2 is 0.5327 (Fig. 6.2).

The risk contributions are calculated at quantiles 50, 90, 95, 99, 99.9 and 99.98%.

Figure 6.3 shows the total Expected Shortfall Contributions allocated to the industries normalized with respect to automobile industry risk contributions and ordered by $ESC_{99\%}$.

In order to capture all risks of the portfolio a risk measure, which combines few quantile levels, is needed. As one can see, Hardware and Materials have mainly tail exposure (largest consumption of ESC at the 99.98%-quantile), where Transportation, Diversified Finance and Sovereign have the second to fourth largest consumption of ESC at the 50%-quantile, i.e. are considerable more exposed to events happening roughly every second year as Hardware and Materials.

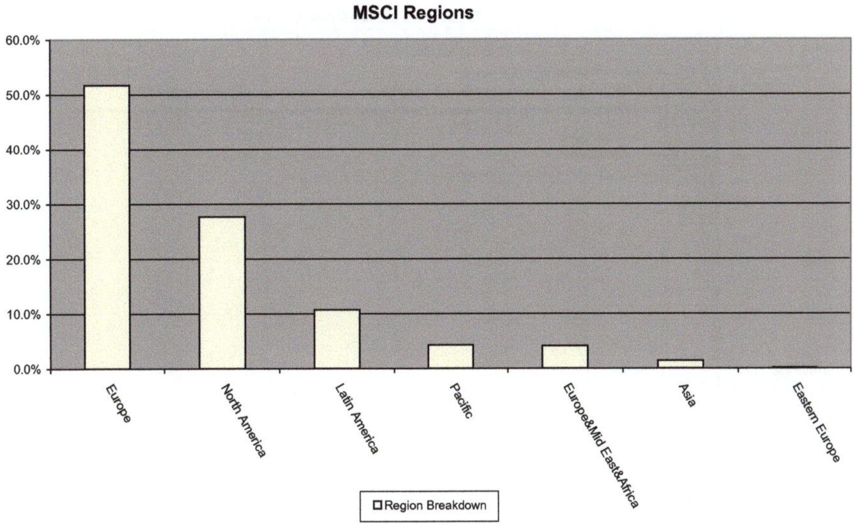

Fig. 6.1 MSCI region breakdown. XFGRegionsBreakdown

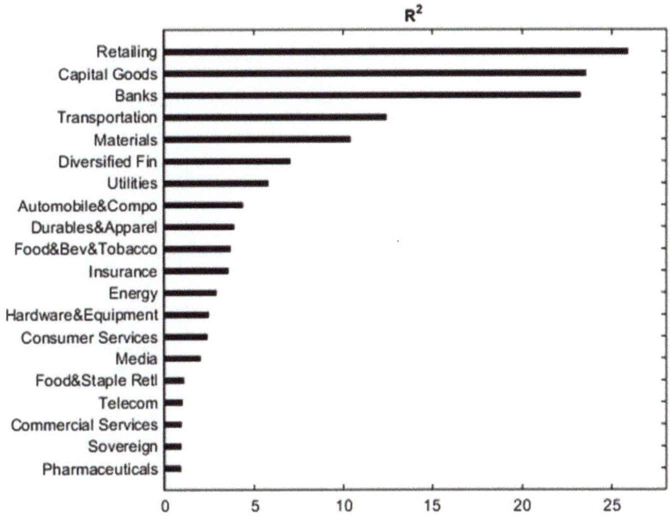

Fig. 6.2 R^2 values of different MSCI industries. XFGRsquared

The spectral risk measure as a convex combination of Expected Shortfall risk measures at the following quantiles 50, 90, 95, 99, 99.9 and 99.98% can capture both effects, at the tail and at the median of the loss distribution.

Four spectral risk measures are calculated. The first three are calibrated in terms of increase of the risk aversion function at each considered quantile as in Fig. 6.4.

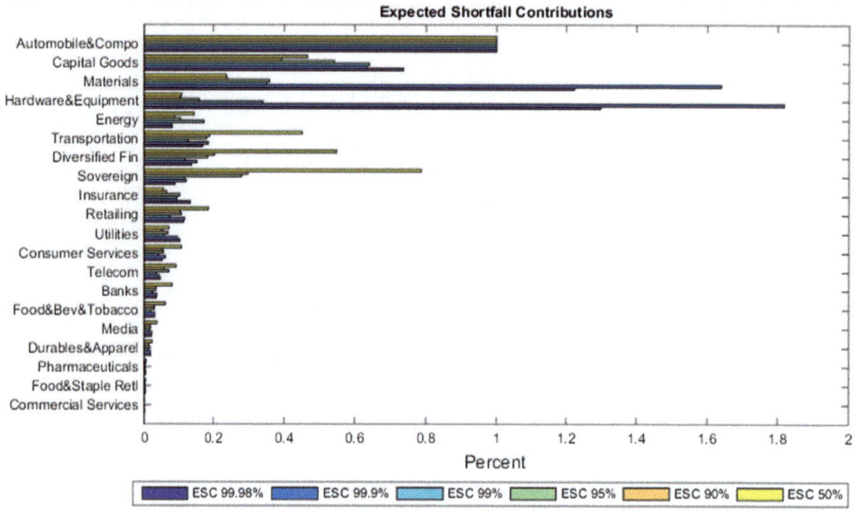

Fig. 6.3 Expected shortfall contributions for different industries at different quantiles

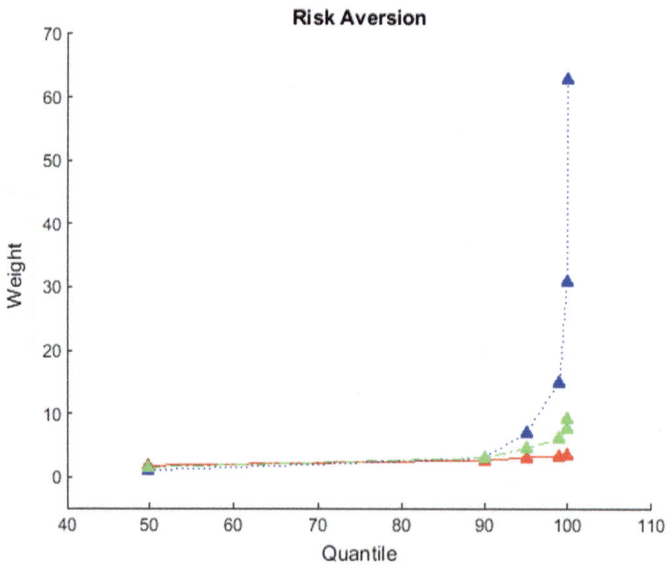

Fig. 6.4 Risk aversion calculated with respect to different methods. The *dotted blue*, *dashed-dotted* and *solid lines* represent "SCA - decreasing steps", "SCA - equal steps" and "SCA - increasing steps" correspondingly. XFGriskaversion

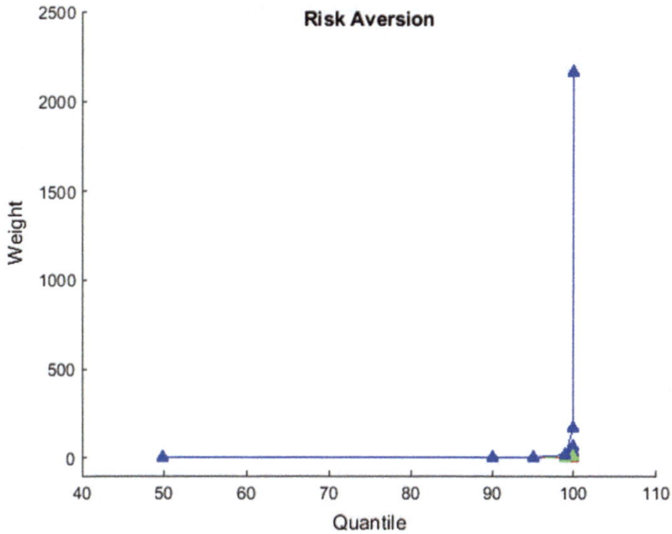

Fig. 6.5 Risk aversion when the weights are directly set to 0.1 at the 50, 90, 95%-quantiles, 0.15 at the 99 and 99.9%-quantiles and 0.4 at the 99.98%-quantile. XFGriskaversion2

The least conservative one is "SCA - decreasing steps" in which the risk aversion increases at each quantile by half the size it has increased at the quantile before. "SCA -equal steps" increases in risk aversion by the same amount at each quantile, "SCA -increasing steps" increases in risk aversion at each quantile by doubling the increase at each quantile. The last most conservative one is SCA - 0.1/0.1/0.1/0.15/0.15/0.4, in which the weights of $\tilde{\mu}$ are directly set to 0.1 at the 50, 90, 95%-quantiles, 0.15 at the 99 and 99.9%-quantiles and 0.4 at the 99.98%-quantile as in Fig. 6.5. The last one has a very steep increase in the risk aversion at the extreme quantiles.

As a comparison to the expected shortfall, the chart below shows the Spectral risk allocation allocated to industries ordered by SCA - equal steps and normalized with respect to automobile industry SCA as in Fig. 6.6.

All tables so far were based on the risk allocated to the industries. Much of the displayed effects are just driven by exposure, i.e. "Automotive" is by far the largest exposure in that portfolio and all sensible risk measure should mirror this concentration. Interestingly enough the most tail emphasizing measures are the exceptions. There the largest contributors Hardware and Materials have actually less than 10% of the entire exposure.

Usually one uses as well percentage figures and risk return figures for portfolio management. On the chart "RC/TRC" the percentage of total risk (TRC) allocated to the specific industries is displayed in Fig. 6.7.

For the risk management Fig. 6.8 showing allocated risk capital per exposure is very useful. It compares the riskiness of the industry normalized by their exposure. Intuitively it means that if you increase the exposure in "transportation" by a small amount like 100.000 Euro than the additionally capital measured by SCA-increasing

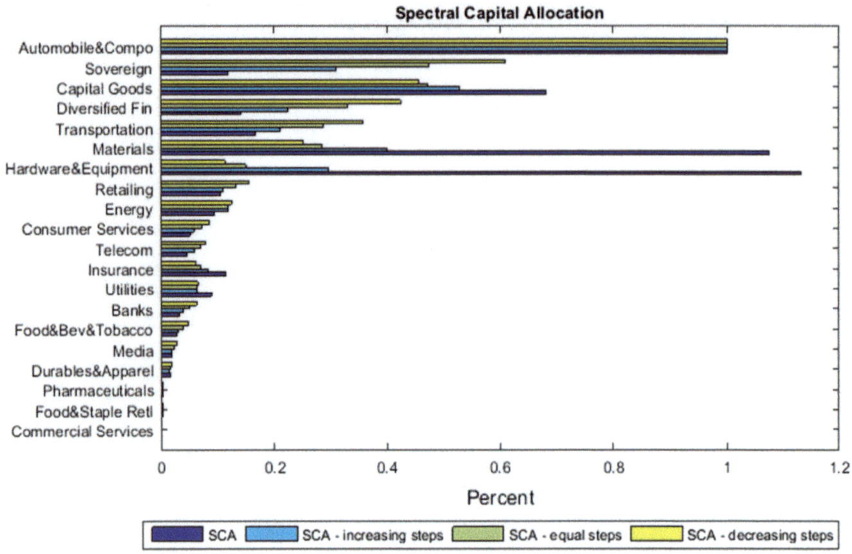

Fig. 6.6 Different risk contributions with respect to different SCA methods

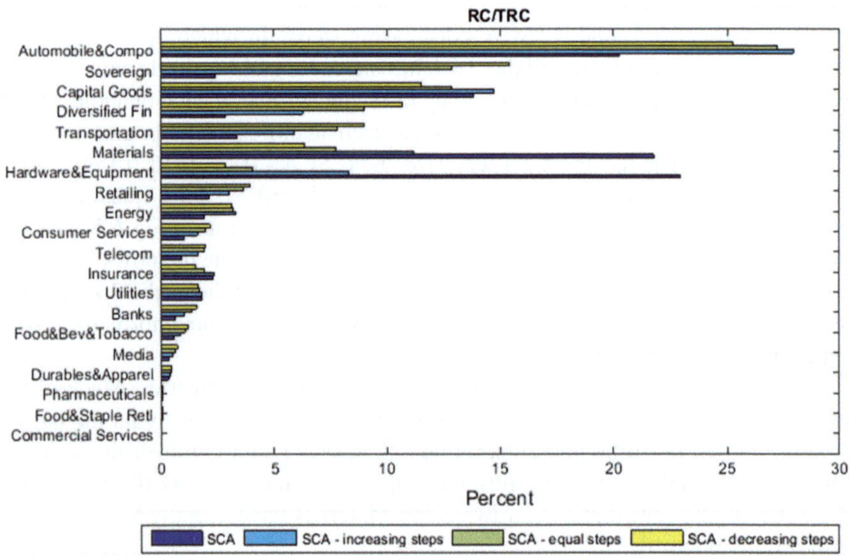

Fig. 6.7 Total risk contributions with respect to different SCA methods

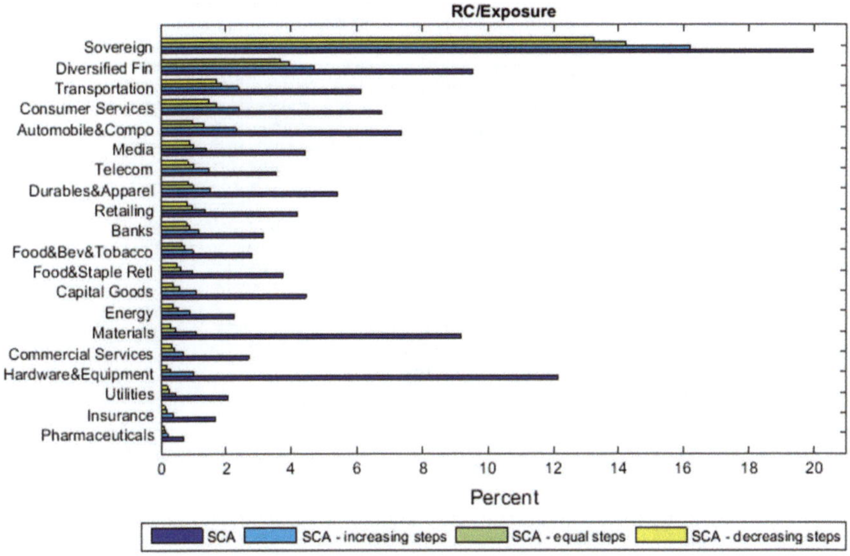

Fig. 6.8 Total risk contributions with respect to different SCA methods

steps will increase by 2.5%, i.e. by 2.5000 Euro. In that sense it gives the marginal capital rate in each industry class. Here the sovereign class is the most risky one. In that portfolio the sovereign exposure was a single transaction with a low rated country and it is therefore no surprise that "sovereign" performance worst in all risk measures (Fig. 6.8).

With that information one should now be in the position to judge about the possible choice of the most sensible spectral risk measure among the four presented. The measure denoted by SCA based on the weights 0.1, 0.1, 0.1, 0.15, 0.15, 0.4, overemphasis tail risk and ignores volatility risk like the 50%-quantile. From the other three spectral risk measures, also the risk aversion function of the one with increasing steps, does emphasis too much the higher quantiles. SCA decreasing steps seems to punished counterparties with a low rating very much, it seems to a large extend expected loss driven, which can be also seen in the following table on the RAROC-type Figs. 6.9. On that table "decreasing steps" does not show much dispersion. One could in summary therefore recommend SCA-equal steps.

For information purpose we have also displayed the Expected Loss/Risk Ratio for the Expected Shortfall Contribution in Fig. 6.10. Here the dispersion for the ESC at the 50% quantile is even lower as for the SCA-decreasing steps.

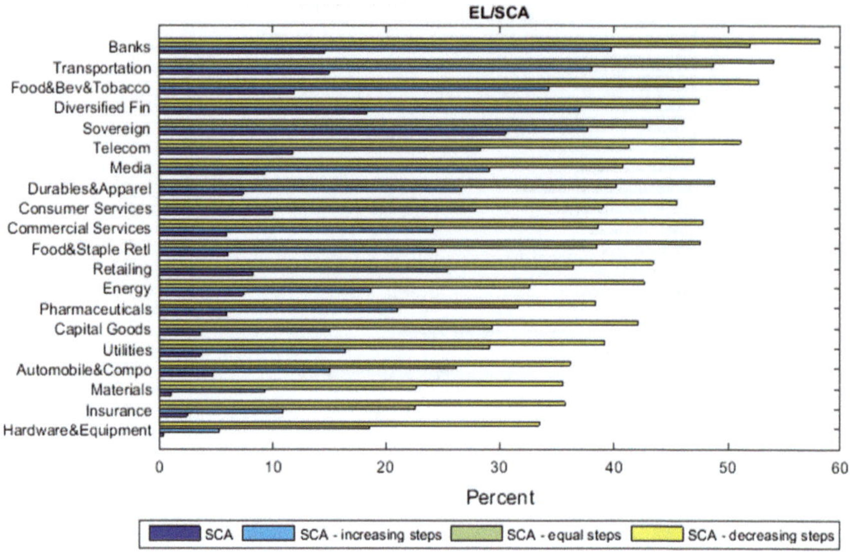

Fig. 6.9 EL/SCA with respect to different SCA methods. XFGELESC

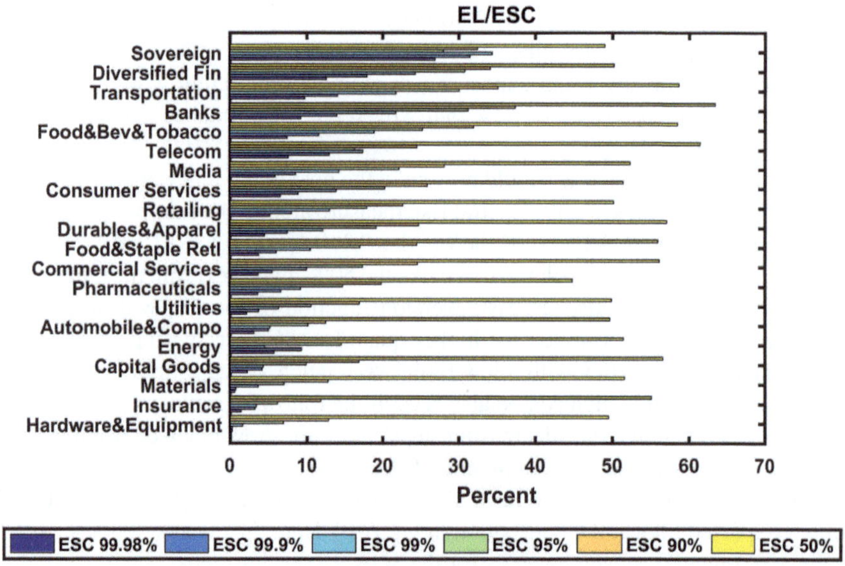

Fig. 6.10 Expected loss/risk ratio for the expected shortfall contribution at different quantiles. XFGELESC

6.8 Summary

In order to combine different loss levels in one risk measure spectral risk measures provide a sensible tool. Weighting of the quantiles is usually be done by the risk aversion function. Starting from an implementation point of view it looks more convenient to write a spectral risk measure as a convex combination of expected shortfall measures. However one has to be careful in the effects on the risk aversion function. All this holds true and become even more important if capital allocation is considered, which finally serves as a decision tool to differentiate sub-portfolios with respect to their riskiness. We analyze an example portfolio with respect to the risk impact of the industries invested in. Our main focus are the different specification of the spectral risk measure and we argue in favour for the spectral risk measure based on a risk aversion which has the same magnitude of increase at each considered quantile, namely the 50, 90, 95, 99, 99.9, and 99.98% quantile. This risk measure exhibits a proper balance between tail risk and more volatile risk.

References

Acerbi, C. (2002). Spectral measures of risk: A coherent representation of subjective risk aversion. *Journal of Banking and Finance*, *26*, 1505–1518.

Acerbi, C., & Tasche, D. (2002). On the coherence of expected shortfall. *Journal of Banking and Finance*, *26*, 1487–1503.

Artzner, P., Delbaen, F., Eber, J.-M., & Heath, D. (1997). Thinking coherently. *RISK*, 68–71.

Artzner, P., Delbaen, F., Eber, J.-M., & Heath, D. (1999). Coherent measures of risk. *Mathematical Finance*, *9*, 203–228.

Bluhmn C., Overbeck, L., Wagner, C. (2002). *An introduction to credit risk modeling*. Boca Raton: CRC Press/Chapman & Hall.

Crouhy, M., Galai, D., Mark, R. (1999). A comparative analysis of current credit risk models. *Journal of Banking and Finance*, *24*.

Delbaen, F. (2000). *Coherent risk measures*. Lecture notes, Scuola Normale Superiore di Pisa.

Denault, M. s. (2001). Coherent allocation of risk capital. *Journal of Risk*, *4*, 1.

Kalkbrener, M. (2002). *An axiomatic approach to capital allocation*. Technical Document, Deutsche Bank AG, Frankfurt.

Kalkbrener, M., Lotter, H. & Overbeck, L. (2004). Sensible and efficient capital allocation. *Risk*.

Kurth, A., & Tasche, D. (2003). Contributions to credit risk. *Risk*, 84–88.

Overbeck, L.(2000). Allocation of economic capital in loan portfolios. In Stahl/Härdle (Eds.), *Proceedings "Measuring Risk in Complex Stochastic Systems"*. Lecture Notes in Statistics, 1999. Berlin: Springer.

Overbeck, L., & Stahl. G. (2003). Stochastic essentials for the risk management of credit portfolios. *Kredit und Kapital*, 1.

Overbeck, L. (2004). *Spectral Capital Allocation*. London: RISK Books.

Rockafellar, R. T., & Uryasev, S. (2000). Optimization of conditional value-at-risk. *Journal of Risk*, 2, 21–41.

Chapter 7
Market Based Credit Rating and Its Applications

R.S. Tsay and H. Zhu

Abstract Credit rating plays a critical role in financial risk management. It is like a name tag of a firm indicating its health condition. Generally, ratings involve a lot of firm-specific information which is hard to obtain or only available quarterly. In this chapter, we propose a two-step algorithm involving ARIMA-GARCH modelling and clustering to obtain a market based credit rating utilizing easily obtained public information. The algorithm is applied to 3-year CDS spreads of 247 publicly listed firms. Empirical result of the application and comparisons between the obtained ratings with the ratings given by agencies show that such a market based credit rating performs quite well.

7.1 Introduction

Credit rating is a reflection of a firm's creditworthiness, traditionally provided by professional rating agencies. It is widely used to measure the credit risk of a company, i.e. the firm's ability to meet its debt servicing obligations, and hence plays a significant role in the financial market. Investors can use credit ratings to aid their investment decisions, e.g., Erlenmaier (2011), while an issuer may use the rating to determine the optimal amount of debt outgoing or signal its low investment risk, e.g., Nordberg (2010). Some investment funds may restrict investing only on firms whose credit ratings exceed certain level.

In the past decades, more and more researchers are interested in credit ratings, especially after the 2008 subprime financial crisis. Some are interested in the effectiveness of agency's ratings. For example, Kliger and Sarig (2000) showed that the credit rating can provide better assessment of default risk than publicly-available information alone. Hull et al. (2004) discussed the relationship between bond yields,

R.S. Tsay (✉)
University of Chicago Booth School of Business, Chicago, IL 60637, USA
e-mail: ruey.tsay@chicagobooth.edu

H. Zhu
Department of Statistics, University of Chicago, Chicago, IL, USA
e-mail: hzhuad@uchicago.edu

W.K. Härdle et al. (eds.), *Applied Quantitative Finance*, Statistics and Computing,
DOI 10.1007/978-3-662-54486-0_7

Credit Default Swap spreads, and credit rating announcements. Others are interested in proposition or replication of the ratings by the agencies. Altman (1968) used five financial ratios to predict bankruptcy, and many researchers employed the same financial variables based method to quantify credit risk, such as Kaplan and Urwitz (1979), Ederington (1985) and Kamstra et al. (2001). This approach often involves substantial firm-specific information which is hard to obtain or only available quarterly. Recently, Creal et al. (2014) proposed a market-based credit rating which makes direct use of the prices of traded assets. The basic idea of market-based credit rating is that asset prices of traded firms should reflects timely the publicly-available firm-specific information. Following the same idea, we propose a market-based credit rating method using CDS spreads and/or their robustified values. The proposed method is easy to understand and use. As a matter of fact, the ratings are easily reproducible.

Credit Default Swap (CDS) is a financial agreement between a buyer and a seller in which the buyer makes periodic payments to the seller and receives a payoff from the seller in exchange if the reference entity defaults before the CDS contract expires. CDS is widely used with other financial derivatives to hedge the risk or to speculate on price movements. The periodic payment the buyer makes, which is also known as the price of CDS, is quoted in spread. Higher spread means the referred entity has a higher possibility to default from market's perspective, indicating its lower creditworthiness. Ericsson et al. (2009) shows that firm leverage, which is closely related to default risk, plays a significant role in determining its CDS spread. Micu et al. (2004) also find that rating changes can cause dynamic shifts on CDS markets. Therefore, there should be a close relation between credit rating and CDS spread. In this chapter, we leverage this close relationship and show that the proposed credit rating based on CDS spreads works well in comparison with the results provided by rating agencies.

The rest of the chapter proceeds as follows. In the next section, we introduce the methodology used. In Sect. 7.3, we consider empirical analysis and provide some discussions. The concluding remarks are presented in Sect. 7.4.

7.2 Methodology

Different from the method of Creal et al. (2014), the proposed method uses a two-stage procedure: forecasting and clustering. Our goal is to make the market-based credit rating easy to follow and use. In particular, no special program is needed. The proposed method can be easily reproduced. On the other hand, unlike Creal et al. (2014), we do not consider ratings of firms that have no CDS data.

7.2.1 Modeling and Forecasting

Assume that we have time series of daily CDS spreads of N firms. Denote the data by $\{y_{it}|i = 1, \ldots, N; t = 1, \ldots, T\}$. These series have the same maturity. In our empirical analysis, we use 3-year CDS spreads.

Instead of using y_{it} directly, we use predictions in the proposed credit rating method. Rating is necessarily concerning future performance of a firm. Thus, it makes sense to use predictions. In our empirical analysis, we use 1-step ahead predictions. If preferred, multi-step predictions can be used. Another reason for using predictions is to mitigate the impact of outliers. Since firm's creditworthiness typically does not change overnight, an abrupt change in CDS spread might be caused by reasons not related to the fundamentals of a firm. Using predictions can mitigate the impacts of such isolated outlying observations.

The proposed rating method uses predictions of the level and volatility of a CDS time series. To obtain the predictions, we apply ARIMA-GARCH models to each CDS time series. The model entertained can be written as

$$z_{it} = (1 - B)^{d_i} y_{it}, \tag{7.1}$$

$$z_{it} = \sum_{j=1}^{p_i} \phi_j z_{i,t-j} + a_{it} + \sum_{j=1}^{q_i} \theta_j a_{i,t-j}, \tag{7.2}$$

$$a_{it} = \sigma_{it} \epsilon_{it}, \tag{7.3}$$

$$\sigma_{it}^2 = \alpha_{i,0} + \sum_{j=1}^{r_i} \alpha_{i,j} a_{i,t-j}^2 + \sum_{j=1}^{s_i} \beta_{i,j} \sigma_{i,t-j}^2, \tag{7.4}$$

where d_i is a nonnegative integer denoting the order of differencing, p_i and q_i are nonnegative integers representing the autoregressive (AR) and moving-average (MA) order of the differenced series z_{it}, respectively, $\{\epsilon_t\}$ is a sequence of independently and identically distributed random variates with mean zero and variance 1, r_i and s_i are also nonnegative integers indicating the autoregressive conditional heteroscedastic (ARCH) order and the generalized ARCH order, respectively. The distribution of ϵ_t can be Gaussian or standardized Student-t or some skewed distributions with heavy tails. Equations (7.1) and (7.2) are referred to as the *mean equations* for y_{it} whereas Eqs. (7.3) and (7.4) are the *volatility equation.* This class of model is general and applicable to the CDS time series. The parameters of the model in Eqs. (7.2) and (7.4) are estimated by the maximum likelihood method.

There are several R packages available for building an ARIMA(p, d, q)-GARCH(r, s) model for a given financial time series. See, for instance, the `fGarch` and `rugarch` packages. The latter package allows for fractional differencing, i.e., d_i of Eq. (7.1) may assume nonnegative real values.

The modeling steps used in this chapter are as follows:

1. Mean equation: For given maximum values of p, d and q, we use the Akaike information criterion (AIC) to select the order (p_i, d_i, q_i) for the time series y_{it}.

As a matter of fact, one can even apply the automatic model selection procedure `auto.arima` of the R package `forecast` to select ARIMA model.

2. ARCH test: Let $\hat{a}_{it}$ be the residual series of the mean equation. We apply Ljung-Box $Q(m)$ statistics to the squared series a_{it}^2 to detect the existence of conditional heteroscedasticity, also known as the ARCH effect. Under the null hypothesis of no conditional heteroscedasticity, the test statistic is distributed asymptotically as χ_m^2.
3. Volatility equation: If the ARCH effect is statistically significant, we entertain ARIMA(p_i, d_i, q_i)-GARCH(r_i, s_i) models with given maximum values r and s for the GARCH model. Again, AIC is used to select the GARCH order and the distribution of ϵ_{it}. If the ARIMA order can be reduced as a result of the joint estimation, we further simplify the mean equation. Again, the modification is carried out using the AIC.

Our choice of AIC is for simplicity. Other information criteria can be used if needed.

Once an ARIMA-GARCH model is built for the CDS time series y_{it}, we use the model to obtain predictions of y_{it} and its volatility. The forecast origin is the sample size T. Denote the h-step ahead forecasts of mean and volatility of y_{it} at the forecast origin T by $\boldsymbol{x}_i(h) = (\hat{y}_{i,T}(h), \hat{\sigma}_{i,T}(h))'$. Let $\boldsymbol{X}_h$ denote the collection of h-step ahead forecasts of mean and volatility at the forecast origin T for all time series. Specifically, the i-row of $\boldsymbol{X}_h$ consists of $\boldsymbol{x}_i(h)$. We use $\boldsymbol{X}_h$ in the proposed credit rating method.

7.2.2 Clustering

Clustering analysis has a long history in the statistical literature. Many methods are available, including agglomerative hierarchical methods, K-means, tree-based methods, and supporting vector machine. In this chapter, we use mainly the K-means for its wide applicability and nonparametric nature. We also apply a tree-based method in our discussion.

Consider the predictions in $\boldsymbol{X}_h$, which contains the mean and volatility of CDS spreads. Intuitively, a high-quality company would have low values in mean and volatility, and higher values in either mean or volatility are indicative of higher default risk. For ease in notation, we shall omit the subscript h and denote the predictions as $\boldsymbol{X}$ with ith row being $\boldsymbol{x}_i$.

Assume that there are k categories in the rating system. The K-means method uses some measurement of similarity between companies. In this chapter, we use the Euclidean distance to measure similarity. The basic idea of the K-means method is that the distances between members of a cluster should be as small as possible, but the total distance between the clusters is large. Let $\boldsymbol{S} = \{S_i | i = 1, \ldots, k\}$ denote the k clusters, and $\boldsymbol{m}_i$ be the mean vector of members in cluster S_i. The K-means method can be described as

$$\arg\min_{S} \sum_{i=1}^{k} \sum_{\boldsymbol{x} \in S_i} \|\boldsymbol{x} - \boldsymbol{m}_i\|^2.$$

A company is assigned to one and only one cluster. There are various algorithms available to achieve K-means clustering. We describe briefly an algorithm below. Randomly select k points from $\boldsymbol{X}$ and assign them to form k clusters. Since each cluster has a single element, we denote the initial mean vector of the clusters as $\boldsymbol{m}_1^{(0)}, \cdots, \boldsymbol{m}_k^{(0)}$. The algorithm then proceeds with the following three steps.

1. Assignment Step: All points $\boldsymbol{x}_i$ in $\boldsymbol{X}$ are assigned to $S_j \in \boldsymbol{S}$ via

$$j = \arg\min_{u} d(\boldsymbol{x}_i, \boldsymbol{m}_u^{(0)})$$

   where d denotes the Euclidean distance. If there are several j satisfying the condition, one randomly assigns the point to one of those S_j.
2. Updating Step: when all points in $\boldsymbol{X}$ are assigned, update the mean vector of each cluster, namely

$$\boldsymbol{m}_j^{(1)} = \frac{1}{|S_j|} \sum_{\boldsymbol{x}_i \in S_j} \boldsymbol{x}_i,$$

   where $|S_j|$ denotes the number of points in S_j.
3. Repeat the Assignment and Updating Steps to obtain $\boldsymbol{m}_j^{(2)}$ and check the condition

$$d(\boldsymbol{m}_j^{(2)}, \boldsymbol{m}_j^{(1)}) = 0, \quad j = 1, \ldots, k.$$

   If the condition fails, repeat Step 3 until it is satisfied.

It is easy to see that the algorithm aims at achieving the stability of the mean vectors. With the stable mean vectors, the clustering is stable too. In theory, the prior algorithm achieves local convergence as the result may depend on the initial assignment. However, one can use different initial assignments to ensure global convergence. In application, some time series may contain outliers that can weaken the accuracy in prediction, leading to inferior clustering analysis. In this case, some data processing might be helpful. For instance, one can apply wavelet smoothing to the observed time series before the modeling. See Nason (2008) for applications of wavelet methods in statistics.

7.3 Empirical Analysis

In this section, we apply the proposed method to a collection of 294 CDS series with 3-year maturity from Markit. The data are from January 2004 to September 2014. A few time series did not start in January 2004. In this case, a shorter time span is

Table 7.1 ARIMA+GARCH Order Combinations

ARIMA order	GARCH order			
	(0,0)	(1,1)	(2,1)	(2,2)
(0,1,0)	0	0	1	9
(0,1,1)	0	0	1	16
(0,1,2)	1	0	0	9
(1,1,1)	4	0	1	21
(1,1,2)	1	0	0	21
(2,1,2)	2	0	0	26
(3,1,2)	0	0	1	10
(3,1,3)	1	0	0	9
(4,1,4)	0	0	3	9
(4,1,5)	0	12	2	5
(5,1,5)	1	18	1	10

used. Since the observed spreads are small, we analyze $y_t = \log(10000s_t)$, where s_t is the observed spreads.

7.3.1 *Modeling and Forecasting*

Following the proposed method, we start the analysis with ARIMA-GARCH modeling. Table 7.1 summarizes the main results of ARIMA-GARCH order selection. The ARIMA orders are shown in row whereas GARCH orders in column. These results are selected by AIC with maximum value 5 for both p and q.

From Table 7.1, a majority of the firms assume the GARCH(2,2) structure. On the other hand, the ARMA orders vary markedly. The need for the first difference in the CDS spreads is not surprising as it is in agreement with most time series of asset prices.

To demonstrate, Fig. 7.1 shows the time plots of observed data, fitted values and 1-step ahead prediction for the 3-year CDS spreads of BestBuy and IBM. The black line, green line, and red point are the observed data, fitted values, and prediction, respectively. From the plots, the fitted models appear to provide good fits.

The plots in Fig. 7.1 also show marked market impacts and difference between companies. Both BestBuy and IBM spreads exhibit substantial increases in default risk during the 2008 financial crisis. On the other hand, the BestBuy spreads show that the company did not do well in 2013. For the IBM series, there was no clear increase in default risk after 2011.

Figure 7.2 shows the time plots of log returns of IBM CDS spreads after wavelet transformation and the associated fitted values. As expected, the model selected by AIC fits the wavelet transformed data well. The main discrepancies between the data

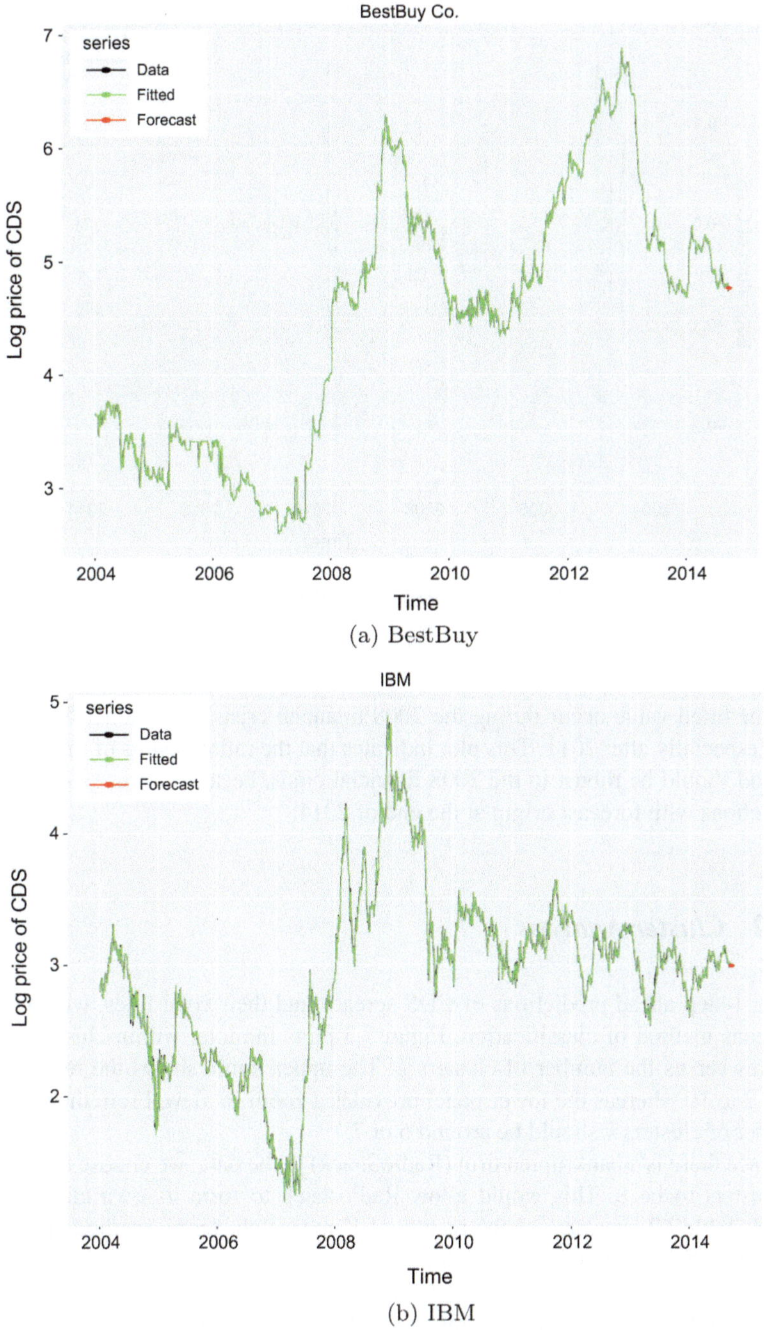

Fig. 7.1 Observed data, fitted values, and a prediction of 3-year CDS spreads of Best Buy and IBM from January 2004 to September 2014. The data are $\log(10000 s_t)$ for the observed spread s_t

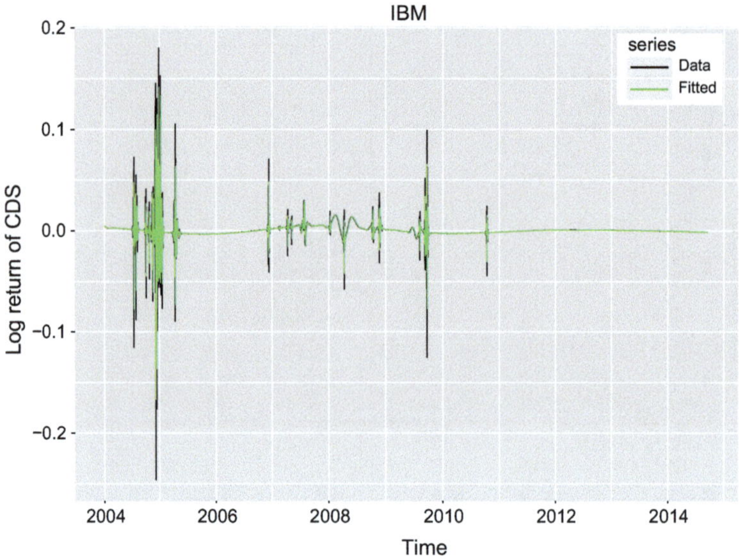

Fig. 7.2 The log return of IBM 3-year CDS spreads after wavelet transformation (in *black*) and the fitted values (in *green*)

and the fitted value occur during the 2008 financial crisis. The model fits the data well, especially after 2011. This plot indicates that the rating results of the proposed method should be robust to the 2008 financial crisis, because we use 1-step ahead predictions with forecast origin at the end of 2014.

7.3.2 Cluster Analysis

Using 1-step ahead predictions of CDS spreads and their volatilities, we apply the K-means method of classification. Figure 7.3 plots the total within cluster sum of squares versus the number of clusters k. The upper figure shows the results for k from 2 to 10 whereas the lower panel provides a zoom-in view. From the plots, the number of clusters k should be around 6 or 7.

Since there is a bankrupted firm (RadioShack) in the data, we choose the number of clusters to be 8. This would allow Radioshack to form its own cluster. With $k = 8$, Table 7.2 summarizes the results of K-means clustering method. To ensure convergence of the K-means method, the results shown are based on 10,000 initial random starts.

From Table 7.2, most of the firms in our data are clustered into Cluster 1, which has lower values in mean and volatility. Thus, as expected, most firms have low default risk. Assuming that the loss recovery rate is 40 %, the expected implied

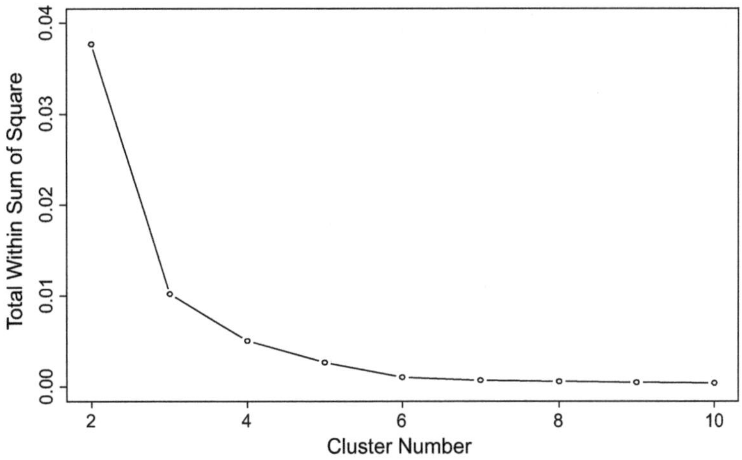

(a) The number of clusters ranges from 2 to 10.

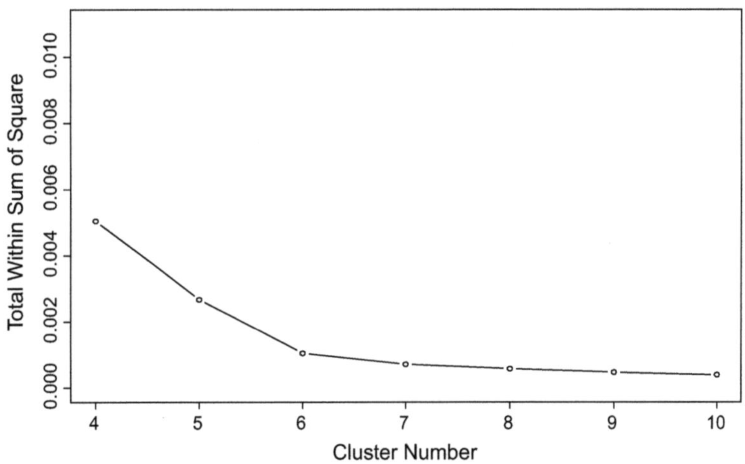

(b) Zoom-in Figure

Fig. 7.3 Total within cluster sum of squares (against the number of clusters)

probability of default (IPD) of the best group is $\frac{0.002545244}{1-0.4} \times 3 \times 100\% = 1.27\%$, which appears to be reasonable. This is understandable because the U.S. economy has largely recovered from the 2008 financial crisis. The default risk of a good company should be low. The outlying firm belongs to the worst cluster with spread being ten or hundred times larger than that of other clusters. Such high spread leads to IPD about 100%, confirming that the firm (RadioShack) is indeed bankrupted. Other firms showing relatively high CDS spreads include Toy"R"US (1630bps) and SHC-Acceptance (1715bps). These firms have been known to be in financial stress in recent years, and they are clustered into the categories 5th to 8th. Note that the

Table 7.2 Results of K-mean clustering method, where μ and σ denote the mean spread and volatility of each cluster

Cluster	μ	σ	Size
1	25.45244	0.8541984	196
2	81.26927	2.8876315	51
3	142.66127	5.5291877	32
4	256.70490	28.0481097	5
5	412.27850	7.8975128	5
6	841.41321	102.7335156	2
7	1622.83575	41.0780870	2
8	13910.92850	147.6544963	1

Table 7.3 S&P rating versus the proposed market-based credit rating

S&P Rating	Market-Based Rating Rank				
	1	2	3	4	5
AA+	1	0	0	0	0
AA	1	0	0	0	0
AA-	5	0	0	0	0
A+	4	0	0	0	0
A	20	0	0	0	0
A-	19	1	0	1	1
BBB+	23	2	1	0	0
BBB	27	4	1	0	0
BBB-	10	6	0	0	0
BB+	2	7	2	0	0
BB	1	2	4	1	0
B+	0	1	0	0	0
B	0	1	2	1	1
B-	0	0	1	0	0
CCC+	0	0	0	0	1

estimated IPD and the distribution of firms across clusters match well with the rating results by ICAP (2013) although they used a different data set.

We also compare results of the proposed rating method with the well-known S&P credit ratings. With a limited subsample of 154 firms whose S&P ratings are gathered, results of the proposed clustering method are directional in line with the S&P ratings. See Table 7.3.

Each cell in Table 7.3 shows the number of firms with S&P rating in row and the clustering result in column. Although the proposed method does not differentiate much between good firms, which might be due to the small number of firms available for the comparison, it is reassuring to see that firms with high ratings by the proposed market-based credit rating procedure also have high S&P ratings.

Finally, Fig. 7.4 shows the time plots of median spreads and volatilities for each cluster obtained by the proposed market-based credit rating method. From the plots, the differences between clusters are clearly seen, indicating that the proposed rating method is capable of ranking firms based on their CDS spreads. For instance, Clusters 1 and 2 have lower spreads and volatilities. The defaulted firm had increasing spreads and volatilities over the data span.

7.3.3 Discussion

Some discussions of the proposed market-based credit rating method are in order. First, as demonstrated by a small subsample, the proposed rating method can produce ratings that are directional in line with those of the S&P rating. This is encouraging as the proposed method only uses the CDS spreads. Indeed, the results show that there exists a close relationship between CDS spreads and the S&P ratings. To demonstrate, we apply a tree-based classification procedure to the S&P rating using the one-step ahead predictions of CDS spreads, indicators of the industrial sectors, and log returns of the spreads as explanatory variables. In other words, we used the subsample of 154 firms mentioned in previous section to build a classification tree with CDS spreads and some additional variables. In a classification tree, branches are determined by relevant explanatory variables with more important variables appearing first and more often.

For detailed explanation of tree procedures and pruned tree classification, see James et al. (2013). The resulting tree is shown in Fig. 7.5a. The first few branches of the tree are obtained by either the spread or the standard error of the spreads. The industrial sector only appears in the high-level branches. In the plot, we use alphabets to represent sectors so that the tree is easier to read. Part (b) of Fig. 7.5 shows a pruned tree which provides a clear relationship between the CDS spreads and the S&P ratings. Consequently, CDS spreads are indeed informative about credit risk of a firm.

Second, there are ways to improve the proposed model-based credit rating. For example, a potential weakness of using CDS spreads alone to perform credit rating is that the method might overlook the variations between industrial sectors. Similar to stock returns, the level and volatility of CDS spreads might depend on the industrial sectors. For instance, healthcare companies tend to have lower volatility as their demands are more robust to the U.S. business cycles. Table 7.4 provides the median end-of-year spreads from 2011 to 2013 and the 1-step ahead predictions of 10 industrial sectors to which the 294 time series belong.

From Table 7.4, we see that sectors whose demands are relatively inelastic like healthcare or industrial sectors have lower spreads all year round while the high-elastic demand sectors, including financial and consumer goods, have higher spreads. This is easy to understand because people will lower consumption or investment

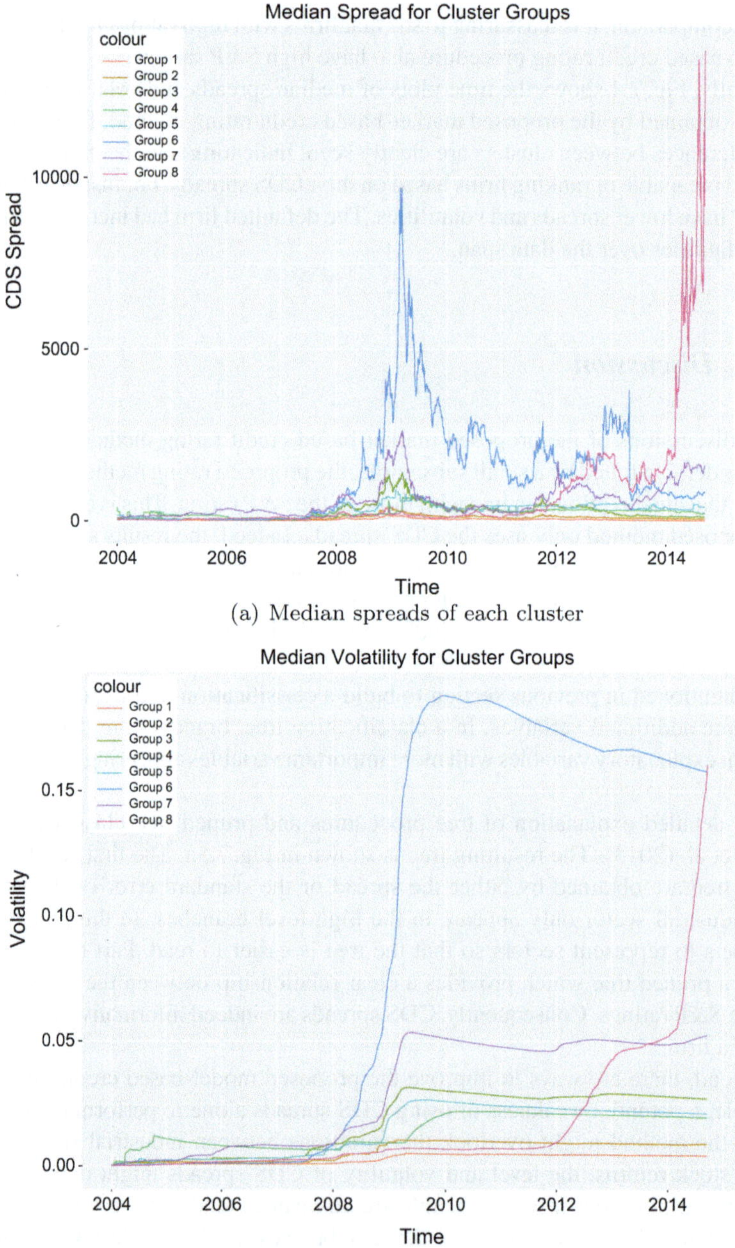

(b) Median volatility of each cluster

Fig. 7.4 Time plots of the median spreads and volatility for each cluster based on results of the proposed market-based credit rating

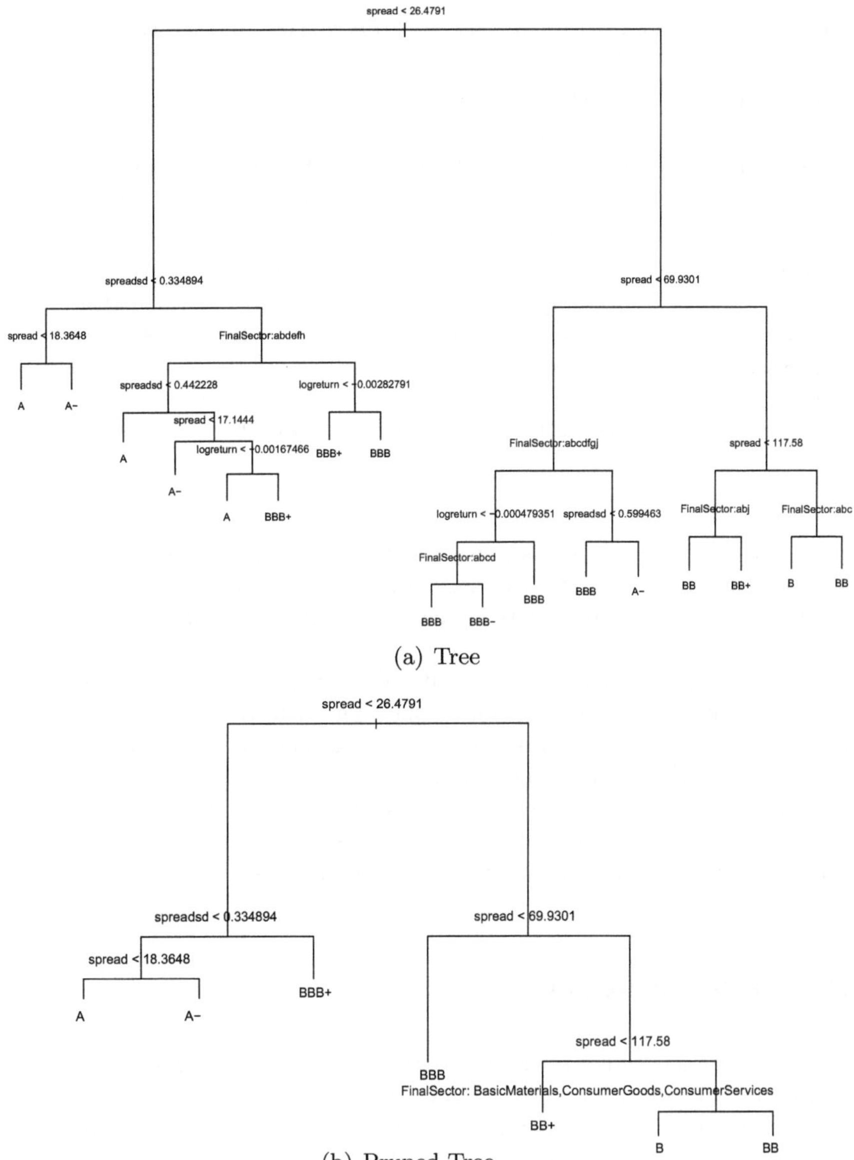

(a) Tree

(b) Pruned Tree

Fig. 7.5 Classification tree on S&P rating

Table 7.4 Median end-of-year CDS spreads from 2011 to 2014 for different industrial sectors

Sector	2011	2012	2013	2014
Basic material	101.32170	66.50631	50.92944	38.05819
Consumer goods	107.78704	68.26233	44.79193	39.74378
Consumer services	109.7089	95.46819	52.94550	39.21939
Energy	70.60976	46.93080	35.96803	33.08610
Financials	164.23591	68.44491	41.62527	32.40109
Healthcare	54.72795	37.95143	21.49869	19.71443
Industrials	64.43414	36.37636	23.81710	23.83260
Technology	103.52434	110.35931	54.64625	41.88988
Telecommunications services	42.88189	42.26581	35.86557	37.04494
Utilities	106.76204	59.46691	34.12895	25.43286

during recession, but will not stop using daily tools or visiting doctors. With the difference between sectors, it seems sector may affect credit rating. However, data from more firms and more sectors are needed to better study the role played by sectors.

Another interesting issue is that volatilities of CDS spreads may vary from sector to sector. Sectors with higher volatilities may be more likely to have lower rating. Since sample variances are sensitive to outliers, we apply wavelet transform to the log returns of CDS spreads. Figure 7.6 shows the scatter plot of sample means and standard deviations of the smoothed log returns for various sectors. The plot confirms that some sectors indeed have higher volatility. Thus, industrial sectors could be used to enhance credit rating. This issue deserves a careful investigation.

7.4 Concluding Remarks

Similar to stock and future prices, CDS spreads reflect the expectation of market participants on credit risk of a firm. Thus, CDS spreads are informative for credit rating. In this chapter, we proposed a market-based credit rating method based on ARIMA-GARCH modeling and prediction of CDS spreads. The proposed method is simple and widely applicable. Limited empirical analysis showed that ratings obtained by the proposed method perform reasonably well. However, further study is needed to improve the results of the proposed rating method. For example, the issue mentioned in the comparison of the proposed method with S&P rating in Sect. 7.3.2 may be solved using additional information. In particular, information concerning industrial sectors, macro-economic factors, and firm size could be helpful.

In the literature, Feng et al. (2008) and Amato and Furfine (2004) argue that there is some effect of business cycle on credit ratings. It's true that macroeconomic factors

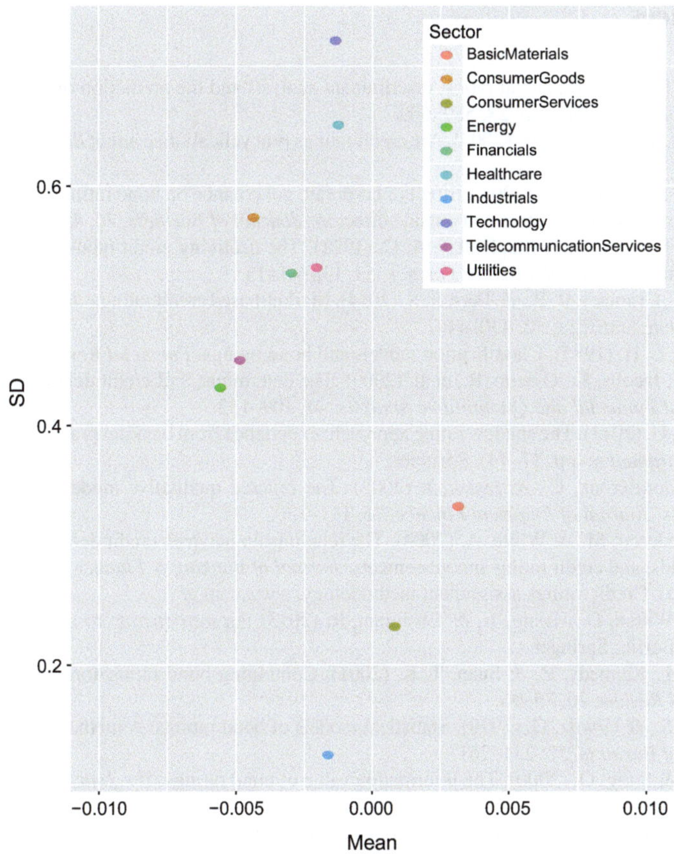

Fig. 7.6 Mean and standard deviation of log returns of CDS spreads across sectors

may affect systematic risk which in turn affect credit ratings. Yet business cycle is still not fully understood or not widely accepted, see Summers (1998). One of such examples is the famous equity premium puzzle in the standard RBC model. Finally, Blume et al. (1998) and Bhojraj and Sengupta (2003) both mention the relationship between credit rating and firm size; thus firm size may be useful in improving credit rating. Intuitively, large firm is less likely to default, or even too big to fail. The issue of firm size also deserves a careful study.

References

Altman, E. I. (1968). Financial ratios, discriminant analysis and the prediction of corporate bankruptcy. *Journal of Finance*, *23*, 589–609.
Amato, J. D., & Furfine, C. H. (2004). Are credit ratings procyclical? *Journal of Banking & Finance*, *28*, 2641–2677.
Bhojraj, S., & Sengupta, P. (2003). Effect of corporate governance on bond ratings and yields: The role of institutional investors and outside directors. *Journal of Business*, *76*, 455–475.
Blume, M. E., Lim, F., & MacKinlay, A. C. (1998). The declining credit quality of us corporate debt: Myth or reality? *Journal of Finance*, *53*, 1389–1413.
Creal, D. D., Gramacy, R. B., & Tsay, R. S. (2014). Market-based credit ratings. *Journal of Business & Economic Statistics*, *32*, 430–444.
Ederington, L. H. (1985). Classification models and bond ratings. *Financial Review*, *20*, 237–262.
Ericsson, J., Jacobs, K., Oviedo, R., et al. (2009). The determinants of credit default swap premia. *Journal of Financial and Quantitative Analysis*, *44*, 109–132.
Erlenmaier, U. (2011). The shadow rating approach: experience from banking practice. In *The Basel II Risk Parameters* (pp. 37–74). Springer.
Feng, D., Gouriéroux, C., & Jasiak, J. (2008). The ordered qualitative model for credit rating transitions. *Journal of Emprical Finance*, *15*, 111–130.
Hull, J., Predescu, M., & White, A. (2004). The relationship between credit default swap spreads, bond yields, and credit rating announcements. *Journal of Banking & Finance*, *28*, 2789–2811.
ICAP (2013). Credit ratings assignment methodology. www.icap.gr.
James, G., Witten, D., Hastie, T., & Tibshirani, R. (2013). *An introduction to statistical learning* (Vol. 6). Berlin: Springer.
Kamstra, M., Kennedy, P., & Suan, T. K. (2001). Combining bond rating forecasts using logit. *Financial Review*, *20*, 75–96.
Kaplan, R. S., & Urwitz, G. (1979). Statistical models of bond ratings: A methodological inquiry. *Journal of Business*, *52*, 231–261.
Kliger, D., & Sarig, O. (2000). The information value of bond ratings. *The Journal of Finance*, *55*, 2879–2902.
Micu, M., Remolona, E., Wooldridge, P., et al. (2004). The price impact of rating announcements: Evidence from the credit default swap market. *BIS Quarterly Review*, *2*, 55–65.
Nason, G. P. (2008). *Wavelet methods in statistics with R*. Berlin: Springer.
Nordberg, D. (2010). *Corporate governance: Principles and issues*. London: Sage.
Summers, L. H. (1998). Some skeptical observations on real business cycle theory. In *Real business cycles: A reader* (p. 97).

Chapter 8
Using Public Information to Predict Corporate Default Risk

C.N. Peng and J.L. Lin

Abstract Corporate defaults are often affected by many factors that are roughly divided into the two types: internal factors and external factors. Internal factors can be measured precisely with firm-specific financial statistics while external factors contain qualitative data, like related news. There are large amount of timely information from news which affects the default probability of corporates. Efficient extraction information contained in the news is the main focus of this study and we propose to use empirical Bayes and Bayesian Networks to achieve this goal. First, we retrieve both macroeconomic and firm-specific news published by major newspapers in Taiwan. Then, word segmentation is applied, keywords are extracted and then the news variables are computed. Instead of adding the news variables to the logistic regression model, we convert them into prior distribution for the parameters in the corporate default model. Finally, we compute the posterior distribution of the model parameters to predict the corporate default. The estimation is performed using the integrated nested Laplace approximations which, to our belief, is better than the traditional Markov Chain Monte Carlo for our model. Empirical analysis using Taiwanese data finds that news has a significant impact on the corporate default rate prediction. Adding the news variable does improve the forecast precision and prove its usefulness.

C.N. Peng (✉) · J.L. Lin
Commerce Development Research Institute, Taiwan, Republic of China
e-mail: Kenal.peng@cdri.org.tw

J.L. Lin
e-mail: jlin@gms.ndhu.edu.tw

W.K. Härdle et al. (eds.), *Applied Quantitative Finance*, Statistics and Computing,
DOI 10.1007/978-3-662-54486-0_8

8.1 Introduction

Due to the rapid development of internet, we can get instant global economic news on all the financial media around the clock. There are basically two kinds of news based on its frequency and involved entity. One is regularly published government economic data and forecast, and the other is occasional occurrence of corporate litigation, financial earning information, personnel changes or industry dynamics. News such as Taiwan HTC's infringement cases sued by the US Apple, US Apple's announcement of its unexpected decrease of sales, or the talk of Morris Chang, TSMC's chairman, will have direct or indirect impacts on business, industry and the overall economic environment. Extracting and interpreting these financial news to forecast corporate default rates have been an important issue. However, since news is mostly qualitative, and is often released irregularly, it is difficult to quantify such information as variables to be included in the econometric models. In practice, credit rating agencies such as S and P and Moody's and other credit rating agencies have taken into account non-quantitative factors to adjust their credit rating results obtained from the statistical models.

Financial information can also be classified as qualitative and quantitative types. News about European debt crisis is qualitative data while credit rating or economic growth rate is quantitative data. Both types of data have significant impacts on corporate earnings and should be included in the corporate default prediction models. For the quantitative data, one can directly feed them into statistical models for empirical analysis. As for extracting information from qualitative data, it would be much more convenient to perform the task using the Bayesian models, which combine prior distribution and likelihood function into posterior distribution. Qualitative information is for prior distribution as data is for the likelihood function. In other words, textual news can be coerced into priori distribution. Yet, there is still one obstacle for this implementation. In traditional Bayesian models, priori distribution is formulated for the model parameters in likelihood functions or in regression models. While we could easily make a statistical inference from the news about its impact on default rate, its implication for models parameters is unclear. For example, the Euro debt crisis will not only increase the potential default probability of the bank, but also slowdown economic growth. Such information is difficult to be converted into priori distribution of model parameters. Therefore, the main purpose of this paper is to quantify financial news and embed it in a Bayesian framework to forecast corporate default rates. The computation and simulation are performed using the Integrated Nested Laplace Approximations (INLA) which is believed to be more efficient than the popular Markov Chain Monte Carlo (MCMC) for our model. It is worth mentioning that our model could be further developed as a real-time and dynamic default prediction model that is very useful for credit risk management.

8.2 Literature Review

Credit rating reflects the soundness of the enterprise and related literatures are voluminous. We shall first examine the influences of credit rating by some major credit rating agencies, followed by evaluating these credit ratings. Then, we discuss papers on modeling corporate default probability and introduce information theory and its application. Finally, we review models containing quantitative and qualitative variables.

Brooksaff et al. (2004) used the Standard & Poor and Fitch's credit ratings to assess their impacts on the global stock market. The empirical analysis confirmed significant effects, especially when the credit rating are downward graded. Yet, it is not the case for newly developing countries. Ferreira and Gama (2007) also found a spillover effect on the stock markets of other countries when a country's rating is downward graded. Kim and Wu (2008) discover some impacts on credit markets when credit rating agencies release long and short term ratings. Orth (2013) applied Bayesian simulation approach to adjust the rating of sovereign debt securities and corporate debt securities. There exist under-estimation of risk for Standard & Poor's credit rating, especially when the rating is downward graded. Literatures on modeling corporate default probability are voluminous and can be roughly divided into two categories: structural model and reduced-form model. Merton's model as in Black and Scholes (1973) and Merton (1974) is the representative structural model. Credit rating agency, Moody, further revised it as Merton-KMV model. In this model, when the market value of a corporate's assets is lower than its liabilities, the company will soon reach default. It uses European option pricing to calculate the default probability. This model has been called firm-value based model. Vasicek (1977) and Shimko (1993) use stochastic interest rates to evaluate the Bond prices. Longstaff and Schwartz (1995) and Hui et al. (2003) relax part of the assumptions and modify Merton model. However, in addition to the internal factors from within the corporate, there are many external factors that could cause corporate default. The changing external environment has gradually made structural model less popular. Reduced-form model, also known as intensity model, mainly explores the linkage among corporate default and the explanatory variables. It was first proposed by Jarrow and Turnbull (1995) and a great deal of related models were developed, including multiple regression analysis (West 1970), multivariate discriminant analysis and Z-score model (Altman 1968), logistic model (Ohlson 1980), Probit model (Zmijewski 1984), order probability model (Gentry et al. 1985; Blume et al. 1998; Guttler and Wahrenburg 2007), fixed proportional hazards model (Cox 1972; Lane et al. 1986; Bharath and Shumway 2008), discrete-time hazard model (Shumway 2001; Chava and Jarrow 2004), credit rating transition matrix (Lando and Skodeberg 2002) and dynamic default intensity model (Duffie et al. 2007). It is worth noting that (Duffie et al. 2007) and its extended models belong to the application of survival models, which use macroeconomic, industry, firm-specific and other variables to estimate the default intensity.

Information arrives in many forms but all affect the corporates performance. While information about corporate earnings and other general information are released on quarterly or monthly based, the daily stock market is often strongly influenced by the news of the day so that the daily close price reflect daily market information rather than corporate real operating conditions. Brown et al. (1988), Braun et al. (1995), Pandher and Currie (2013), Coval and Shumway (2001) and others interpret this phenomenon from different angles. Tetlock (2007) studied medias (Wall Street Journal) impact on investors and found significant impacts of negative news on stock trading volume. Tetlock et al. (2008) show that negative wording will affect corporate revenue and can be used as an important predictor for the stock returns and the corporate revenue. Antweiler and Frank (2004) studied the impact of the web news on stock market. Yet, it is rather difficult to evaluate the composite impacts of news from different sources as their basic characteristics might be different from each other in a fundamental way.

For Bayesian credit risk literature, Czado (1994) derived Bayesian inference of binary regression models with parametric link; Gössl (2005), and McNeil and Wendin (2007) used Bayesian inference method to revise portfolio credit risk calculation; Kiefer (2008, 2009, 2010), Jacobs and Kiefer (2010, 2011), Gössl (2005) and McNeil and Wendin (2007) included outside experts opinions via Bayesian framework to compute the posterior density of underlying parameters in credit risk models. Orth (2013) studied the evaluation of sovereign and corporate credit risk, and calculated credit rating transition matrix. Lock and Gelman (2010) transforms the poll results into a priori distribution and then combine it with the general regression model to predict the US presidential election results. Ben-Gal (2007) and Fernandez and Salmeron (2008) show that Bayesian network model could be represented by directed acyclic graph, which describes the relationship between two or more nodes, and the node strength was expressed by probability. Yet, this approach requires clear definitions of all nodes with real data that limited its applicability. Among few related researches, Alexander (2000) used Bayesian belief networks (BBNs) to design work insurance policy. Pourret et al. (2008a), mentioned that Denmark's largest financial services company (Nykredit) applied BBNs to predict the default probability of large corporates. It is worth noting that Bayesian network model is mainly applied in computational biology and bioinformatics gene regulatory networks, gene expression analysis, document classification, information retrieval, decision support systems and so on.

Furthermore, both Back et al. (2001) and Kloptchenko et al. (2004) combined firm-specific variables with news processed using text mining methods to evaluate the impact of the news on the corporation. However, this approach is limited to specific event and is difficult to generalize to general cases. Only few studies combine quantitative and qualitative data into a single model to predict corporate default rates and Lu et al. (2012) is one exception. He retrieved keywords from news, classified these keywords into crisis and non-crisis categories, use chi-square test to screen proper keywords and then assign weights to construct Intensity of Default-Corpus (ITDC) which latter is fed into a logistic regression model for corporate default probability prediction. The empirical results showed that the closer to the crisis point the better estimation of default probability.

8.3 Econometric Models

We shall discuss our econometric models in two parts. The conventional corporate default model is first introduced and then the news variables are added.

8.3.1 *Logistic Models for Default Rate*

Among existing models, we select Shumway (2001)'s model as our base model because it is a dynamic discrete-time hazard model. Let T denote the time of default and the firm starts at $t = 1$. Then, the survival probability at Δt, is

$$\varphi(t|x) \equiv p(t \in [t, t + \Delta t | T \geq t, x]) \tag{8.1}$$

$$= \frac{1}{1 + e^{-\theta_1 g(t) - \theta_2 X}} \tag{8.2}$$

The multi-period logistic model for empirical analysis. Equation (8.2) now becomes

$$\lambda(t|x) \equiv ln(\varphi(t|x)) = \theta_1 g(t) + \theta_2 X \tag{8.3}$$

where $g(t) = ln(t)$ is a function of t, θ_1, θ_2 are estimated parameters, and x could be firm-specific earnings or macroeconomic variables. By plugging-in estimated parameters into the model, we get the strength of default, the higher the value the higher the default probability. Note that model defined in (8.3) will be reduced to standard logistic model if the term $g(t)(= ln(t))$ is removed.

8.3.2 *Default Models Including News Information*

In Bayesian models, past data can be used to specify priori distribution (Robbins (1985), Brandel (2004)). Assume that $p(x|\theta)$ is the likelihood function of x, and θ is the unknown parameter of interest. Let $g(\theta|\eta)$ be the prior distribution of θ, where η is called hyper-parameters vector. Brandel (2004) applied Bayes theory and obtained posterior distribution as

$$p(\theta|x, \eta) = \frac{p(x|\theta)g(\theta|\eta)}{m(x|\eta)} = \frac{p(x|\theta)g(\theta|\eta)}{\int p(x|\theta)g(\theta|\eta)d\theta}$$

where $m(x|\eta) = \int p(x|\theta)g(\theta|\eta)d\theta$ is the marginal distribution of x. Then the expectation of posterior density is

$$E[\theta|x] = \frac{\int \theta p(x|\theta)g(\theta|\eta)d\theta}{\int p(x|\theta)g(\theta|\eta)d\theta} \tag{8.4}$$

In (8.4), the estimation result will be affected by the hyper-parameters vector, η. Estimation is straightforward if η is known but η is usually unknown in practice. In turn, Marginal Maximum Likelihood Estimation (MMLE) can be applied and the resulting marginal distribution $m(x|\eta)$ of x is then used to estimate η. This process is called empirical Bayes method.

Obviously, for default probability model, the dependent variable is 0 (event does not occur) or 1 (event occurs), the estimated default probability is within (0,1] and the explanatory variables are macroeconomic or firm-specific financial variables. This explains why Kleinman (1973) Wilhelmsen et al. (2009), Kiefer (2009, 2010), and Jacobs and Kiefer (2010, 2011) all choose Beta-Binomial model. Assume that variable Y_{it} represents the default status of i-th corporate at time t. $Y_{it} = 1$ when it defaults or $Y_{it} = 0$ when it does not default. Y_{it} has Bernoulli(π_i) distribution, where π_i is default probability of corporate i. Assume the default status of corporate i is independent over time. Let X_i be the default status up to time n_i, we have the following formula

$$X_i = \sum_{t=1}^{n_i} Y_{it} \sim B(n_i, \pi_i)$$

where X_i has binomial distribution and variable X_i will vary with π_i. The maximum likelihood function of corporate i with default probability at time n_i is

$$p(X_i = x_i|\pi_i) = C_{x_i}^{n_i} \pi_i^{x_i} (1 - \pi_i)^{n_i - x_i}$$

Through dynamic default probability model, we can solve for π_i. Assume π_i has $Beta(r, s)$ distribution and is re-parameterized as $Beta_{rep}(\mu, M)$ where

$$\mu = \frac{\gamma}{\gamma + s}, M = \gamma + s$$

Put (8.5) into $Beta(\mu, M)$, the joint probability density function is

$$g(\Pi = \pi|\mu, M) = \frac{\Gamma(M)}{\Gamma(M\mu)\Gamma(M(1-\mu))} \pi^{M\mu - 1} (1 - \pi)^{M(1-\mu)-1}$$

Thus the marginal probability function is

$$m(X = x|\mu, M) = C_x^n \frac{\Gamma(M)}{\Gamma(M\mu)\Gamma(M(1-\mu))} \frac{\Gamma(x + M\mu)\Gamma(n - x + M(1-\mu))}{\Gamma(n + M)}$$

Finally, the posterior distribution is $Beta(r_{EB}, s_{EB})$ where

$$\gamma_{EB} = x + M\mu, S_{EB} = n - x + M(1 - \mu)$$

In the estimation process, the relationship of hyper-parameters requires simulation estimation. While there exist a great of simulation estimation methods, Markov Chain Monte Carlo (MCMC) or EM-algorithm are commonly used. This paper adopts more efficient Integrated Nested Laplace Approximations (INLA). Wilhelmsen et al. (2009) compared the difference between MCMC and INLA, and found that the efficiency and accuracy of INLA are better than that of MCMC. See Rue et al. (2009) for details.

8.3.3 Bayesian Network Model

Ben-Gal (2007) pointed out that the main structure of Bayesian network model is non-circulate probability graphical model where there exist sequential causal relationships among various events. In this paper, we shall estimate $\lambda(t|x)$ in the discrete-time hazard model. As it is affected not only by firm-specific variables at time t, but also by the news information at time $t - 1$. Hence, we specify the default probability function as

$$f(Y_t|X_{i,t-1}), i = 1, 2, ..., n$$

where $X_{i,t-1}$ is the news information factor at time $t - 1$. News information will be retrieved, quantified and its probability distribution will be simulated. Finally, using Bayesian network method, we can get revised default probability as

$$f(Y_t, X_{i,t-1}), i = 1, 2, ..., n$$

From above, assume there are two news X_1 and X_2 then

$$f(Y, X_1, X_2) = f(Y|X_1, X_2)f(X_2|X_1)f(X_1)$$

where $f(Y|X_1, X_2)$ is the default probability from the corporate default model, and $f(X_2|X_1)$ is mutual impact between news events. This, in principle, can be used to estimate the impact of sequent news events on default probability but it is difficult to implement in practice. Thus, we follow Fernandez and Salmeron (2008) and Rijmen (2008) and apply regression analysis. Since Bayesian network is non-circulate directed, each news event can be treated as an explanatory variable, and the dependent variable is the corporate default variable. Under multiple news events, we need to consider whether they are related with each other. Its mathematical formula is

$$f(Y|X) = \alpha X + \varepsilon$$

were Y represent default of the corporate, X is news event and ϵ is random error. Obviously, we have

$$f(Y|X) = \frac{f(Y,X)}{f(X)} = f(X|Y)\frac{f(Y)}{f(X)}$$

and

$$f(X|Y) \propto f(Y)f(X|Y)$$

It is called Naive Bayes (NB) when each news event is independent and Tree Augmented Naive Bayes (TAN) when news are dependent. Rijmen (2008) adopts logistic regression in Bayesian Network model where the weight of each segmented word is estimated with the logistic regression model. Wilhelmsen et al. (2009) assumed the prior distribution of logistic regression coefficients is

$$\beta_j \sim \pi(\beta_j|\theta_j), \quad j = 0, 1, ..., M$$

where $\pi(\cdot|\theta)$ denotes all possible distributional function, and θ_j is a scalar or vector parameter. In this paper, θ_j is assumed as a scalar from news information, we obtain posterior distribution as

$$\pi(\beta, \theta|y) = \frac{\pi(\beta, \theta, y)}{\pi(y)} \propto \pi(y|\beta, \theta)\pi(\beta|\theta)\pi(\theta) \tag{8.5}$$

$$= \prod_i \pi(y_i|\beta, \theta)\pi(\beta|\theta)\pi(\theta) \tag{8.6}$$

Solved by INLA, we obtain $\pi(\beta, \theta|y)$ where news information is included.

Rue et al. (2009) derive the test for parameters. Let $y = (y_1, y_2, \cdots, y_n)$ be the observed variable, its probability function be $\pi(\beta|\theta)$, and the model for unknown parameter β be $\pi(\beta|\theta)$, and θ is hyper-parameter. $\pi(\theta)$ is distribution function of hyper-parameter, and through Bayesian theory we get marginal posterior distribution as

$$\pi(\beta_i|y) = \int_\theta \pi(\beta_i|\theta, y)\pi(\theta|y)d\theta \tag{8.7}$$

$$\pi(\theta_j|y) = \int \pi(\theta|y)d\theta_{-j} \tag{8.8}$$

Through INLA, we get the approximation of marginal posterior distribution as

$$\tilde{\pi}(\beta_i|y) = \int_\theta \tilde{\pi}(\beta_i|\theta, y)\tilde{\pi}(\theta|y)d\theta \tag{8.9}$$

$$\tilde{\pi}(\theta_j|y) = \int \tilde{\pi}(\theta|y)d\theta_{-j} \tag{8.10}$$

where $\int \pi(\theta|y)d\theta_{-j}$ represents integration over all but the j-th parameter. In other words, to obtain the estimated value of the parameters, we have to integrate over all hyper-parameters. As the parameter vector θ is multi-dimensional, we must use the Laplace estimate. In order to improve accuracy, latent Gaussian models are applied. To obtain the estimation of $\tilde{\pi}(\beta_i|y)$, we need to get an approximation of $\tilde{\pi}(\beta_i|\theta, y)$ and $\tilde{\pi}(\theta|y)$, which are assumed as Gaussian distribution. We use Kullback–Leibler Divergence (KLD) test which is defined as below:

$$D_{KL}(P||Q) = \int_{-\infty}^{\infty} ln(\frac{p(x)}{q(x)})p(x)dx$$
$$D_{KL}(P||Q) = \sum_{x} ln(\frac{p(x)}{q(x)})p(x)$$

where P, Q are respective two cumulative probability distribution for continuous and discrete random variables. Let Q be normal distribution, when $D_K L(P||Q) \approx 0$, P is also normally distributed.

For model selection, we shall use two methods: in-sample Receiver Operating Characteristic Curve (ROC) and out-of-sample forecasting error (Lin and Tsay 2007). Altman and Bland (1994) proposed ROC as a method of diagnostic test, which is widely used by biometrics. Within a 2×2 table, P denotes positive and N negative.

Diag	P	N	Total
Truth			
P	TP	FN	
N	FP	TN	
Total			Nobs

The True Positive(TP) and True Negative (TN) are the cells for the right diagnostics. Let Nobs denote total number of samples, then accuracy ratio, sensitivity and specificity are respectively defined as (TP+TN)/Nobs and TP/(TP+FN). ROC is based upon sensitivity and specificity, and can be used for model comparison.

As for out of sample forecasting error, we can calculate its Root Mean Square of Error (RMSE)

$$RMSE_t = \sqrt{\sum_{i=t+1}^{T} \frac{(\hat{y}_i - y_i)^2}{n_i}}$$

To summarize, these news frequencies are used to obtain the prior distribution of the regression parameters β in Shumways model.

8.4 Extracting News Information

Chinese characters can be divided into three types: classical, vernacular and other dialects. Their usages and the structures are all different from each other. Vernacular is currently used, which might vary due to the geographical environment and social backgrounds, but in general follows certain syntax. Tsay (2008) pointed out that a sentence is constituted by two basic components, subject and predicate. Subject is the major part of the sentence, either the perpetrators of the action, or the objects being interpreted, clarified or depicted. The predicate is the statement to clarify the subject. In this paper, news from various media also follow a set of rules. For example, editorial manual of Central News Agency depicts the main structure and term usage. We further classify economic news in Taiwan into two categories. One is economic news containing economic data, business cycle indicators, or economic policy announcement released by the government official or agencies, which does not make judgment of any corporate. The other one is public talks or comments on specific corporate. In addition to Taiwan's local news, foreign financial news also has a considerable impact. We must distinguish their impacts.

8.4.1 *News Keywords*

Keyword is set in accordance with the commonly used terms and categorized by subject, verb and adjective. Six main structures of the subject are set including raw materials, European debt crisis, people and institutions, economic data release, as well as business and policy agreements. Within each main structure, at least eight keywords are selected, which can be different words with same meaning. The predicate is mainly verbs, such as recovery, recess, rise, fall, up, down, strength and the like. Default keywords defined by Taiwan Economic Journal (TEJ) are also included. There are 10 categories: bankruptcies, restructuring, bounced checks, bail out, take over, CPAs doubt on continuous operation, net worth is negative, unlist, tight budget, negative worth, and shut down. Finally, these keywords are classified as positive, neutral and negative.

8.4.2 *Keyword Conversion*

Segmented keywords from all news items (documents) are compiled into the document-term matrix where columns are news items, and rows are keywords. For each cell, 0 and 1 indicate if there is such keyword. For each keyword, summing over all news items during any specific quarter will produce frequency of keywords. This process is repeated separately for positive and negative keywords and their ratios are then computed.

8.5 Empirical Analysis and Results

This paper uses quarterly firm-specific data of all listed companies in Taiwan from 2000 to 2012. The data is taken from TEJ, excluding incomplete data entries, financial firms and news media corporations. There are 908 corporates where 805 are still listed at the end of the sample period and 103 are unlisted. As for news, there are mainly two sources: newspaper and networks news. Yet, as the latter is only available for one month after posting, we only use newspapers news. The major four newspapers in Taiwan are China Times, United Daily News, Free News and Apple Daily. The data is collected daily from the first quarter of 2008 to the fourth quarter of 2012, amounting to about 270,000 news items.

8.5.1 Empirical Models

This empirical analysis is illustrated in two parts. First, we follow previous research in selecting firm-specific quantitative variables under the constraint that the resulting ROC curve is above 90%. Second, as for news variables, we employ empirical Bayes and Bayesian networks to convert as quantitative variables and then feed them into the base default model as is introduced previously. We compare the performance of the following six models:

1. Model I: Earnings model
   This is the conventional default model only based upon firm-specific financial variables and $ln(t)$. Standard logistic regression estimation will suffice.
2. Model II: Earnings-macroeconomic model
   In additional to firm-specific financial variables and $ln(t)$, macroeconomic variables are also included in the model for default prediction. Again, the model is estimated using standard logistic regression.
3. Model III: Bayesian earnings model
   Earnings models are formulated under Bayesian framework and is used to predict corporate defaults. Empirical Bayes is used for model estimation. To be specific, default variable is first regressed against firm-specific financial variables and the estimation results are then converted into prior distribution of the associated parameters using INLA algorithm. Finally, the posterior distribution are derived with prior and likelihood function.
4. Model IV: Bayesian earnings-macroeconomic model
   Both firm-specific financial variables and macroeconomic variables are included in the model under Bayesian framework. Estimation procedure is the same as Bayesian earnings model except that macroeconomic variables are added.
5. Model V: Bayesian news-earnings model
   News variables are added to the Bayesian earning model via empirical Bayes method and INLA. To be specific, firm-specific news are classified as good news, and bad news and their relative frequencies to all news are computed.

For macroeconomic news, only those containing the five most and least frequent keywords, such as *price, monetary policy* are counted. These news are further classified as good news or bad news. Next, regress the firm default variable against *ln*(*t*), firm-specific good news and firm-specific bad news for each firm. Then, for each ten macroeconomic key variables, regress firm default variable against macroeconomic good news and bad news for each firm. Summing the predicted probability distribution obtained from five most frequent keywords and firm-specific regressions give rise to model 5(L). Similarly, summing the predicted probability distribution obtained from five least frequent keywords and firm-specific regressions gives rise to model 5(S). It is worth noting that the idea of Bayesian network model is used in this step. Now, we could combine news effects with Shumway's model with firm-specific variable using INLA.

6. Model VI: Bayesian news-earning-macroeconomic model
   News variables are added to the Bayesian earnings-macroeconomic model via empirical Bayes method and INLA. Computation procedure is the same as Bayesian news-earnings model except for the added macroeconomic variables.

8.5.2 *Variable Selection*

In the discrete-time hazard model, explanatory variables must be included to predict corporate default probability. Altman (1968), Ohlson (1980) and Zmijewski (1984) used three to nine financial ratio variables. Shumway (2001) included two financial ratios and three market-driven variables. Chava and Jarrow (2004) added industrial variables to those in Altman (1968) and Zmijewski (1984). Lee and Yeh (2004) focused on the relationship between corporate governance and financial distress. Duffie et al. (2007) added macroeconomic variables to the dynamic intensity model. Campbell et al. (2008) added two firm-specific financial ratios and stock return to the list of variables compiled by Shumway. Standard & Poor consider eighteen variables on liquidity, terms of profitability, capital structure, cash flow and ability to repay interest etc. in corporate's credit rating.

After taking all these literatures into consideration, we select seven variables: assets-liabilities ratio, quick ratio, ratio of retained earnings to total assets, earnings per share, operating expense ratio, unemployment rate, and TAIEX (Taiwan Stock Exchange Capitalization Weighted Stock Index) return. The definitions of the selected variables are reported in Table 8.1. In additional to the variable definition and type of variables, their expected sings are also listed. Table 8.2 summarizes basic statistics of the variables. Except for unemployment rate and the stock market return, extremely large skewness and kurtosis of firm-specific financial variables indicate obvious departure from normal distribution assumption. Table 8.3 reports the parameter estimates for Model I and II. As can be seen from the table, except for the ratio of retained earnings to total assets, all variables are significant and their signs are consistent with prediction from finance theory. The Bayesian estimates for Model III

Table 8.1 Variable definitions

Category	Name	Variable definition	Sign
Financial structure	Asset-liability ratio	Total asset/total liability	Negative
Solvency	Quick ratio	(Liquid asset-inventory)/liquid liability	Negative
Profitability	Ratio of retained earnings to total assets	Retained earning/total assets	Negative
	Earning per share	Earning/number of shares	Negative
Operating capacity	Operating expense ratio	Operating expense/net revenue	Positive
Macro variables	Unemployment rate		Positive
	Stock market return		Negative

Table 8.2 Summary statistics for explanatory variables

Variable	Mean	Std	Median	Skewness	Kurtosis
Assets-liabilities ratio	3.52	5.11	2.61	25.81	1083.35
Quick ratio	1.65	4.08	1.06	26.86	1139.92
Retained earnings to total assets ratio	0.06	0.66	0.10	−49.58	3273.97
Earnings per share	1.15	3.56	0.66	54.42	5886.05
Operating expense ratio	0.26	6.49	0.10	110.16	14210.83
Unemployment rate	4.48	0.72	4.32	0.19	2.77
Stock market return	3.80	26.47	6.89	0.13	3.22

Table 8.3 Parameter estimates for Model I and II. Signif. codes: $p < 0.001$ ****; $p < 0.01$ ***; $p < 0.05$ **; $p < 0.1$ *

	Model I		Model II	
	Est.	t-stat	Est.	t-stat
Intercept	−0.9814	−3.306****	−2.8829	−6.067****
Time trend	−0.2667	−4.151****	−0.4406	−5.145****
Assets-liabilities ratio	−1.0066	−6.734****	−1.0574	−6.427****
Quick ratio	−3.1559	−11.742****	−3.0579	−10.585****
Retained earnings to total assets ratio	0.0776	1.495	0.0767	1.470
Earnings per share	−0.1998	−9.201****	−0.1974	−8.292****
Operating expense ratio	0.0085	3.188***	0.0080	2.457**
Unemployment rate			0.5361	5.726****
Stock market return			−0.0041	−1.668*

Table 8.4 Estimation results using empirical Bayes method

	Model III					Model IV				
	Mean	Std	2.50%	97.50%	KLD	Mean	Std	2.50%	97.50%	KLD
Intercept	−0.923	0.414	−1.729	−0.103	7.07E-14	−2.847	0.618	−4.074	−1.645	1.08E-13
Time trend	−0.264	0.090	−0.437	−0.085	1.23E-12	−0.436	0.111	−0.655	−0.217	0.00E+00
Asset-lib rat	−1.037	0.209	−1.464	−0.644	1.17E-11	−1.087	0.214	−1.523	−0.682	8.67E-12
Quick ratio	−3.169	0.375	−3.932	−2.459	8.80E-12	−3.066	0.376	−3.830	−2.353	8.13E-12
Rtn earnings/Total asset	0.081	0.072	−0.037	0.245	2.67E-10	0.080	0.068	−0.031	0.234	2.39E-10
Earn per share	−0.202	0.030	−0.260	−0.142	1.55E-13	−0.199	0.031	−0.259	−0.138	1.79E-13
Oper. exp. rat	0.009	0.004	0.002	0.017	9.15E-11	0.008	0.004	0.001	0.018	2.37E-10
Unemp rat						0.542	0.122	0.303	0.781	4.67E-15
Stock mkt rtn						−0.004	0.003	−0.010	0.002	1.50E-14

Table 8.5 Estimation results of logistic model with news variables Signif. codes: p < 0.001****; p < 0.01***; p < 0.05**; p < 0.1*

	Pooled news		Category news	
	est.	t-stat	est.	t-stat
Intercept	2.19453	0.464	0.8633	0.164
Time trend	−2.18977	−1.711	−1.9608	−1.387
Pooled news	−0.01627	−1.247		
Positive news			−1.688	−2.234**
Negative news			1.6849	2.627***

and IV are summarized in Table 8.4. In additional to mean, standard deviation, 2.50 and 97.5% quantiles, we also compute Kullback-Leibler Divergence (KLD) statistics which measures divergence from normal distribution. KLD values of all parameters are very small, indicating little divergence of the posterior distribution from normal distribution. Furthermore, we also find that except for the ratio of retained earnings to total assets and TAIEX return, the 95% confidence interval of all parameters do not include 0.

8.5.3 Adding News Variables

For the purpose of comparison, we perform a logistic regression of corporate default indicator directly against news variables and put the results in Table 8.5. On the left panel of the table all news are pooled together while on the right panel positive and negative news are separated. As is expected, pooled news variable is not significant while negative news has stronger effect than positive news on corporate default rate though both estimates are significant. Similar findings were found in Lu et al. (2012).

Now we turn to models V and VI where news variables are added to Shumway's model on the Bayesian framework. Empirical results are reported in Table 8.6. A detailed comparison of estimation results, we make the following observations. First, estimation results of Shumway model without news variables are similar whether it is estimated within classical logistic model or empirical Bayesian model. Second, the results of Model V and VI are similar to those of models III and IV that except for the ratio of retained earnings to total assets and TAIEX return, the 95% confidence interval of all parameters do not include 0 and all KLDs are close to 0. Third, adding news variables to the Bayesian model would change the parameter estimates a great deal. For example, the impacts on quick ratio double in Models V and VI. Fourth, as is in Fig. 8.1 where RMSE for out-of-sample forecast over time are graphed, model II with macroeconomic variables consistently outperform the base model I with only firm-specific variable. Fifth, as is shown in Fig. 8.2, ROC curves of all six models are all above 90%, but the difference is small among models.

Table 8.6 Estimation results using INLA with news variables

	Model V					Model VI				
	Mean	Std	2.50%	97.50%	KLD	Mean	Std	2.50%	97.50%	KLD
Intercept	5.370	4.966	−4.330	15.157	7.56E-15	−3.244	7.574	−19.120	10.672	2.56E-11
Time trend	−1.751	1.323	−4.373	0.822	7.71E-14	−0.256	1.803	−3.648	3.444	1.29E-11
Asset-lib rat	−1.810	0.478	−2.810	−0.930	1.94E-11	−1.831	0.476	−2.831	−0.957	2.43E-11
Quick ratio	−1.577	0.545	−2.709	−0.565	2.44E-11	−1.504	0.542	−2.636	−0.503	2.20E-11
Rtn earnings to total asset rat	0.312	0.239	−0.079	0.851	1.97E-10	0.305	0.240	−0.085	0.847	2.47E-10
Earn per share	−0.149	0.047	−0.238	−0.055	4.28E-12	−0.152	0.046	−0.238	−0.058	1.46E-11
Oper. exp. ratio	0.011	0.006	0.002	0.024	2.05E-09	0.011	0.006	0.001	0.025	2.75E-09
Unemp rat					0.673	0.291	0.126	1.272	1.38E-11	
Stock mkt rtn					−0.006	0.006	−0.019	0.006	9.48E-12	

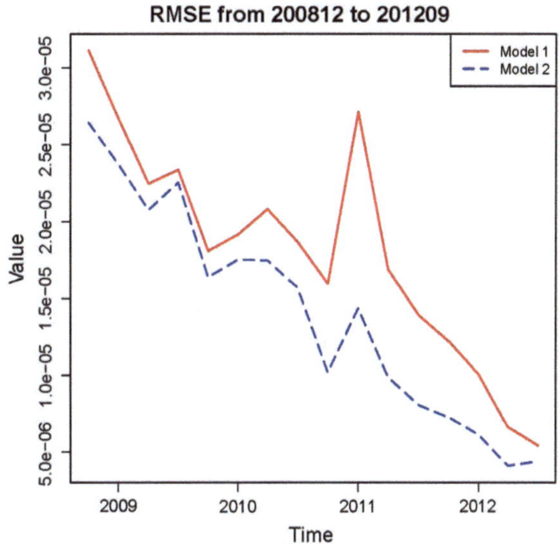

Fig. 8.1 RMSEs for Model I and II

Figure 8.3 is the time series graph of average corporate default rate where annual default rates are put in the left panel whereas quarterly default rates are put in the right panel. The upper panel are based on ratio of number of unlisting stocks to total stocks while the bottom panel is computed following the definition of default in TEJ. As is obvious from the figures, the peaks and troughs of default rate defined by TEJ leads those defined by unlisting.

Figure 8.4 displays the average corporate default intensity of all six models which is the simple average of each corporate's default intensity in each model respectively. Comparing the resulting intensity figures of paired models will highlight their differences. Models I, III and V do not contain macroeconomic variables and are put in left panel of the figure while Model II, IV and VI include macroeconomic variables and are put in the right panel. From the figure, we make the following findings. First, the estimated default intensity of empirical Bayesian model (model III/IV) are smaller than those from Shumway model (model I/II). Both estimates differ from each other by a big margin from 2002 to 2008 when the subprime mortgage crisis broke out. Yet both estimates converge after 2008 crisis. The patterns are similar for both paired models with and without macroeconomic variables. Second, as news variables are collected from Jan 1, 2008 to Dec 31, 2012, comparing estimation results of two sub-periods with and without news variable would reveal the impacts of new variables. Considering that each keyword might have different impact on corporates default probability, we add one more step. We first perform a logistic regression again each macroeconomic keyword, compute the squared root of residual sum of squares, RSS, and then sort them in ascending order. Next, we select the keywords with the 5 largest

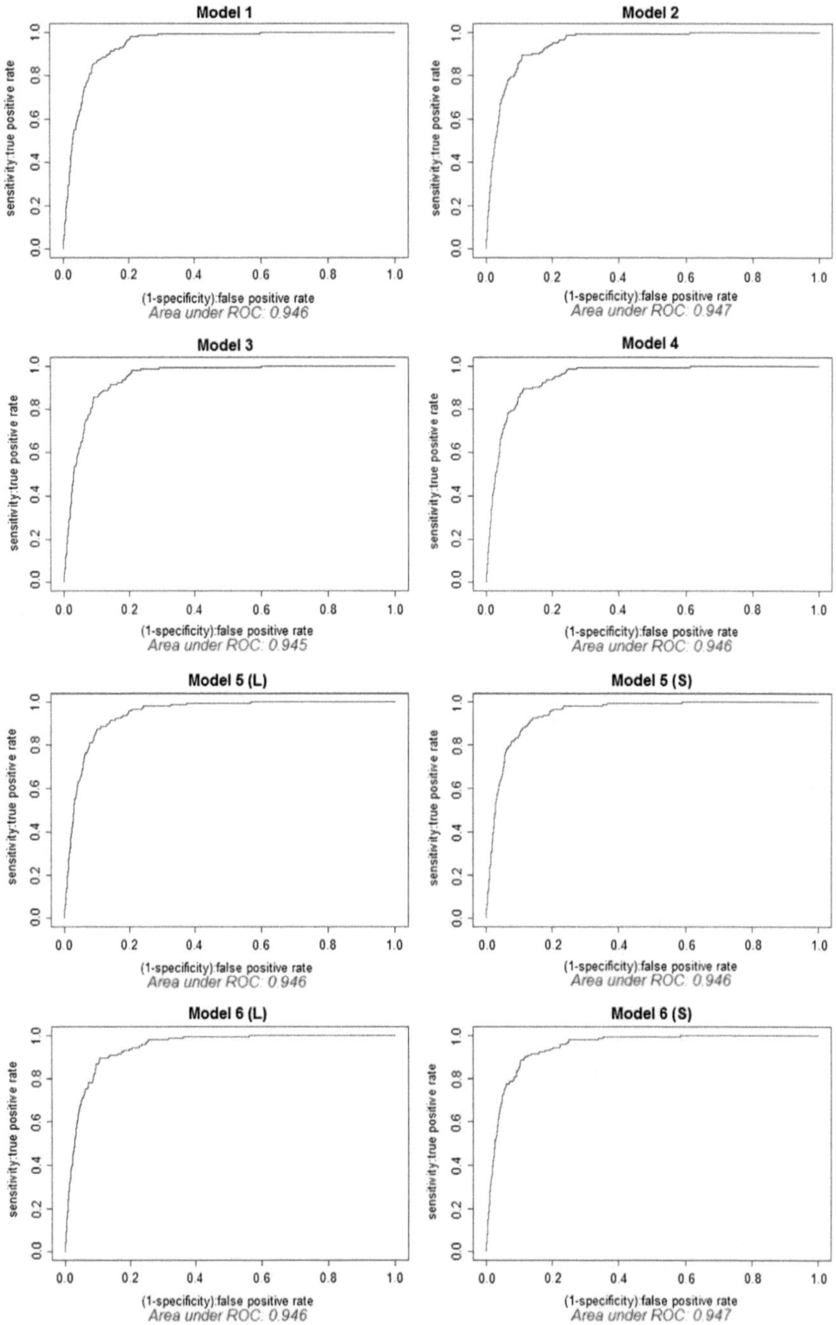

Fig. 8.2 ROC curves for all six models

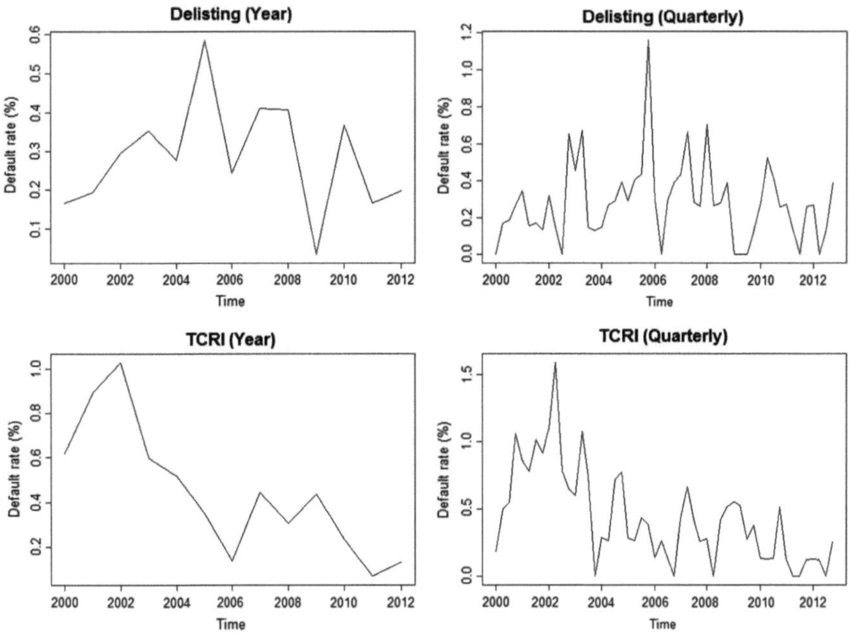

Fig. 8.3 Time series plot of average default rate

RSSs (denoted as L-keywords) and keywords with 5 smallest RSSs (denoted as S-keywords). These L- and S-keywords are then respectively combined with keywords for each corporate, fed into the Bayesian models and estimated using INLA the algorithm. The results are put in the middle and bottom panels of Fig. 8.4. From the figure, we observe that without macroeconomic variables, adding S-keywords produces a sharp increase of corporate default intensity in early 2008 while the impact of S-keyword are much smaller. The situation is reversed when macroeconomic variables are included in the model where L-keywords has a stronger impact on default intensity than S-keywords. It deserves further investigation to explain this phenomenon. Finally, the ROC curves for all six models are reported in Fig. 8.2. They are all above 90% but adding news variables does not significantly increase the ROC curve.

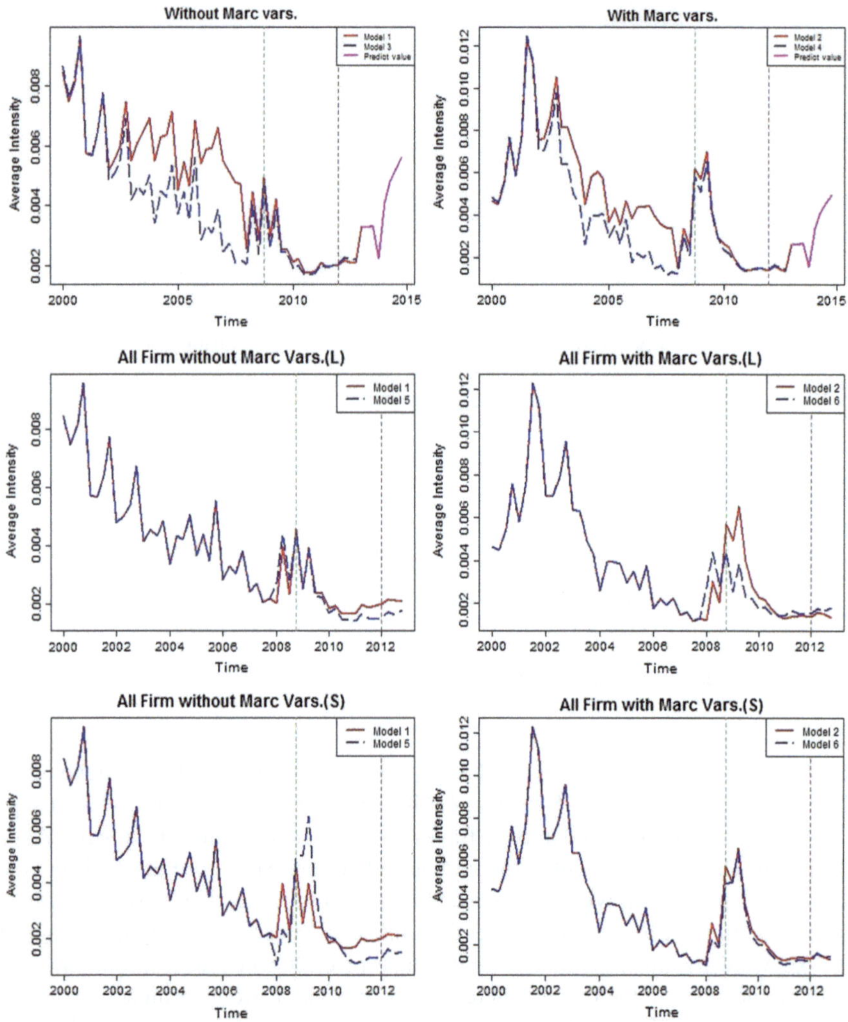

Fig. 8.4 Time series plot of default intensity of all six models

8.6 Conclusions

While corporates' financial reports are released on a quarterly basis, daily economic or financial news could provide timely and useful information about the corporate default probability. This paper provides a framework to extract information from text-based news to improve corporate default prediction. Instead of converting news as a new variable in a standard logistic regression model, we employ the complicated INLA method to transform news into prior information of corporate default and then estimate its impact within a Bayesian model. The conversion is completed using the

INLA. Empirical analysis confirms usefulness of the proposed method though there are rooms for improvement. For example, each keyword might have different weight and the timing of the news within each quarter might be important. These issues deserve further investigation in the future.

References

Alexander, C. (2000). Bayesian Methods for Measuring Operational Risks. *Discussion papers in Finance*.

Altman, D. G, & Bland, J M. (1994). Diagnostic tests. 1: Sensitivity and specificity. *BMJ: British Medical Journal, 308*(6943), 1552

Altman, E. I. (1968). Financial ratios, discriminant analysis and the prediction of corporate bankruptcy. *The Journal of Finance, 23*, 589–609.

Antweiler, W., & Frank, M. Z. (2004). Is all that talk just noise? the information content of internet stock message boards. *The Journal of Finance, 59*, 1259–1294.

Ben-Gal, I. (2007). Bayesian networks. In F. Ruggeri, R. Kenett, & F. Faltin (Eds.), *Encyclopedia of statistics in quality and reliability*. New York: Wiley.

Back, B., Toivonen, J., Vanharanta, H., & Visa, A. (2001). Comparing numerical data and text information from annual reports using self-organizing maps. *International Journal of Accounting Information Systems, 2*, 249–269.

Bharath, S. T., & Shumway, T. (2008). Forecasting default with the merton distance to default model. *Review of Financial Studies, 21*, 1339–1369.

Black, F., & Scholes, M. (1973). The pricing of options and corporate liabilities. *The Journal of Political Economy, 81*, 637–654.

Blume, M. E., Lim, F., & MacKinlay, A. C. (1998). The declining credit quality of U.S. corporate debt: Myth or reality? *Journal of Finance, 53*, 1389–1414.

Brandel, J. (2004). Empirical Bayes methods for missing data analysis. *Department of Mathematics Uppsala University, Project Report.*

Braun, P. A., Nelson, D. B., & Sunier, A. M. (1995). Good news, bad news, volatility, and betas. *The Journal of Finance, 50*, 1575–1603.

Brooks, R., Faff, R. W., Hillier, D., & Hillier, J. (2004). The national market impact of sovereign rating changes. *Journal of Banking and Finance, 28*, 233–250.

Brown, K. C., Harlow, W. V., & Tinic, S. M. (1988). Risk aversion, uncertain information, and market efficiency. *Journal of Financial Economics, 22*, 355–385.

Campbell, J. Y., Hilscher, J., & Szilagyi, J. (2008). In search of distress risk. *Journal of Finance, 63*, 2899–2939.

Chava, S., & Jarrow, R. A. (2004). Bankruptcy prediction with industry effects. *Review of Finance, 8*, 537–569.

Coval, J. D., & Shumway, T. (2001). Is sound just noise? *The Journal of Finance, 56*, 1887–1910.

Cox, D. R. (1972). Regression models and life-tables. *Journal of the Royal Statistical Society Series B, 34*, 187–220.

Czado, C. (1994). Bayesian inference of binary regression models with parametric link. *Journal of Statistical Planning and Inference, 41*, 121–140.

Duffie, D., Saita, L., & Wang, K. (2007). Multi-period corporate default prediction with stochastic covariates. *Journal of Financial Economics, 83*, 635–665.

Fernández, A., & Salmerón, A. (2008). Extension of Bayesian network classifiers to regression problems. In H. Geffner, R. Prada, I. M. Alexandre, & N. David (Eds.), *Advances in artificial intelligence - IBERAMIA 2008* (Vol. 5290, pp. 83–92), Lecture notes in artificial intelligence. Berlin: Springer.

Ferreira, M. A., & Gama, P. M. (2007). Does sovereign debt ratings news spill over to international stock markets? *Journal of Banking and Finance, 31*, 3162–3182.

Gentry, J. A., Newbold, P., & Whitford, D. T. (1985). Predicting bankruptcy: If cash flow's not the bottom line, what is? *Financial Analyst's Journal, 41*, 47–56.

Gössl, C. (2005). Predictions based on certain uncertainties-a Bayesian credit portfolio approach. *Working Paper*, Hypo Vereinsbank AG.

Güttler, A., & Wahrenburg, M. (2007). The adjustment of credit ratings in advance of defaults. *Journal of Banking and Finance, 31*, 751–767.

Hui, C. H., Lo, C. F., & Tsang, S. W. (2003). Pricing corporate bonds with dynamic default barriers. *Journal of Risk, 5*(3), 17–37.

Jacobs, M, Jr., & Kiefer, N. M. (2010). *The bayesian approach to default risk: A guide* (2nd ed.). Ithaca: Center for Analytical Economics.

Jacobs Jr, M., & Kiefer, N. M. (2011). The bayesian approach to default risk analysis and the prediction of default rates. *Discussion paper.*

Jarrow, R. A., & Turnbull, S. M. (1995). Pricing derivatives on financial securities subject to credit risk. *The Journal of Finance, 50*, 53–85.

Kiefer, N. M. (2008). Default estimation for low-default portfolios. *Journal of Empirical Finance, 16*, 164–173.

Kiefer, N. M. (2009). Correlated defaults, temporal correlation, expert information and predictability of default rates. *CAE Working Paper.*

Kiefer, N. M. (2010). Default estimation and expert information. *Journal of Business and Economic Statistics, 28*, 320–328.

Kim, S. J., & Wu, E. (2008). Sovereign credit ratings, capital flows and financial sector development in emerging markets. *Emerging Markets Review, 9*, 17–39.

Kleinman, J. C. (1973). Proportions with extraneous variance: Single and independent sample. *Journal of the American Statistical Association, 68*, 46–54.

Kloptchenko, A., Eklund, T., Karlsson, J., Back, B., Vanharanta, H., & Visa, A. (2004). Combining data and text mining techniques for analysing financial reports. *Intelligent systems in accounting, finance and management, 12*, 29–41.

Lando, D., & Skodeberg, T. (2002). Analyzing ratings transitions and rating drift with continuous observations. *Journal of Banking and Finance, 26*, 423–444.

Lane, W. R., Looney, S. W., & Wansley, J. W. (1986). An application of the cox proportional hazards model to bank failure. *Journal of Banking and Finance, 10*, 511–531.

Lee, T. S., & Yeh, Y. H. (2004). Corporate governance and financial distress: Evidence from Taiwan. *Corporate Governance, 12*(3), 378–388.

Lin, J. L., & Tsay, R. S. (2007). Comparisons of forecasting methods with many predictors. Working paper, Department of Finance, National DongHwa University.

Lock, K., & Gelman, A. (2010). Bayesian combination of state polls and election forecasts. *Political Analysis, 18*, 337–348.

Longstaff, F. A., & Schwartz, E. S. (1995). A Simple Approach to Valuing Risky Fixed and Floating Rate Debt. *The Journal of Finance, 50*(3), 789–819.

Lu, Y. C., Wei, Y. C., Chang, T. Y., & Liao, W. J. (2012). Does soft information from news improve the forecasting performance of habitually used Taiwan corporate credit risk index? *Taiwan Banking and Finance Quarterly, 13*(4), 27–53. (in Chinese).

McNeil, A. J., & Wendin, J. P. (2007). Bayesian inference for generalized linear mixed models of portfolio credit risk. *Journal of Empirical Finance, 14*, 131–149.

Merton, R. C. (1974). On the pricing of corporate debt: The risk structure of interest rates. *The Journal of Finance, 29*, 449–470.

Ohlson, J. A. (1980). Financial ratios and the probabilistic prediction of bankruptcy. *Journal of Accounting Research, 18*, 109–31.

Orth, W. (2013). Default probability estimation in small samples-with an application to sovereign bonds. *Quantitative Finance, 13*, 1891–1902.

Pandher, G., & Currie, R. (2013). CEO compensation: A resource advantage and stakeholder-bargaining perspective. *Strategic Management Journal*, *34*, 22–41.

Pourret, O., Naim, P., & Marcot, B. (Eds.). (2008a). *Bayesian networks: a practical guide to applications* (Vol. 73). New Jersey: Wiley.

Rijmen, F. (2008). Bayesian networks with a logistic regression model for the conditional probabilities. *International Journal of Approximate Reasoning*, *48*, 659–666.

Robbins, H. (1985). An empirical bayes approach to statistics. In Samuel Kotz & Norman L. Johnson (Eds.), *Breakthroughs in Statistics* (pp. 388–394). New York: Springer.

Rue, H., Martino, S., & Chopin, N. (2009). Approximate Bayesian inference for latent Gaussian models by using integrated nested Laplace approximations. *Journal of the royal statistical society: Series B*, *71*, 319–392.

Shimko, D. C. (1993). Bounds of probability. *Risk Magazine*, *6*(4), 33–37.

Shumway, T. (2001). Forecasting bankruptcy more accurately: A simple hazard model. *The Journal of Business*, *74*, 101–124.

Tetlock, P. C. (2007). Giving content to investor sentiment: The role of media in the stock market. *The Journal of Finance*, *62*, 1139–1168.

Tetlock, P. C., Saar-Tsechansky, M., & Macskassy, S. (2008). More than words: Quantifying language to measure firms' fundamentals. *The Journal of Finance*, *63*, 1437–1467.

Tsay, Z. Y. (2008). *Chinese grammer*. Taipei: Wanjuan. (in Chinese).

Vasicek, O. (1977). An equilibrium characterization of the term structure. *Journal of Financial Economics.*, *5*, 177–188.

West, R. (1970). An alternative approach to predicting corporate bond ratings. *Journal of Accounting Research*, *7*, 118–127.

Wilhelmsen, M., Dimakos, X. K., Husebo, T., & Fiskaaen, M. (2009). Bayesian modelling of credit risk using integrated nested laplace approximations. Working paper. Oslo, Norway: Norwegian Computing Center.

Zmijewski, M. E. (1984). Methodological issues related to the estimation of financial distress prediction models. *Journal of Accounting Research*, *22*, 59–82.

Chapter 9
Stress Testing in Credit Portfolio Models

M. Kalkbrener and L. Overbeck

Abstract As, in light of the recent financial crises, stress tests have become an integral part of risk management and banking supervision, the analysis and understanding of risk model behaviour under stress has become ever more important. In this paper, we present a general approach to implementing stress scenarios in a multi-factor credit portfolio model and analyse asset correlations, default probabilities and default correlations under stress. We use our results to study the implications for credit reserves and capital requirements and illustrate the proposed methodology by stressing a large investment banking portfolio. Although our stress testing approach is developed in a particular credit portfolio model, the main concept - stressing risk factors through a truncation of their distributions - is independent of the model specification and can be applied to other risk types as well.

9.1 Introduction

Stress testing has been adopted as a generic term describing various techniques used by financial firms to analyze their potential vulnerability to extreme yet plausible events, see para 718 in Basel Committee on Banking Supervision (2006) for specific requirements on banks' stress testing programs. Stress scenarios have long been used in risk management to supplement risk measures like value-at-risk (VaR) and economic capital (EC), e.g. Kupiec (1998) and Berkowitz (2000), but stress testing has gained new prominence in the aftermath of the subprime crisis and the European sovereign debt crisis. In particular, it has become an integral part of banking supervision, which is reflected in regulatory stress testing programs such as the annual Comprehensive Capital Assessment Review (CCAR) performed by the FED since

M. Kalkbrener (✉)
Deutsche Bank AG, Risk Methodology, Frankfurt, Germany
e-mail: michael.kalkbrener@db.com

L. Overbeck
Department of Mathematics, University of Gießen, Arndtstraße 2,
35392 Giessen, Germany
e-mail: Ludger.Overbeck@math.uni-giessen.de

W.K. Härdle et al. (eds.), *Applied Quantitative Finance*, Statistics and Computing,
DOI 10.1007/978-3-662-54486-0_9

2010 (Board of Governors of the Federal Reserve (2012)) and the EU-Wide Stress Tests, see European Banking Authority (2011). Principles of sound stress testing practices have been laid down by the Basel Committee on Banking Supervision (2009), analysis and surveys of macroeconomic stress testing can be found in Čihák (2007), Alfaro and Drehmann (2009), Drehmann (2009), Quagliariello (2009) and Borio et al. (2012).

An important challenge in designing effective stress tests is the selection of scenarios that are both severe and plausible. One approach frequently used by risk managers is the application of historical scenarios such as the 1987 stock market crash or the subprime crisis. By their very nature, historical scenarios are plausible and provide useful information on the sensitivity of a portfolio to specific market shocks but they restrict attention to prior stress episodes. Hypothetical scenarios, in contrast, are not constrained to replicate specific past incidents and can therefore cover a broader spectrum of potential risks. However, depending on the choice of the hypothetical scenarios, stress test results might misrepresent risks either because the most dangerous scenarios are not considered or because the selected scenarios are too implausible. In order to overcome this problem, systematic approaches to scenario selection have been investigated for more than 15 years, e.g. Studer (1999). More recent work on that subject includes Breuer et al. (2009), Breuer and Csiszár (2013), Flood and Korenko (2015) and Glasserman et al. (2015).

In this paper, we present an alternative approach to the specification of stress scenarios, which has initially been introduced in Bonti et al. (2006) for analyzing credit concentrations. Duellmann and Erdelmeier (2009) use the same methodology for stressing credit portfolios of German banks. In this approach, statistical EC or VaR models serve as quantitative framework for the specification of stress scenarios. More precisely, stress scenarios are defined through constraints on the risk factors of the model. These constraints are then used to truncate the distribution of the stressed risk factors or - in other words - restrict the state space of the model, where each state represents values of the risk factors. The response of the peripheral (or unstressed) risk factors is specified by the dependence structure of the model. As an example, consider an economic downturn in the automotive sector. In a structural credit portfolio model with industry and country factors this scenario can be implemented by truncating the systematic risk factor for the automotive industry. The severity of the downturn scenario is reflected through the truncation threshold, so that a lower threshold implies more severe stress. Since the automotive industry is positively correlated to most industry and country factors non-automotive exposures are affected as well.

The specification of stress scenarios through constraints on risk factors of VaR or EC models has a number of advantages:

1. Stress scenarios are implemented in a way that is consistent with the existing quantitative framework. This implies that the relationships between (unrestricted) risk factors remain intact and the experience gained in the day-to-day use of the model can be utilized in the interpretation of stress testing results. It has to be analyzed, however, whether historical correlation patterns, which are typically used for calibrating (unstressed) risk capital models, provide an appropriate depen-

dence structure for stress testing, see Sect. 9.4 for a sensitivity analysis of model correlations under stress.

2. In a given stress scenario, risk factors are not set to deterministic values but remain stochastic variables, i.e., stressed as well as unstressed factors follow a joint distribution conditional on the truncation thresholds that define the stress scenario. This feature distinguishes our approach from standard stress tests, which are typically based on deterministic stress scenarios. As a consequence, stressed risk measures, e.g. expected loss, value-at-risk or economic capital, can be calculated in each stress scenario.
3. The probability of each stress scenario, e.g. the probability that the risk factors satisfy all the constraints under non-stress conditions, can be easily calculated in the statistical model. This is a good indicator for the severity of a stress scenario.

Our stress testing methodology is developed in a multi-factor credit portfolio model. We provide details on the implementation of stress scenarios and discuss practical issues such as the calculation of truncation thresholds in multi-factor stress scenarios. Another objective of this paper is to review recent results on stressed asset correlations, default probabilities and default correlations presented in Kalkbrener and Packham (2015a) and Packham et al. (2014). In these papers, the analysis is performed in a factor model that follows a normal variance mixture distribution, which covers a wide range of light-tailed to heavy-tailed distributions. Aside from analysing the behaviour under stress for given stress levels or stress probabilities, the asymptotic behaviour, that is, the behaviour under stress as the stress level becomes arbitrarily high, is investigated. Contrary to popular belief, it is shown that the impact of stress on the asymptotic behaviour is greater in light-tailed models than in heavy-tailed models. More specifically,

- asset correlations under stress are less sensitive for heavy-tailed models than light-tailed models;
- default correlations under stress converge to 0 for light-tailed models and to a number strictly greater than 0 for heavy-tailed models;
- default probabilities converge to 1 for light-tailed models and to a number strictly smaller than 1 for heavy-tailed models.

However, the asymptotic behaviour of stresses PDs is not representative for ordinary stress tests: only for rather extreme stress severities, stressed PD's become higher in light-tailed than in heavy-tailed models. Finally, these results are used to study the implications for risk measures, credit reserves and capital requirements under stress.

The paper is structured in the following way. The second section introduces the quantitative framework we will work in. The third section describes our approach to implementing stress scenarios in a multi-factor credit portfolio model. In addition, results from stressing a sample portfolio are presented. In Sect. 9.4, the impact of stress on asset correlations, default probabilities and default correlations is analyzed. Section 9.5 concludes.

9.2 Quantitative Framework for Stress Testing

The objective of this section is the introduction of a class of multi-factor credit portfolio models that serve as the formal framework for the implementation of stress scenarios.

In a typical bank, the economic as well as regulatory capital charge for credit risk far outweighs capital for any other risk class. Key drivers of credit risk capital are concentrations in a bank's credit portfolio, either caused by material concentrations of exposure to individual names or large exposures to a single sector or to several highly correlated sectors. As a consequence, the stress testing methodology for credit risk has to be implemented in a credit portfolio model that provides sufficient flexibility for modeling risk concentrations.

The IRB approach in Basel Committee on Banking Supervision (2006) does not provide an appropriate quantitative framework. It is based on a credit portfolio model that was originally designed to produce portfolio-invariant capital charges. However, it is only applicable under the assumptions that (cf. Gordy 2003)

1. bank portfolios are perfectly fine-grained and
2. there is only a single source of systematic risk.

The simplicity of the model ensures its analytical tractability. However, it makes it impossible to model risk concentrations in a reasonable way.

In order to develop meaningful stress tests, we need to generalize the IRB approach to a multi-factor credit portfolio model that takes into account individual exposures and has a richer correlation structure. In this paper, we use a structural model (Merton 1974), which links the default of a firm to the relationship between its assets and the liabilities that it faces at the end of a given time period $[0, T]$ [1].

More generally, in a structural credit portfolio model the j-th obligor defaults if its ability-to-pay variable A_j falls below a default threshold c_j: the default event at time T is defined as $\{A_j \leq c_j\} \subseteq \Omega$, where A_j is a real-valued random variable on the probability space $(\Omega, \mathcal{A}, \mathbb{P})$ and $c_j \in \mathbb{R}$. We denote the default indicator $\mathbf{1}_{\{A_j \leq c_j\}}$ of the j-th obligor and its default probability $\mathbb{P}(\{A_j \leq c_j\})$ by I_j and p_j respectively. The portfolio loss variable is defined by

$$L := \sum_{j=1}^{n} l_j \cdot I_j, \tag{9.1}$$

where n denotes the number of obligors and l_j is the loss-at-default of the j-th obligor. In order to reflect risk concentrations, a joint distribution of the A_j has to be specified that captures the dependence between defaults of different obligors. This is done via the introduction of a factor model consisting of systematic and idiosyncratic factors. More precisely, each ability-to-pay variable A_j is decomposed into a sum of systematic factors $\Psi_1, \ldots, \Psi_m$ and an idiosyncratic [or specific] factor ε_j, that is

[1] A survey on credit portfolio modeling can be found in Bluhm et al. 2002 and McNeil et al. 2005

$$A_j = \sqrt{R_j^2} \sum_{i=1}^{m} w_{ji} \Psi_i + \sqrt{1 - R_j^2} \varepsilon_j. \tag{9.2}$$

It is usually assumed that the vector of systematic factors $\Psi = (\Psi_1, \ldots, \Psi_m)$ follows an m-dimensional normal distribution with mean $\mathbf{0} = (0, \ldots, 0)$ and covariance matrix $\Sigma = (\Sigma_{kl})$. The systematic weights $w_{j1}, \ldots, w_{jm} \in \mathbb{R}$ determine the impact of each systematic factor on the ability-to-pay variable A_j. The systematic weights are scaled such that the systematic component

$$\phi_j := \sum_{i=1}^{m} w_{ji} \Psi_i \tag{9.3}$$

is a standardized normally distributed variable, i.e., ϕ_j has mean 0 and variance 1. The idiosyncratic factors $\varepsilon_1, \ldots, \varepsilon_n$ are standardized normally distributed variables, they are independent of each other as well as independent of the systematic factors. Each R_j^2 is an element of the unit interval $[0, 1]$. It determines the impact of the systematic component on A_j and therefore the correlation between A_j and ϕ_j: it immediately follows from (9.2) that

$$R_j^2 = \mathrm{Corr}(A_j, \phi_j)^2. \tag{9.4}$$

In order to quantify portfolio risk, measures of risk are applied to the portfolio loss distribution (9.1). The most widely used risk measures in banking are value-at-risk and expected shortfall: value-at-risk $\mathrm{VaR}_\alpha(L)$ of L at level $\alpha \in (0, 1)$ is simply an α-quantile of L whereas expected shortfall of L at level α is defined by

$$\mathrm{ES}_\alpha(L) := (1 - \alpha)^{-1} \int_\alpha^1 \mathrm{VaR}_u(L) du.$$

For most practical applications the average of all losses above the α-quantile is a good approximation of $\mathrm{ES}_\alpha(L)$: for $c := \mathrm{VaR}_\alpha(L)$ we have

$$\mathrm{ES}_\alpha(L) \approx \mathrm{E}(L|L > c) = (1 - \alpha)^{-1} \int L \cdot \mathbf{1}_{\{L>c\}} d\mathbb{P}.$$

These risk measures are used to determine the economic capital, which is designed to state with a high degree of certainty the amount of capital needed to absorb unexpected losses. Economic capital $\mathrm{EC}(L)$ is usually defined as value-at-risk $\mathrm{VaR}_\alpha(L)$ at a high level α, e.g., $\alpha = 0.9998$, minus the expected loss $\mathrm{E}(L)$ of L:

$$\mathrm{EC}(L) := \mathrm{VaR}_\alpha(L) - \mathrm{E}(L),$$

where the subtraction of the expected loss reflects the fact that only unexpected losses are covered by economic capital.

9.2.1 Definition of Asset and Default Correlations

The critical quantities entering the risk measures defined above are the default probabilities and the risk concentrations of the default indicators I_j, either specified by default or asset correlations. In this subsection, we provide a formal definition of these quantities, an analysis of default or asset correlations under stress is performed in Sect. 9.4.

The default or event correlation ρ_{ij}^D of obligors i and j, with $i \neq j$, is defined as the correlation $\mathrm{Corr}(I_i, I_j)$ of the corresponding default indicators. Because

$$\mathrm{Var}(I_j) = \mathrm{E}(I_j^2) - p_j^2 = p_j - p_j^2,$$

the default correlation equals

$$\rho_{ij}^D = \mathrm{Corr}(I_i, I_j) = \frac{\mathrm{E}(I_i I_j) - p_i p_j}{\sqrt{(p_i - p_i^2)(p_j - p_j^2)}}. \tag{9.5}$$

The indicator variables I_j are defined in terms of ability-to-pay variables A_j, which are typically interpreted as log-returns of asset value processes. The correlation $\mathrm{Corr}(A_i, A_j)$ is therefore called the asset correlation ρ_{ij}^A of obligors $i \neq j$. As an immediate consequence of (9.2), the correlation as well as the covariance of the ability-to-pay variables of the counterparties i and j are given by

$$\mathrm{Corr}(A_i, A_j) = \mathrm{Cov}(A_i, A_j) = \sqrt{R_i^2}\sqrt{R_j^2} \sum_{k,l=1}^{m} \omega_{ik}\omega_{jl}\mathrm{Cov}(\psi_k, \psi_l). \tag{9.6}$$

There exists an obvious link between default and asset correlations. For given default probabilities, the default correlation ρ_{ij}^D is determined by $\mathrm{E}(I_i I_j)$ according to (9.5), and

$$\mathrm{E}(I_i I_j) = \mathbb{P}(A_i \leq c_i, A_j \leq c_j) = \int_{-\infty}^{c_i} \int_{-\infty}^{c_j} f_{ij}(u, v) du dv,$$

where $f_{ij}(u, v)$ is the 2-dimensional joint density function of A_i and A_j. Hence, default correlations depend on the joint distribution of A_i and A_j. If (A_i, A_j) is bivariate normal the correlation of A_i and A_j determines the copula of their joint distribution and hence the default correlation:

$$\mathrm{E}(I_i I_j) = \frac{1}{2\pi\sqrt{1 - {\rho_{ij}^A}^2}} \int_{-\infty}^{c_i} \int_{-\infty}^{c_j} \exp(-\frac{1}{2(1 - {\rho_{ij}^A}^2)}(u^2 - 2\rho_{ij}^A uv + v^2)) du dv. \tag{9.7}$$

Note, however, that for general ability-to-pay variables outside the multivariate normal class, the asset correlations do not fully determine the default correlations.

9.3 Factor Stress Methodology

In this section, we describe each of the steps of the stress testing process:

1. Specification of an economic stress scenario or scenario based on the characteristics of the portfolio
2. Translation of the scenario into constraints on the systematic factors of the credit portfolio model
3. Quantification of the impact of the stress scenario by calculating the conditional expected loss and other statistics of the portfolio

9.3.1 Specification of Stress Scenarios

The following classification should serve as a rough guide and distinguish different types of stress scenarios.

1. *Macroeconomic scenarios.* A macroeconomic scenario usually requires the use of a macroeconomic model. It specifies an exogenous shock to the whole economy that is propagated over time and may impact the banking system in various ways. This type of stress scenario is used by financial regulators or central banks in order to gain an understanding of the resilience of financial markets or the banking system as a whole.
2. *Market shocks.* These scenarios specify shocks to financial markets. This category also includes certain shocks of a "systemic" nature affecting credit risk (such as a sudden flight to liquidity), or sectoral shocks, for instance the deterioration in credit spreads in the TMT (Technology Media-Telecommunications) sector. Historical scenarios are frequently used for this type of shocks in order to increase the plausibility of these stress scenarios.
3. *Portfolio specific worst case scenarios.* The objective of this worst case analysis is to identify scenarios that are most adverse for a given portfolio. The specification of worst case scenarios can either be based on expert judgement or quantitative techniques.

These scenario types serve different purposes. Economic stress scenarios and market shocks are usually specified by risk management. The objective is to quantify the impact of a plausible economic downturn or a market shock on a credit portfolio.

The aggregated loss of portfolio specific worst case scenarios, on the other hand, serves more as a benchmark to create some awareness of the current market situation. The construction of these scenarios is driven by portfolio characteristics instead of economic considerations.

Regardless of the motivation for considering a particular scenario, there exist a number of criteria that characterize useful stress scenarios:

1. *Plausible.* Stress scenarios must be realistic, e.g. have a certain probability of actually occurring. Risk management will not take any actions based on scenarios that are regarded as implausible.
2. *Consistent.* One objective is to implement stress scenarios in a way that is consistent with the existing quantitative framework. This has the advantage that the relationships between risk factors remain intact and the experience gained in the day-to-day use of the model can be utilized in the interpretation of stress testing results.
3. *Adapted.* Stress tests should include scenarios that are specifically designed for the portfolio at hand. They should reflect certain portfolio characteristics and particular concerns in order to give a complete picture of the risks inherent in the portfolio.
4. *Reportable.* Stress scenarios should provide useful information for risk management purposes, which can be translated into concrete actions. For reporting purposes, it is crucial that the stress scenario is characterized by a clearly identifiable set of stressed risk factors, sometimes called the "core" factors. The remaining "peripheral" factors should then move in a consistent way with those "core" factors.

When designing specific stress scenarios, we usually focus on a small number of directly stressed factors, e.g. those factors that correspond to the sectors of interest. In addition, a small number of stressed factors makes it easier to transform the stress results into concrete management actions. The response of the other risk factors is specified by the dependence structure of the model. This approach is also a superior way to identify risk concentrations compared to just aggregating exposures per sector, because there it can happen that concentrations in distinct but highly correlated sectors remain undetected.

9.3.2 Implementation of Stress Scenarios in Credit Portfolio Models

In order to translate a given stress scenario into model constraints, a precise meaning has to be given to the systematic factors of the portfolio model. Recall that each ability-to-pay variable

$$A_j = \sqrt{R_j^2} \sum_{i=1}^{m} w_{ji} \Psi_i + \sqrt{1 - R_j^2} \varepsilon_j$$

is a weighted sum of m systematic factors $\Psi_1, \dots, \Psi_m$ and one specific factor ε_j. The systematic factors often correspond either to countries (or geographic regions) and industries. Equity data is frequently used to construct time-series for the systematic factors. Statistical techniques are then applied to these time-series to derive the joint distribution of the systematic factors. The systematic weights w_{ji} are chosen according to the relative importance of the corresponding factors for the given counterparty. They are either based on economic information or calculated via statistical techniques such as linear regression.

The economic interpretation of the systematic factors is essential for implementing stress scenarios in the model. The actual translation of a scenario into model constraints is done in two steps:

1. Identification of the appropriate risk factors based on their economic interpretation
2. Truncation of their distributions by specifying upper bounds that determine the severity of the stress scenario

Using the credit portfolio model introduced in Sect. 4.2 as quantitative framework, the specification of the model constraints is formalized as follows. A subset $S \subseteq \{1, \ldots, m\}$ is defined, which identifies the stressed factors Ψ_i, $i \in S$. For each of these factors a cap $C_i \in \mathbb{R}$ is specified. The purpose of the thresholds C_i, $i \in S$, is to restrict the sample space of the model. More formally, the restricted sample space $\bar{\Omega} \subseteq \Omega$ is defined by

$$\bar{\Omega} := \{\omega \in \Omega \mid \Psi_i(\omega) \leq C_i \text{ for all } i \in S\}. \tag{9.8}$$

In other words, $\omega \in \Omega$ is an element of the restricted sample space $\bar{\Omega}$ if none of the stressed factors exceeds its threshold in the event ω. Note that the probability $\mathbb{P}(\bar{\Omega})$ of the restricted sample space $\bar{\Omega}$ under the original probability measure $\mathbb{P}$ provides information on the likelihood of the stress scenario.

Although the formal framework for implementing stress scenarios is simple the actual translation of scenarios into model constraints can be rather complex depending on the specification of the scenario. If a scenario is defined in terms of constraints on the existing systematic country and industry factors the implementation is straightforward. However, even the identification of systematic risk factors is a difficult problem if the given scenario specification involves economic variables that cannot easily be mapped to the country and industry classification used in the model, e.g. the implementation of a drop in US house prices would require an analysis of the potential impact on different countries and industries before the scenario can be translated into model constraints. A more transparent approach, however, is

1. to add a US house price index to the set of systematic factors,
2. to extend the joint distribution of systematic factors in order to capture the dependence between US house prices and the country and industry factors of the model and
3. to implement this stress scenario through a constraint on the new factor.

It is important to note that the new macroeconomic factor - in the present example the US house price index - is not included in the decomposition of the ability-to-pay variable in (9.2), i.e. the US house price index has a weight of zero in all ability-to-pay variables. As a consequence, the behaviour of the unstressed model is not affected.

However, the dependence between new macroeconomic factors, denoted by $\Xi_1, \ldots, \Xi_k$, and the industry and country factors $\Psi_1, \ldots \Psi_m$ is captured in the extended covariance matrix of the larger factor model $(\Psi_1, \ldots \Psi_m, \Xi_1, \ldots, \Xi_k)$. In a stress scenario, the conditional distribution $\mathcal{L}((\Psi_1, \ldots \Psi_m)|\Xi_1 \leq C_1, \ldots, \Xi_k \leq C_k)$ of the country and industry factors given the constraint on the macroeconomic factors is used in (9.2) to obtain the stressed ability-to-pay variables. Therefore, the constraints on macroeconomic factors have an impact on the distribution of the country and industry factors in a stress scenario and, consequently, also on the ability-to-pay variables of all counterparties.

The above example illustrates that, in principle, the initial set of country and industry factors can be extended by a large number of macroeconomic and market factors in order to provide a comprehensive model for stress testing. However, the specification of the joint distribution of these different factors $(\Psi_1, \ldots \Psi_m, \Xi_1, \ldots, \Xi_k)$ is a challenging problem due to differences in the data frequency, e.g. quarterly GDP data versus daily market data, potential time lags between market and macroeconomic variables, etc.

Stress tests are frequently specified by setting the respective risk factors to specific values, e.g. a 10% drop in US house prices in a stress scenario compared to a 2% increase in the baseline scenario. In order to implement this scenario in our model the 10% drop has to be translated into a truncation threshold:

1. using historic house price volatility together with the baseline scenario we calibrate a distribution of US house price changes and
2. based on that distribution, we specify the truncation threshold C such that the conditional mean, i.e., the average of US house price changes below C, equals the 10% drop.

This technique can be generalized to a multi-factor stress scenario. However, if a stress scenario is not consistent with the correlation structure of the model, e.g. if two factors behave differently in the stress scenario although they are almost perfectly correlated in the underlying model, it will not be possible to precisely replicate the specified stress values through multi-dimensional thresholds. In this case, an optimization problem has to be solved instead that results in thresholds that provide the best possible replication but not a perfect match.

Restricting the state space through constraints on systematic factors is a flexible technique to incorporate stress scenarios into the portfolio model. So far, we have only considered stress scenarios that are defined by truncating factor distributions. Alternatively, stress scenarios could be defined via defining more complex constraints than simple caps on individual factors. One possibility is to restrict the state space of the model in such a way that the dependence of particular risk factors is increased. This technique provides an interesting alternative to simply changing correlation parameters of the model. By keeping the original model parameters intact, consistency problems are avoided such as maintaining the positive semi-definiteness of the correlation matrix of the systematic factors.

9.3.3 Calculation of Stressed Risk Capital

The actual calculation of the stressed loss distribution of the portfolio is done through Monte Carlo simulation on the restricted model space $(\bar{\Omega}, \bar{\mathcal{A}}, \bar{\mathbb{P}})$, see (9.8). It is therefore straightforward to calculate risk measures like expected loss, value-at-risk or expected shortfall for the loss distribution under stress and to use statistical techniques such as QQ-plots to study its behavior.

It depends on the particular purpose of a stress test which of those risk measures is used to quantify the impact of a stress test on the credit portfolio. One possibility is to analyze whether current capital requirements cover realized losses in stress scenarios and to use stress tests for the calculation of the conditional expected loss. Another application of stress tests is the analysis of future capital requirements, e.g. the bank wishes to satisfy its EC constraint one year into the future. If the stress event arrives within the one year horizon, then the bank will need capital sufficient to meet its EC requirement conditional on that stress event. This type of analysis requires the calculation of the VaR of the stressed portfolio. Finally, the future regulatory capital requirements in stress scenarios can be assessed by recalculating the Basel II formula with the stressed PDs from the multi-factor model. Since regulatory capital requirements are essential for capital management and strategic planning we regard this impact analysis as an important component of the stress testing methodology in a financial institution.

In the following, we will describe our approach by means of a specific scenario. As an example, consider a downturn scenario for the automotive industry. The simplest implementation in the portfolio model is the following restriction of the state space of the model: only those samples are considered in the Monte Carlo simulation where the automotive industry factor decreases by a certain percentage, say at least 2%. In other words, the distribution of the automotive industry factor is truncated from above at –2%. More precisely, the steps in the calculation of stressed EL and EC are:

- simulate risk factors under their original (non-stress) joint distribution,
- dismiss any simulation not satisfying the scenario constraints,
- derive EL, EC and other statistics from the loss distribution specified by the MC scenarios that satisfy the constraints.

Note that the automotive downturn scenario does not only have an impact on the automotive industry factor: because of correlations, other country factors as well as industry factors are also affected. Figure 9.1 shows the stressed distribution of the automotive industry factor (left) and the impact on the factor for the chemical industry (right): the distribution of the automobile factor has been truncated, while the distribution of the chemical industry factor is no longer centered but has moved to the left.[2]

[2]The distributions in Fig. 9.1 can be represented in a simple way: if $F_{auto}(x)$ denotes the (Gaussian) distribution of the automobile factor, its truncated distribution is given by $F_{auto}(x)/F_{auto}(-2\%)$ for $x \leq -2\%$. The factor for the chemical industry is called an incidentally truncated variable. Its marginal distribution is given by $F_{auto,chem}(-2\%, y)/F_{auto}(-2\%)$, where $F_{auto,chem}$ denotes the joint distribution of the two industry factors.

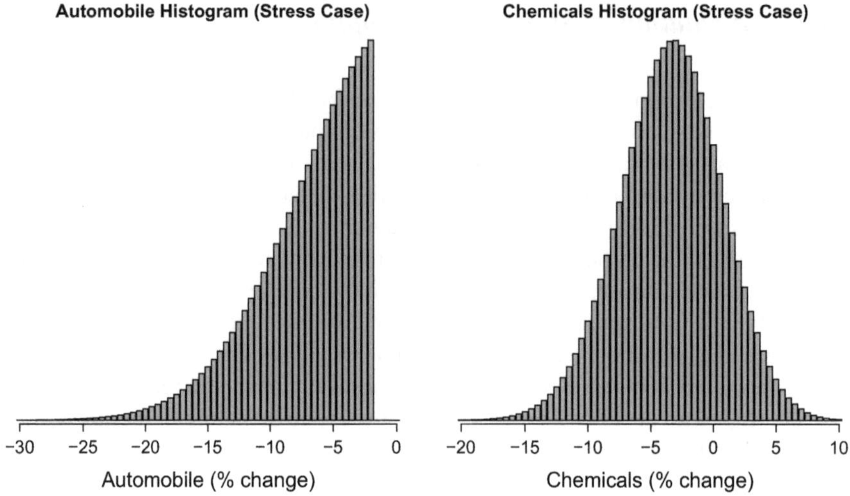

Fig. 9.1 Histogram of simulated factor changes (stress case)

9.3.4 Case Study

We consider the following downturn scenario for the automotive industry: the industry production is forecast to drop by 8% during next year. Using the methodology presented in Sect. 9.3.2 this forecast is translated into a cap on the distribution of the automobile factor.

In this case study, the stress is applied to a sample investment banking portfolio, which consists of 25,000 loans with an inhomogeneous exposure and default probability distribution. Its total exposure is 1000 mn EUR, average exposure size is 0.004% of the total exposure and the standard deviation of the exposure size is 0.026%. Default probabilities vary between 0.02 and 27%. Figure 9.2 exhibits the portfolio's exposure by rating class both for automotive companies and all other borrowers.

Application of the downturn scenario yields the risk estimates shown in Table 9.1.

These key statistics provide important information on the impact of the stress scenario. The 99.98% confidence interval has been chosen because we use the corresponding value-at-risk for the EC calculation. Note that the relative EL increase of 55.6% is significantly higher than the 19% increase of the 99.98% VaR. This results in a 16.3% increase of economic capital defined as 99.98% VaR minus EL.

Figure 9.2 exhibits the portfolio's exposure by rating class both in the non-stress and stress case. The analysis is done separately for automotive companies and all other borrowers. Figure 9.2 clearly shows that exposure is shifted from investments grades (BBB or above) to non-investment grades. As expected, the deterioration of ratings is more pronounced for the automotive industry. Note, however, that due to

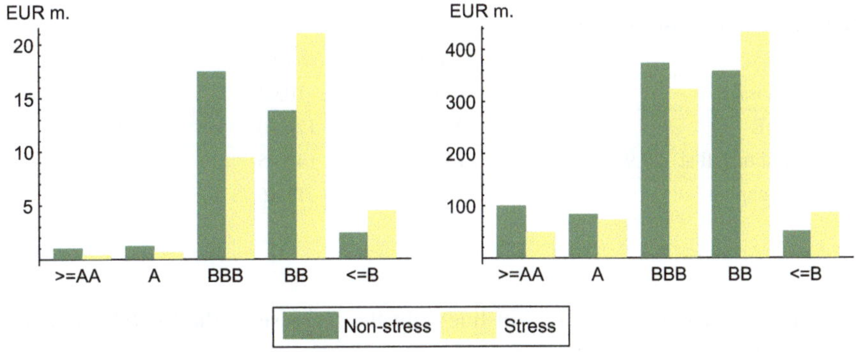

Fig. 9.2 Exposure by rating class for automotive companies (*left*) and all other borrowers (*right*)

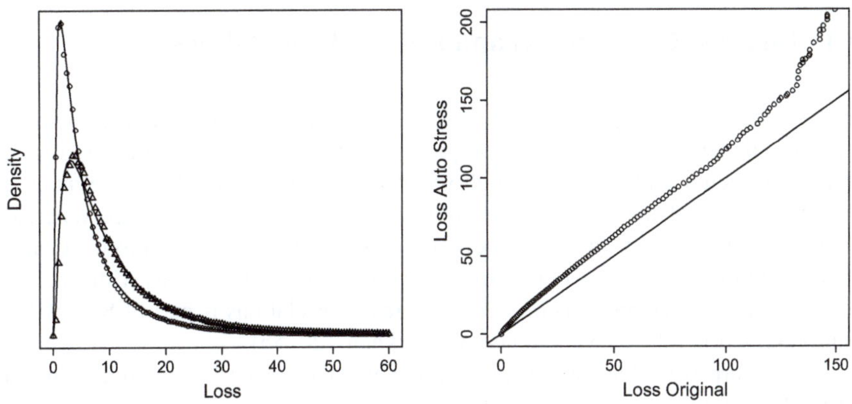

Fig. 9.3 *Left graph* Density plots of original (*circles*) and stressed (*triangles*) loss distributions, together with fitted Vasicek curves. *Right graph* QQ plot of original against stressed loss distribution

the dependence structure of the portfolio this stress scenario also has a significant impact on other borrowers.

Rather than just looking at certain quantiles or other summary statistics, we can get a better understanding of the impact of a stress scenario by studying the whole loss distribution before and after the stress. In order to see the effect of the automotive stress scenario on the portfolio loss, the left graph of Fig. 9.3 shows the original (circles) and the stressed (triangles) loss densities, together with fitted Vasicek distributions (curves). The corresponding QQ-plot, i.e., the quantiles of the two distributions plotted against each other, is shown in the right graph.

The final step in this case study is the calculation of the regulatory capital requirements conditional on the stress event: recalculating the Basel II formula with the stressed PDs increases the regulatory capital from 131.41 to 156.48 mn. In this

Table 9.1 Portfolio risk estimates

	Non-stress	Stress	% chg.
Expected loss	7.03	10.94	55.6
99.98% VaR	103.23	122.80	19.0
Expected shortfall at 99.98%	119.68	145.45	21.5
Economic capital	96.20	111.86	16.3

example, the increase of 19% is in line with the increase of the 99.98% quantile (see Table 9.1).

9.4 Stressed Correlations and Default Probabilities

In the above case study, the expected loss of the portfolio is increased by more than 50% under stress whereas the proportional EC increase is significantly lower. In order to better understand the high sensitivity of the expected loss we analyse the behaviour of default probabilities in stress scenarios, see Sect. 9.4.3. Whereas default probabilities are the only relevant component for the EL, stressed EC also depends on the correlations in the stressed model. Section 9.4.2 deals with stressed asset correlations, an analysis of stressed default correlations is part of Sect. 9.4.3. Our presentation follows Kalkbrener and Packham (2015b).

It is not surprising that the joint distribution of risk factors has a significant impact on the behaviour of default probabilities and correlations under stress. In order to cover a wide range of light-tailed to heavy-tailed distributions we perform our analysis in factor models that follow a normal variance mixture distribution, which is introduced in Sect. 9.4.1.

9.4.1 Distribution of Model Variables

The standard approach in credit risk management is to model the risk factors and ability-to-pay variables through a joint multi-variate normal (aka Gaussian) distribution. In order to specify a more flexible dependence structure we introduce an additional random variable W, the so-called *mixing variable*, which is strictly positive and independent of the systematic and idiosyncratic factors. The definition of the ability-to-pay variables is generalized to

$$A_j = W(\sqrt{R_j^2}\sum_{i=1}^{m} w_{ji}\Psi_i + \sqrt{1-R_j^2}\varepsilon_j) \tag{9.9}$$

$$= \sqrt{R_j^2}\sum_{i=1}^{m} w_{ji}W\Psi_i + \sqrt{1-R_j^2}W\varepsilon_j \tag{9.10}$$

and the systematic and idiosyncratic risk factors now have the form $W\Psi_i$ and $W\varepsilon_j$ respectively. The ability-to-pay variables and risk factors specified in this way follow a so-called multivariate *normal variance mixture (NVM) distribution*. The most important distribution classes covered in this general model are the multivariate normal distribution, in which case the variable W equals 1, and the multivariate Student-t distribution, where W^2 follows an inverse gamma distribution. The Student-t distribution allows for more extreme events than the normal distribution and is therefore a commonly used alternative in financial modelling. Compared to the normal distribution, it takes one additional parameter, the so-called *degrees of freedom*, denoted by ν, that controls the heaviness of the tails. For more details we refer to McNeil et al. (2005).

In general, the tail behaviour of the risk factor W determines the so-called *heaviness* of the tails of the A_j: If the tail function $\mathbb{P}(W \geq x)$ follows a power law, e.g. $\mathbb{P}(W \geq x) \approx x^{-\nu}$ for a $\nu > 0$ and large x, then the ability-to-pay variables are said to have heavy tails. If W is bounded or its tail function decays exponentially, e.g. $\mathbb{P}(W \geq x) \approx e^{-x}$ for large x, then $A_1, \ldots, A_n$ are light-tailed.[3] The normally distributed model and the Student-t distributed model are examples of light-tailed and heavy-tailed models, respectively.

For the sake of simplicity, it will always be assumed that the first risk factor $W\Psi_1$ is truncated. We denote this factor by $V := W\Psi_1$.

9.4.2 Asset Correlations Under Stress

For ability-to-pay variables (or asset returns) A_i and A_j we denote their (unconditional) correlation by ρ_{ij}, the correlation of A_i with risk factor V will be denoted by ρ_i.

It turns out that asset correlations are less sensitive to stress in heavy-tailed models than in light-tailed models. For illustration, we assume that A_1 and A_2 are normally distributed and set $\rho_{12} = 0.4$. Figure 9.4 shows the impact on asset correlations when risk factor V is truncated: The left plot shows a scatter plot of 5000 simulated samples of A_1 and A_2. All simulated scenarios are relevant in the unstressed model. In the right plot only those scenarios are shown where the stressed risk factor V does not exceed a threshold C, where C is chosen such that the stress probability $\mathbb{P}(V \leq C)$

[3]The precise definition is based on the theory of regular variations, see McNeil et al. (2005). Heavy-tailed models correspond to a regularly varying tail function of W, whereas a model is light-tailed if W is bounded or its tail function is rapidly varying.

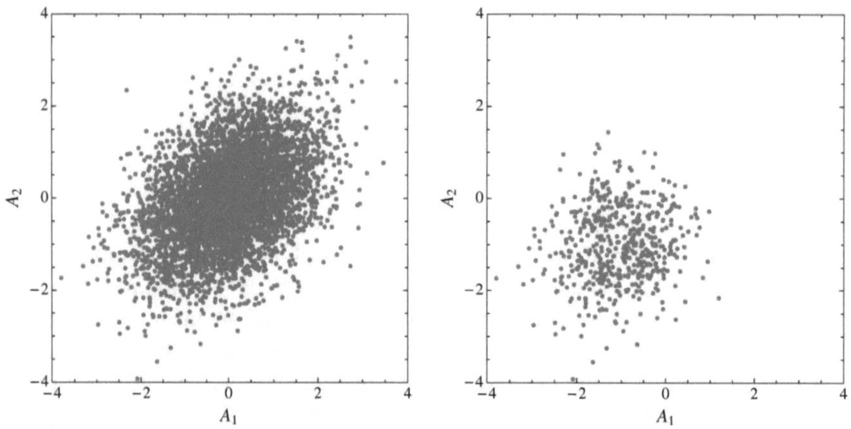

Fig. 9.4 *Left* Simulated normally distributed asset returns A_1 and A_2 with correlation 0.4; A_1 and A_2 are correlated to the joint driving risk factor V with correlation 0.6. *Right* Samples conditional on $V \leq -1.28$ which corresponds to a stress event with probability 10%; the correlation of the sample is 0.1, which is far smaller than the original correlation of 0.4

equals 10%. As a consequence, only approximately 500 of the 5000 scenarios are considered under stress. Since the A_i and V have a positive correlation of 0.6 the average value of A_i in the stressed model is negative, which results in a higher number of defaults. It can also be observed that the asset correlation of 0.4 is significantly reduced under stress, i.e., the correlation of A_1 and A_2 drops to 0.1.

For comparison, we now repeat the calculation for heavy-tailed t-distributed A_i using the same correlation assumptions as in Fig. 9.4. The left graph of Fig. 9.5 shows stressed asset correlations, where instead of the stress level C, stress is expressed by stress probabilities, which are just the probabilities associated with the stress event, $\mathbb{P}(V \leq C)$. For instance, values at 10^{-1} correspond to a stress scenario with probability 10%. Stressed asset correlations are shown for normally distributed and t-distributed assets with degrees of freedom $\nu = 10$ and $\nu = 4$.

Stressed asset correlations may be either greater or smaller than the unconditional asset correlation depending largely on the correlations between the risk factor and the respective asset returns. As illustrated in Fig. 9.5, when the assets in question are sufficiently correlated with the risk factor, the stressed correlation is typically smaller than the unstressed correlation. Loosely speaking, in such a case systematic risk is reduced by conditioning on the risk factor, whereas unsystematic risk remains.

The stressed correlations in the left graph of Fig. 9.5 are calculated with analytic formulas derived in Kalkbrener and Packham (2015a). For normally distributed asset returns A_i, A_j their asset correlations conditional on stress level C are given by

$$\mathrm{Corr}^C(A_i, A_j) = \frac{\rho_i \rho_j \mathrm{Var}^C(V) + \rho_{ij} - \rho_i \rho_j}{\sqrt{(\rho_i^2 \mathrm{Var}^C(V) + 1 - \rho_i^2)(\rho_j^2 \mathrm{Var}^C(V) + 1 - \rho_j^2)}}, \tag{9.11}$$

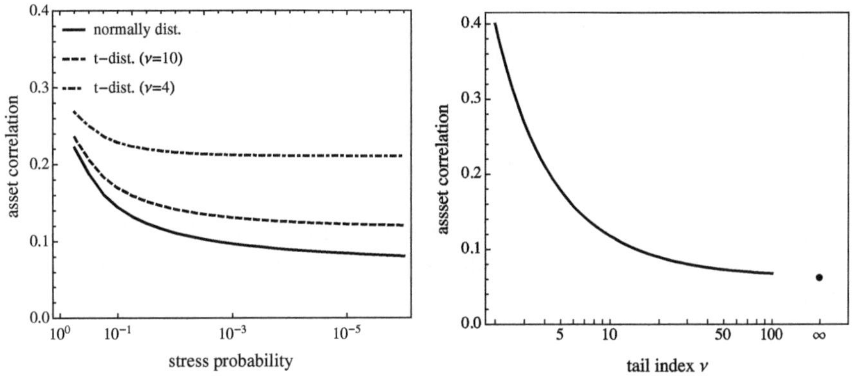

Fig. 9.5 *Left* Stressed asset correlation for different distribution assumptions as a function of the stress probability. *Right* Stressed asset correlation as a function of the tail index when the stress event is taken to the limit $-\infty$. Correlations are as in Fig. 9.4

with

$$\mathrm{Var}^C(V) = 1 - \frac{C\,\phi(C)}{\mathrm{N}(C)} - \frac{(\phi(C))^2}{(\mathrm{N}(C))^2},$$

where ϕ denotes the standard normal density function and N denotes the standard normal distribution function. A corresponding, but more involved, formula is also derived for the Student t-distribution.

The severity of the stress is increased by setting the stress level C to higher negative values, or equivalently, reducing the probability of the stress scenario specified by $\mathbb{P}(V \leq C)$. By letting $\mathbb{P}(V \leq C)$ converge to 0, e.g. by moving to the right in the left graph of Fig. 9.5, we arrive at the asymptotic limit, which is of particular importance for understanding the model behaviour under stress. The right-hand side graph of Fig. 9.5 shows the asymptotic limit of stressed asset correlations for t-distributed assets with different values for ν, where $\nu = \infty$ corresponds to the normally distributed case. The asymptotic analysis confirms the higher sensitivity of light-tailed asset variables under stress.

We have also derived concrete formulas for the asymptotic case, see Kalkbrener and Packham (2015a). These formulas hold in the more general setup of normal variance mixture models. For heavy-tailed NVM models the asymptotic limit of the stressed correlation of A_i and A_j equals

$$\frac{\rho_i\,\rho_j + (\rho_{ij} - \rho_i\,\rho_j)\,(\nu-1)}{\sqrt{(\rho_i^2 + (1-\rho_i^2)\,(\alpha-1))\,(\rho_j^2 + (1-\rho_j^2)\,(\nu-1))}}, \quad \nu > 2, \tag{9.12}$$

if the risk factor is stressed asymptotically, i.e., if V is truncated at a threshold C, and C converges to $-\infty$. The parameter ν specifies the tail index of the asset returns and the risk factor in the heavy-tailed case and corresponds just to the degrees of freedom defined for t-distributions. The case when the variables are light-tailed corresponds to the limit as $\nu \to \infty$, in which case the asymptotic limit of the conditional correlation between A_i and A_j is

$$\frac{\rho_{ij} - \rho_i \rho_j}{\sqrt{(1-\rho_i^2)(1-\rho_j^2)}}. \tag{9.13}$$

Finally, note that the analysis in this section is not restricted to credit portfolio models but holds for any portfolio model with asset variables and risk factors that follow a normal variance mixture distribution.

9.4.3 Default Probabilities and Default Correlations Under Stress

The credit-specific quantities entering credit portfolio models are the default probabilities and default correlations. Just as for asset correlations, their asymptotic behaviour depends on whether the credit portfolio model follows a light- or heavy-tailed NVM distribution. In the light-tailed case, default probabilities converge to 1 under extreme stress and default correlations converge to 0.[4] In other words, default of the entire portfolio becomes a sure event under extreme stress and correlations between default indicators become irrelevant.

In contrast, asymptotic default probabilities and asymptotic default correlations are in (0, 1) in the heavy-tailed case. Both quantities depend on the tail index ν and can be expressed in terms of the Student t-distribution function. More specifically, the asymptotic default probability under stress for a model with tail index ν is given by Abdous et al. (2005) and Packham et al. (2014):

$$\lim_{C \to -\infty} \mathbb{P}(A_1 \le D_1 | V \le C) = t_{\nu+1}\left(\frac{\sqrt{\nu+1}\,\rho_1}{\sqrt{1-\rho_1^2}}\right) \in [1/2, 1), \tag{9.14}$$

where t_ν is the distribution function of the Student-t distribution with parameter ν. A formula for bivariate default probabilities – albeit more involved – and an integral representation for multivariate default probabilities that can be calculated numerically, are derived in Packham et al. (2014).

In all models – whether heavy-tailed or light-tailed – the asymptotic limit of stressed default probabilities and default correlations does not depend on the unstressed default probabilities. For the heavy-tailed case, the tail index and the unstressed correlations enter the asymptotic results.

---

[4]In this subsection, we assume that the unconditional correlations between asset returns $A_1, \ldots, A_n$ and the risk factor V are positive and less than 1, i.e., $\rho_i, \rho_{ij} \in (0, 1)$ for $i, j \in \{1, \ldots, n\}$.

In summary, the impact of stress on the asymptotic limit of default probabilities and correlations is greater in light-tailed models than in heavy-tailed models. This is a remarkable observation since light-tailed models, in particular normally distributed models, are usually considered less sensitive to extreme stress than heavy-tailed models: a popular measure in finance to assess the ability of a bivariate distribution to generate joint extreme events – the tail dependence – is zero in light-tailed models, whereas it is a positive number in heavy-tailed models. In order to better understand this phenomenon we now compare the behaviour of limiting default probabilities to tail dependence.

The tail dependence, or more precisely, the *coefficient of (lower) tail dependence* of the identically distributed variables V and A_1 is defined as[5]

$$\lambda_l(V, A_1) := \lim_{C \to -\infty} \mathbb{P}(A_1 \leq C | V \leq C). \tag{9.15}$$

Hence, the tail dependence of V and A_1 measures the probability $\mathbb{P}(A_1 \leq C)$ conditional on the event $\{V \leq C\}$ for stress levels C converging to $-\infty$. If the NVM distributed random variables V, A_1 are heavy-tailed with tail index ν, the tail dependence coefficient is given by

$$\lambda_l(V, A_1) = 2t_{\nu+1}\left(-\sqrt{\frac{(\nu+1)(1-\rho_1)}{1+\rho_1}}\right),$$

see McNeil et al. (2005). It follows that the tail dependence is strictly positive for heavy-tailed models, provided that $\rho_1 > -1$. For light-tailed NVM distributions, the tail dependence is zero. This includes, of course, the normal distribution, which is still the de-facto standard for modelling risk factors and asset log-returns in structural credit portfolio models, such as CreditMetrics™(Gupton et al. 1997) and Moody's KMV Portfolio Manager™(Crosbie and Bohn 2002).

The zero tail dependence is in contrast to the asymptotic default probability in the light-tailed case, where default is a sure event. Similarly, tail dependence and asymptotically stressed default probabilities disagree in the heavy-tailed case. The left graph of Fig. 9.6 illustrates the difference between tail dependence and asymptotic stressed PD's as a function of the tail index ν.

To make the relation between tail dependence and asymptotic stressed PD's more precise, we introduce an additional parameter $x \in \mathbb{R}$ and measure the probability $\mathbb{P}(A_1 \leq x \cdot C)$ conditional on the event $\{V \leq C\}$ for stress levels C converging to $-\infty$. More formally, we consider the function

$$\lambda(V, A_1, x) := \lim_{C \to -\infty} \mathbb{P}(A_1 \leq x \cdot C | V \leq C), \quad x \in \mathbb{R},$$

[5]In the general case, when V and A_1 are not identically distributed, the tail dependence coefficient is defined via quantiles, see McNeil et al. (2005).

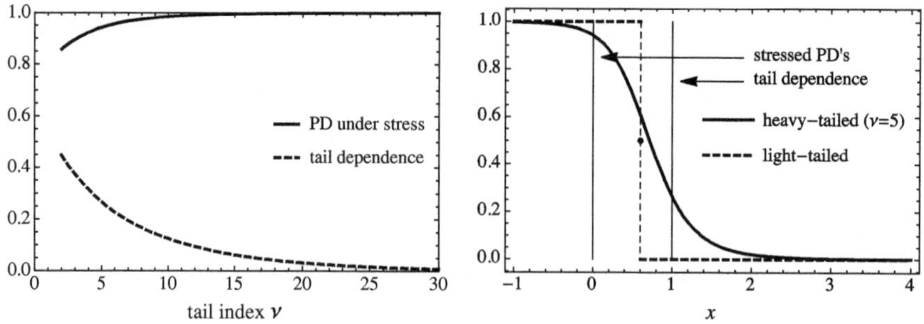

Fig. 9.6 *Left* Tail dependence coefficient and asymptotic PD under stress as a function of the tail index ν. *Right* Tail dependence function $\lambda(V, A_1, x)$ for light- and heavy-tailed variables; special cases arise at $x = 0$ (stressed PD's) at $x = 1$ (tail dependence). The initial correlation between the ability-to-pay variable and the risk factor is 0.6 in both cases

which provides an elegant generalization of both concepts: the tail dependence coefficient of V and A_1 equals $\lambda(V, A_1, 1)$, whereas the asymptotic stressed PD corresponds to $\lambda(V, A_1, 0)$. A closed-form expression for $\lambda(V, A_1, x)$ can be obtained via elementary transformations from Abdous et al. (2005), see also Packham et al. (2014). The tail dependence function is illustrated in the right graph of Fig. 9.6.

The analysis of the function $\lambda(V, A_1, x)$ illustrates the fundamentally different behaviour of the tail dependence coefficient and asymptotic stressed PD's in light-tailed and heavy-tailed credit portfolio models. In the light-tailed case, the asset variable A_1 converges to $-\infty$, more specifically, it is concentrated at $\rho_1 \cdot C$ when $V \leq C$ and $C \to -\infty$. In the heavy-tailed case, however, A_1 does not show the same uniform asymptotic behaviour: $0 < \lambda(V, A_1, x) < 1$ holds for all $x \in \mathbb{R}$ and, in particular, tail dependence as well as stressed default probabilities are in (0, 1).

In summary, this analysis clearly shows that the tail dependence coefficient only provides partial information on a model's ability to produce extreme (joint) events. A more comprehensive picture is given by function $\lambda(V, A_1, x)$, which also explains the observed differences between tail dependence and asymptotic stressed PD's.

So far, our analysis has focused on asymptotic stressed default probabilities. For practical purposes, the model behaviour at smaller and therefore more realistic stress levels is even more important. Hence, we now take a closer look at PD's under stress for various stress levels C and compare them in light- and heavy-tailed models. Figure 9.7 shows PD's under stress for both normally distributed and t-distributed ($\nu = 3$) models as a function of the stress probabilities. The unconditional correlation between the ability-to-pay variable and the risk factor is 0.6. Despite converging to a value smaller than 1, PD's under stress in the t-distributed model dominate the normally distributed case unless the stress probability is very small: If the unconditional PD is 10%, then for stress probabilities greater than approximately $10^{-3.5}$, the PD under stress in the t-distributed model is greater than the respective PD in the normal model. If the unconditional PD is 1%, then the threshold lies beyond 10^{-8}.

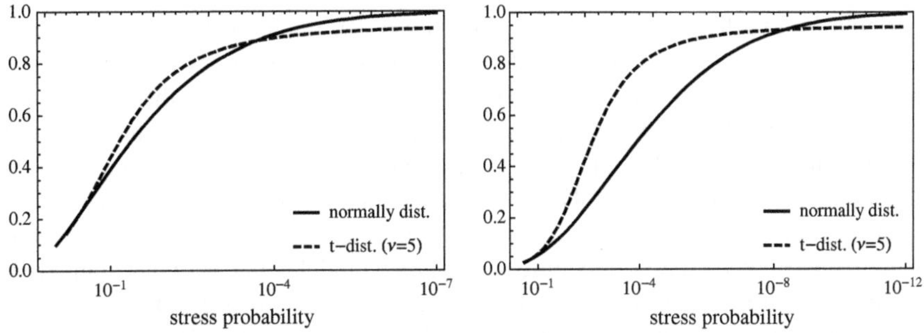

Fig. 9.7 PD's under stress as a function of the stress probability. Models considered are the normal distribution and the t-distribution with parameter $\nu = 5$. Correlations are 0.6. *Left* unconditional PD is 0.1. *Right* unconditional PD is 0.01

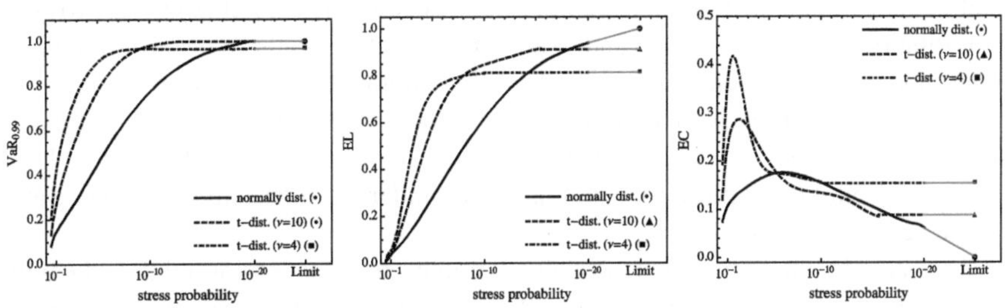

Fig. 9.8 Risk measures for portfolio consisting of 60 homogeneous counterparties, each with a PD of 1%. *Left* Value-at-risk at 99% confidence level; *middle* Expected loss; *right* Economic capital

This example shows that for realistic stress tests the impact on PD's is usually greater in heavy-tailed models. Only for rather extreme stress severities, stressed PD's become higher in light-tailed models and eventually converge to 1.

9.5 Risk Measures

The different behaviour of light-tailed and heavy-tailed models has implications on the credit reserves and capital requirements in stress scenarios, as demonstrated by the following stylized example. Consider a homogeneous portfolio consisting of 60 counterparties. Each counterparty has notional and loss-at-default of $1/60$ and defaults with a probability of 1%. The asset variables of the counterparties are correlated through one risk factor, with $\rho = 0.4$ the correlation between any one counterparty and the risk factor. This implies that the counterparties are correlated with $\rho^2 = 0.16$.

Figure 9.8 shows the value-at-risk, the expected loss and the resulting economic capital for the portfolio under different distribution assumptions, i.e., under a normal distribution and t-distributions with $\nu = 4$ and $\nu = 10$, and for different stress levels. As before, the stress level C is translated into a stress probability, which denotes the probability that a certain stress event occurs. The left graph shows the 99%-value-at-risk of the portfolio. Despite being lower under moderate stress, the VaR in a normally distributed model converges to 1, whereas the VaR for a very heavy-tailed model (t-distribution with $\nu = 4$) converges to a number strictly smaller than 1, see Packham et al. (2014) for the calculation of the asymptotic results. When comparing the two t-distributed models, the more heavy-tailed model with $\nu = 4$ has higher risk for moderate stress levels, but lower risk for less probable stress events.

Similar observations hold for the expected loss (middle graph). The expected loss under stress corresponds just to the probability of default under stress, since the recovery rate is 0 in this example. The asymptotic results, Eq. (9.14), confirm that the EL converges to 1 in the light-tailed case, whereas it converges to a number strictly smaller than 1 in the heavy-tailed cases. Finally, economic capital converges to zero for normally distributed models and to a number strictly greater than zero for heavy-tailed models (a confidence level of 99% for economic capital may not be realistic in practice, but serves well to illustrate some key characteristics of the stressed portfolios). Because stress has different impact on value-at-risk and expected loss, economic capital is not monotone, but increases under moderate stress and decreases for greater stress levels.

To conclude, in light-tailed models, extreme stress scenarios tend to heavily increase the credit reserves specified by the expected loss whereas economic capital, which defines capital requirements, converges to 0. The impact of extreme stress on expected loss and economic capital is more balanced in heavy-tailed models, whose asymptotic limit retains a richer dependence structure.

9.6 Conclusion

In this paper, we have presented a general approach to implementing stress scenarios in a multi-factor credit portfolio model. The general philosophy behind this type of stress test is that stress scenarios are implemented through a restriction of the probability space of the model or, in other words, certain future scenarios are no longer considered possible. The calculation of the stressed portfolio loss distribution is done under a probability measure that contains additional information. The scenarios are then implemented in a way that is consistent with the quantitative framework, i.e., without destroying the dependence structure of risk factors in the model. This is achieved by translating the economic stress scenarios into constraints on the systematic factors. The main prerequisite here is that the systematic factors of the credit portfolio model can be linked to economic variables.

Although the methodology has been developed in a particular factor model, the main concept - implementing stress scenarios through a truncation of the distribution of the risk factors - is completely independent of the model specification and the way that default dependencies are parameterized, e.g. whether asset or default correlations are used. In fact, it can be applied to factor models for market and operational risk as well. However, the model choice has significant implications for the behavior of correlations under stress. In ordinary stress tests, stressed PD's are usually higher in heayy-tailed models. Contrary to popular belief, however, the impact of stress on the asymptotic behaviour is greater in light-tailed models than in heavy-tailed models.

Disclaimer

The views expressed in this paper are those of the author and do not necessarily reflect the position of Deutsche Bank AG.

References

Abdous, B., Fougères, A.-L., & Ghoudi, K. (2005). Extreme behaviour for bivariate elliptical distributions. *Canadian Journal of Statistics*, *33*(3), 317–334.

Alfaro, R., & Drehmann, M. (2009). Macro stress tests and crises: what can we learn? *BIS Quarterly Review* (December 2009), 29–41.

Basel Committee on Banking Supervision. (2006). International Convergence of Capital Measurement and Capital Standards: A Revised Framework, Comprehensive Version. Technical report, Bank for International Settlements.

Basel Committee on Banking Supervision. (2009). Principles for sound stress testing practices and supervision. Technical report, Bank for International Settlements.

Berkowitz, J. (2000). A coherent framework for stress-testing. *Journal of Risk*, *2*(2), 5–15.

Bluhm, C., Overbeck, L., & Wagner, C. (2002). *An Introduction to Credit Risk Modeling*. Boca Raton: CRC Press/Chapman & Hall.

Board of Governors of the Federal Reserve (2012). Comprehensive Capital Analysis and Review 2012: Methodology and Results for Stress Scenario Projections. Technical report, March 2012.

Bonti, G., Kalkbrener, M., Lotz, C., & Stahl, G. (2006). Credit risk concentrations under stress. *Journal of Credit Risk*, *2*(3), 115–136.

Borio, C., Drehmann, M., & Tsatsaronis, K. (2012). Stress Testing Macro Stress Testing: Does It Live up to Expectations? BIS Working Papers no. 369, Bank for International Settlements.

Breuer, T., & Csiszár, I. (2013). Systematic stress tests with entropic plausibility constraints. *Journal of Banking and Finance*, *37*(5), 1552–1559.

Breuer, T., Jandačka, M., Rheinberger, K., & Summer, M. (2009). How to find plausible, severe, and useful stress scenarios. *International Journal of Central Banking*, *5*(3), 205–224.

Čihák, M. (2007). Introduction to Applied Stress Testing. IMF Working Paper No 07/59, International Monetary Fund, Washington DC.

Crosbie, B., & Bohn, J. (2002). Modelling Default Risk. Working paper, KMV Corporation. www.kmv.com.

Drehmann, M. (2009). Macroeconomic stress testing banks: A survey of methodologies. In Quagliariello, M. (ed.): *Stress Testing the Banking System: Methodologies And Applications* (pp. 37–67). Cambridge: Cambridge University Press.

Duellmann, K., & Erdelmeier, M. (2009). Crash testing German banks. *International Journal of Central Banking*, *5*(3), 139–175.

European Banking Authority. (2011). 2011 EU-Wide Stress Test Aggregate Report. Technical report, London, 2011.

Flood, M., & Korenko, G. (2015). Systematic scenario selection: stress testing and the nature of uncertainty. *Quantitative Finance*, *15*(1), 43–59.

Glasserman, P., Kang, C., & Kang, W. (2015). Stress scenario selection by empirical likelihood. *Quantitative Finance*, *15*(1), 25–41.

Gordy, M. B. (2003). A risk-factor model foundation for ratings-based bank capital rules. *Journal of Financial Intermediation*, *12*(3), 199–232.
Gupton, G., Finger, C., & Bhatia, M. (1997). CreditMetrics: Technical Document. Technical report, JP Morgan & Co.
Kalkbrener, M., & Packham, N. (2015a). Correlation under stress in normal variance mixture models. *Mathematical Finance*, *25*(2), 426–456.
Kalkbrener, M., & Packham, N. (2015b). Stress testing of credit portfolios in light- and heavy-tailed models. *Journal of Risk Management and Financial Institutions*, *8*(1), 34–44.
Kupiec, P. (1998). Stress testing in a value at risk framework. *Journal of Derivatives*, *24*, 7–24.
McNeil, A. J., Frey, R., & Embrechts, P. (2005). *Quantitative Risk Management*. Princeton: Princeton University Press.
Merton, R. C. (1974). On the pricing of corporate debt: the risk structure of interest rates. *Journal of Finance*, *29*, 449–470.
Packham, N., Kalkbrener, M., & Overbeck, L. (2014). Multivariate default probabilities and default correlations under stress. *Journal of Applied Probability*, forthcoming.
Quagliariello, M. (2009). Macroeconomic stress-testing: Definitions and main components. In Quagliariello, M. (ed.): *Stress Testing the Banking System: Methodologies and Applications* (pp. 18–36). Cambridge: Cambridge University Press.
Studer, G. (1999). Market risk computation for nonlinear portfolios. *Journal of Risk* *1*(4), 33–53.

Chapter 10
Penalized Independent Factor

Y. Chen, R.B. Chen and Q. He

Abstract We propose a penalized independent factor (PIF) method to extract independent factors via a sparse estimation. Compared to the conventional independent component analysis, each PIF only depends on a subset of the measured variables and is assumed to follow a realistic distribution. Our main theoretical result claims that the sparse loading matrix is consistent. We detail the algorithm of PIF, investigate its finite sample performance and illustrate its possible application in risk management. We implement the PIF to the daily probability of default data from 1999 to 2013. The proposed method provides good interpretation of the dynamic structure of 14 economies' global default probability from pre-Dot Com bubble to post-Sub Prime crisis.

10.1 Introduction

Sovereign default probability reflects financial vulnerability and sovereign financing or refinancing difficulties or default of advanced and emerging market economies. It is considered as a fundamental early warning indicator of financial crises and contagions of global financial markets. Thus, sovereign credit ratings and the associated sovereign default rates continue to be a major concern of international financial markets and economic policy makers. According to the current version of Basel Capital Accord 3, financial institutions will be allowed to use credit ratings and the corresponding default rates to determine the amount of regulatory capital they have to

Y. Chen (✉) · Q. He
Department of Statistics and Applied Probability, National University of Singapore, Singapore, Singapore
e-mail: stacheny@nus.edu.sg

Q. He
e-mail: hq19861027@gmail.com

R.B. Chen
Department of Statistics, National Cheng Kung University, Tainan, Taiwan
e-mail: rbchen@mail.ncku.edu.tw

W.K. Härdle et al. (eds.), *Applied Quantitative Finance*, Statistics and Computing,
DOI 10.1007/978-3-662-54486-0_10

reserve against their credit risks. It prompts the booming research interests on the determinants and co-movements of sovereign defaults.

While the large amount of information containing in the sovereign default data makes it possible to understand the dependence among economies, the massive sample size, high dimensionality and complex dependence structure of the data create computational and statistical challenges. It turns out that data analysis in a reduced space often accompanies with improved interpretability and estimation accuracy. This possibly explains the wide adoption of factor models in literature.

Factor models try to decipher complex phenomena of large dimensional data through a small number of basic causes or factors. Though the factors are often supposed to be macroeconomic and financial determinants, our study intends to launch a new investigation into the identification of factors of sovereign default probabilities in a data-driven way. From a statistical viewpoint, understanding the dependence among these sovereign default probabilities relies on the estimation of the joint probability distribution of the multiple variables. The conventional methods such as Principal Component Analysis (PCA) and Factor Analysis (FA) extract a set of uncorrelated factors from the multivariate and dependent data within a linear framework. Under Gaussianity, non-correlation is identical to independence. With the aid of Jacobian transformation, the complex joint distribution can be obtained by using the marginal distributions of each factor in a closed form. Thus, the high dimensional statistical problem is converted to univariate cases. Independence however does not hold, if the measured variables e.g. the sovereign default probabilities are not Gaussian distributed, which is most likely in practice. In this case, the joint distribution estimation cannot be easily solved with the help of the conventional methods.

The recently developed Independent Component Analysis (ICA) method sheds lights on possible solutions. Similar to the PCA and FA methods, the ICA identifies essential factors via a linear transformation. Instead of projecting onto the eigenvectors of the covariance matrix as PCA does, the ICA directly extracts statistical independent factors from the original complex data via solving an optimization problem on statistical cross-independence. Depending on the definition of independence, various estimation methods have been proposed, including the maximization of nongaussianity (Jones and Sibson 1987; Cardoso and Souloumiac 1993; Hyvärinen and Oja 1997), the minimization of mutual information (Comon 1994; Hyvärinen 1998, 1999a), the maximum likelihood estimation (Pham and Garat 1997; Bell 1995; Hyvärinen 1999b), and the local parametric estimation with time varying loading (Chen et al. 2014).

In high dimensional space, however, ICA leads to redundant dependence by assuming each factor is associated with all the measured variables. The overparametrization is solvable by either reducing the number of factors or simplifying the structure of the loading matrix. Wu et al. (2006) proposed an ordering approach based on the mean-square-error criterion to identify the number of ICs. This dimension reduction eventually accompanies with loss of information. On the other hand, the dependence between the measured sovereign default probabilities and the factors can be sparse. A possibly more realistic situation is that each measured variable is only driven by a few factors, while others depend on a possibly different set of factors. It suggests necessity to reduce dimensionality in parameter space, with a sparse loading matrix.

Sparse estimation has been widely used especially in the regularized regression analysis. Under the sparsity assumption, unnecessary dependence is penalized and insignificant coefficients are pushed to zeros, see e.g. Lasso (Tibshirani 1996), Ridge (Frank and Friedman 1993) and the smoothly clipped absolute deviation (SCAD) penalty (Fan and Li 2001) and so on. The adoption of sparsity in independent component analysis is still new. Hyvärinen and Raju (2002) proposed sparse Bayesian ICA, where the loading matrix is assumed to be random and a conjugate sparse prior is imposed to the loading matrix. Zhang et al. (2009) incorporated adaptive Lasso in the maximum likelihood estimation method to obtain sparse loading matrix, where the statistical independent factors are assumed to follow a simple distribution family with one parameter. Theoretical properties of the estimators are unknown in the above works.

We are motivated to propose a penalized independent component analysis method, named PIF, to extract statistical independent factors via a sparse linear transformation. The sparse loading matrix is estimated under normal inverse Gaussian distributional assumption with SCAD penalty. Our main theoretical result claims that the sparse loading matrix estimator is consistent. The proposed PIF method displays appealing performance in simulation study. We implement the PIF to the daily probability of default data of Corporate Vulnerability Index from 1999 to 2013. The proposed method shows superior interpretation of the dynamic structure of 14 economies' global default probability from the pre-Dot Com bubble period to the post-Sub Prime crisis period.

The remainder of the paper is structured as follows. Section 10.2 details the sovereign default probability data. Section 10.3 presents the penalized independent factor method, the estimation procedure and statistical prosperity of the estimator. Its finite sample performance is investigated along with simulation study in Sect. 10.4. Section 10.5 implements the PIF method to the sovereign default probabilities. Section 10.6 concludes.

10.2 Data

We consider the sovereign default probabilities of 14 economies from 1st April 1999 to 31st December 2013. The data are the equally-weighted Corporate Vulnerability Index (CVI), proxies of sovereign default probability, maintained in the Credit Research Initiative, Risk Management Institute at National University of Singapore. The CVI of each economy is constructed by averaging of all the listed firms' probability of default (PD) in the corresponding exchange. It is worth mentioning that the number of firms considered over the time horizon is not fixed, given the happening of default events and IPOs. For example, on 1st Apr 1999, there were 717 firms listed in the stock exchange of China, and on 31st Dec 2013, the number of listed firms went up to 3017. The PDs were computed using the forward intensity approach in Duan

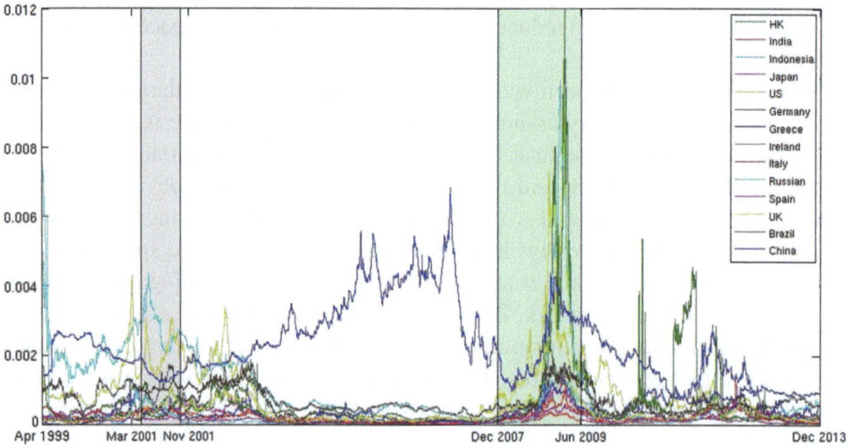

Fig. 10.1 Time series plot of the 14 economies CVI data. *Gray shadow* is the Dot Com bubble period and *light greed shadow* is the Sub Prime crisis

et al. (2012) with input variables of common economic factors including e.g. stock index returns and 3-month interest rates, and firm specific factors of e.g. distance to default, ratio of cash (equivalent) to total assets, return on assets, market to book ratio and 1-year idiosyncratic volatility. The 14 economies include 9 advanced economies of Hong Kong, Japan, US, Germany, Greece, Ireland, Italy, Spain, and UK, and 5 emerging ones of China, India, Indonesia, Russian and Brazil.

Figure 10.1 displays the movements of the 14 CVIs from 1999 to 2013. To understand the dynamic structure of CVIs over time, we divide the time horizon of the 15 years into five sub-periods according to the business cycles announced by the National Bureau of Economic Research, including two recessions occurred from 1st March 2001 to 30th November 2001 (Dot Com bubble) and from 1st December 2007 to 30th June 2009 (US Sub Prime crisis). During the two recessions, the level of CVI increases on average 26 and 53% respectively. The relatively high level of the sovereign default probabilities continues after the recessions for a while and then drops to low value. China, however, behaves distinctively from the rest. The CVI of China is much larger than the others during 2002–2007, i.e. the post-Dot Com bubble period. For example, China's CVI is 3 times of the second highest value of Indonesia. Table 10.1 reports the CVI summary statistics of each economy over the 15 years. China and US have the highest level (mean) of CVI. The level of the US' CVI is high mainly during the two recessions, the Doc Com bubble and Sub Prime crisis. China, on the other hand, though immune to the Dot Com bubble recession, due to its constantly achieved 2-digits growth during 2003 to 2007, accompanies with high level for the "higher return higher risk" philosophy. In terms of variation, China reaches to the highest CVI variation, with a standard deviation of at least 12% larger than the rest. Moreover, all CVIs are positively skewed with extreme values and the JB statistics are all significant, indicating the deviation from Gaussianity.

Table 10.1 Summary statistics of the CVI data over the time horizon, Apr 1999–Dec 2013

	Mean(10^{-3})	SD(10^{-3})	Skewness	Kurtosis	JB-stats(10^4)
China	2.19	1.21	0.71	3.17	0.06*
Hong Kong	0.46	0.38	2.17	9.53	1.38*
India	0.21	0.11	0.70	2.70	0.05*
Indonesia	0.92	0.95	2.06	9.68	1.39*
Japan	0.26	0.20	1.98	8.54	1.04*
US	1.00	1.08	2.86	14.43	3.67*
Germany	0.53	0.44	1.13	3.23	0.12*
Greece	0.46	0.45	2.02	7.73	0.87*
Ireland	0.58	1.17	4.68	29.57	17.82*
Italy	0.22	0.16	1.83	8.00	0.86*
Russian	0.40	1.07	5.19	32.99	22.61*
Spain	0.18	0.13	1.13	3.72	0.13*
UK	0.41	0.48	3.76	19.83	7.63*
Brazil	0.72	0.33	0.79	2.46	0.06*

The conventional PCA is not able to deliver independent factors. Table 10.2 reports the correlation matrix of the CVI data during the 15 years, which are mostly positive except China. While China has either negative or weak correlations with the other economies, the US remains high positive correlations to most of the advanced economies such as Japan and UK, consistent to its influential role in the global financial markets (Tables 10.3, 10.4, 10.5 and 10.6).

More detailed summary statistics on CVIs over the 5 time periods can be found in Tables 10.7, 10.8, 10.9, 10.10, 10.11, 10.12, 10.13, 10.14, 10.15 and 10.16 in the Appendix.

10.3 Penalized Independent Factor

Consider p-dimension random vector $\mathbf{X} = \left(X_1, \cdots, X_p\right) \in \mathbb{R}^p$. The penalized independent factor analysis is to factorize the variables into a linear combination of latent independent random factors $\mathbf{Z} = \left(Z_1, \cdots, Z_p\right) \in \mathbb{R}^p$:

$$\mathbf{Z} = B\mathbf{X} \tag{10.1}$$

where B refers to a sparse and invertible loading matrix. Given the observed realizations $\mathbf{X}_i = \left(X_{i1}, \cdots, X_{ip}\right)$ with $i = 1, \cdots, n$, the task here is to estimate the sparse loading B as well as to obtain the independent factor $\mathbf{Z}_i$ with $i = 1, \cdots, n$, without any prior knowledge of the sparsity structure of B.

Table 10.2 Correlation matrix of the 14 economies from 1st April 1999 to 31st December 2013

	China	HK	India	Indo	Japan	US	DE	Greece	Ireland	Italy	Russian	Spain	UK	Brazil
China	1	−0.04	−0.35	−0.09	−0.02	−0.05	−0.19	−0.40	0.01	−0.27	0.08	−0.43	0.03	−0.08
HK		1	0.60	0.55	0.76	0.74	0.69	0.24	0.40	0.67	0.54	0.41	0.76	0.79
India			1	0.77	0.51	0.46	0.40	0.18	0.14	0.47	0.32	0.32	0.32	0.70
Indo				1	0.50	0.42	0.28	−0.18	0.02	0.22	0.23	0.00	0.16	0.66
Japan					1	0.88	0.74	0.18	0.66	0.66	0.66	0.42	0.81	0.77
US						1	0.72	0.24	0.68	0.72	0.70	0.49	0.87	0.70
DE							1	0.34	0.49	0.73	0.43	0.51	0.73	0.68
Greece								1	0.34	0.63	0.26	0.68	0.34	0.04
Ireland									1	0.56	0.71	0.46	0.73	0.35
Italy										1	0.57	0.78	0.75	0.54
Russian											1	0.47	0.76	0.44
Spain												1	0.58	0.31
UK													1	0.61
Brazil														1

Table 10.3 Summary statistics of simulation results. The results of the PIF is marked in bold and reported in the first row of each scenario. The results of the NIG-ICA and ICA are reported in the second and third row of each scenario. Detection of zeros is the percentage of zeros correctly estimated, and miss-detection is the percentage of non-zero entries estimated as zero. λ is the optimal penalty parameter obtained via minimizing BIC. Each measurement is given in the form of sample average (sample standard deviation)

	Method	ED	MN	RMSE	Det and Miss-det of zeros %	λ
Non-sparsity	PIF	**4.80(2.65)**	**3.97(2.50)**	**0.09(0.03)**	**−/0**	**0**
	NIG-ICA	4.80(2.65)	3.97(2.50)	0.09(0.03)	−/0	0
	ICA	14.51(8.25)	10.26(5.71)	0.19(0.14)	−/0	–
Sparsity	PIF	**6.67(3.98)**	**5.54(3.65)**	**0.09(0.03)**	**100/0**	**0.04**
	NIG-ICA	6.96(3.99)	5.75(3.65)	0.10(0.03)	0/0	0
	ICA	27.19(17.47)	20.40(13.61)	0.20(0.14)	0/0	–
Highly-sparsity	PIF	**3.17(1.83)**	**2.69(1.62)**	**0.08(0.03)**	**99/0**	**0.07**
	NIG-ICA	3.90(1.86)	3.20(1.64)	0.11(0.04)	0/0	0
	ICA	13.81(10.72)	10.19(7.28)	0.28(0.86)	0/0	–

Table 10.4 True Loading matrix. Zero entries are left blank

17	14	10		−13				21			−15		
	−10	15		13	19		−10	12	17			−14	
	−34	−11	−23	27			7	27	43	37	−24	27	18
		−44		−14	13	−97	50	−12	−127		172	−65	
			11										
	7			−79	−29	146	20	−40	−206		−74	223	9
	33	28	−6	−51	−19	71		182	−32		−99	−89	10
	18	−109		−51		−28		−33	54		70	−62	28
				−77	−20	88	−66	30	−95		−144	308	9
	40	43	−13	−108		−49	−14	−79	10		−51	177	−21
	−32	−92	7	57	6	−21	16	14	40	8	−114	29	−9
		64		90	6	−10	−11	−22			−153	−29	28
	−86	−25	17	−116		−24	26	−9	123		−47	108	10
	7	−73		34	−26	−47	−17	25	−32		163	372	

Table 10.5 Simulation results for large dimension loading matrix. Each measurement is given in the form of mean(std). The penalty parameter is $\lambda = 0.08$ by minimizing BIC. #0s is the percentage of zero elements estimated correctly by the method. Mis-detection is the number of elements that are wrongly pushed to zero

	ED	MN	RMSE	Detection of zeros %	λ
PIF	**88.60(26.11)**	**60.00(24.63)**	**0.20(0.10)**	**99.85**	**0.08**
NIG-ICA	90.23(27.74)	61.50(25.68)	0.22(0.10)	0.00	0
ICA	419.24(56.11)	204.00(36.54)	1.29(0.05)	0.00	–

Table 10.6 Number of factors participated by each economy. Sparsity is reflected by of the percentage of zeros in the loading matrix

Country	China	HK	India	Indo	Japan	US	DE	Greece
1999:4–2001:2	1	6	9	4	12	6	8	6
2001:3–2001:11	3	9	9	5	10	6	9	12
2001:12–2007:11	3	10	12	8	13	10	11	9
2007:12–2009:6	7	11	12	9	12	7	9	11
2009:7–2013:12	6	11	11	10	11	9	6	8
Country	Ireland	Italy	Russian	Spain	UK	Brazil	Total	Sparsity%
1999:4–2001:2	9	10	1	11	12	4	99	49
2001:3–2001:11	9	11	9	12	11	7	122	38
2001:12–2007:11	13	13	12	12	12	11	149	24
2007:12–2009:6	7	13	6	9	11	12	136	31
2009:7–2013:12	2	12	9	11	11	9	126	36

Table 10.7 Summary statistics of the CVI data, Apr 1999–Feb 2001

	Mean(10^{-3})	SD(10^{-3})	Skewness	Kurtosis	JB-stats(10^3)
China	2.08	0.59	−1.94	7.55	1.04*
Hong Kong	0.55	0.33	1.33	4.91	0.31*
India	0.33	0.08	0.61	3.05	0.04*
Indonesia	2.27	1.04	3.18	14.04	4.73*
Japan	0.32	0.05	0.24	2.35	0.02*
US	1.13	0.63	2.62	10.94	2.64*
Germany	0.23	0.13	2.11	6.99	0.98*
Greece	0.25	0.27	1.08	3.14	0.14*
Ireland	0.20	0.06	1.57	4.71	0.37*
Italy	0.15	0.07	1.47	5.07	0.38*
Russian	0.39	0.97	3.51	14.63	5.38*
Spain	0.15	0.04	0.47	1.94	0.06*
UK	0.17	0.07	1.08	3.77	0.15*
Brazil	0.95	0.26	−0.69	2.55	0.06*

Table 10.8 Summary statistics of the CVI data during DOT COM bubble, Mar 2001–Nov 2001

	Mean(10^{-3})	SD(10^{-3})	Skewness	Kurtosis	JB-stats(10^3)
China	1.40	0.27	−2.56	16.33	2.34*
Hong Kong	0.78	0.22	−0.03	2.11	0.01*
India	0.44	0.04	−0.90	3.88	0.05*
Indonesia	2.85	0.60	0.54	2.41	0.02*
Japan	0.40	0.06	0.07	1.64	0.02*
US	2.08	0.45	0.40	2.10	0.02*
Germany	1.07	0.34	0.39	1.96	0.02*
Greece	0.42	0.11	0.62	3.45	0.02*
Ireland	0.36	0.16	1.20	3.31	0.07*
Italy	0.40	0.08	−0.53	2.09	0.02*
Russian	0.44	0.18	0.49	2.25	0.02*
Spain	0.18	0.03	1.09	3.83	0.06*
UK	0.58	0.20	0.78	2.77	0.03*
Brazil	1.05	0.09	0.98	4.00	0.06*

The loading matrix and independent factors are only identifiable up to scale. For any constant $c \neq 0$, one obtains another set of loading matrix cB and independent factors denoted $c\mathbf{Z}$ satisfying (10.1). To avoid the identification problem, we assume that the independent factors have unit variance. Moreover, we set the number of independent factors to p, as the primary goal of our study is to convert the multivariate problem into a number of univariate ones with sparsity such that it eases the

Table 10.9 Summary statistics of the CVI data, Dec 2001–Nov 2007

	Mean(10^{-3})	SD(10^{-3})	Skewness	Kurtosis	JB-stats(10^3)
China	2.96	1.26	0.12	2.80	0.01*
Hong Kong	0.43	0.32	1.20	3.43	0.54*
India	0.17	0.09	1.21	3.60	0.57*
Indonesia	0.80	0.62	1.24	3.33	0.57*
Japan	0.21	0.18	1.21	3.12	0.54*
US	0.62	0.69	1.59	4.44	1.11*
Germany	0.46	0.46	1.27	3.23	0.59*
Greece	0.22	0.14	1.88	8.01	3.58*
Ireland	0.23	0.28	1.98	6.71	2.70*
Italy	0.15	0.09	0.80	2.35	0.27*
Russian	0.05	0.04	8.03	126.92	1425.52*
Spain	0.08	0.05	0.93	2.40	0.35*
UK	0.30	0.24	1.38	3.45	0.71*
Brazil	0.68	0.33	0.90	2.34	0.34*

Table 10.10 Summary statistics of the CVI data during Sub Prime crisis, Dec 2007–Jun 2009

	Mean(10^{-3})	SD(10^{-3})	Skewness	Kurtosis	JB-stats(10^3)
China	2.29	1.02	0.23	1.98	0.03*
Hong Kong	0.87	0.66	1.11	3.38	0.12*
India	0.20	0.10	0.25	1.73	0.04*
Indonesia	0.74	0.39	0.33	1.55	0.06*
Japan	0.52	0.35	0.64	2.19	0.06*
US	2.64	1.97	1.15	3.25	0.13*
Germany	0.92	0.51	0.22	1.38	0.07*
Greece	0.59	0.33	0.34	1.63	0.06*
Ireland	2.12	2.61	1.60	4.71	0.32*
Italy	0.42	0.20	0.54	2.55	0.03*
Russian	1.97	2.54	1.33	3.70	0.18*
Spain	0.36	0.10	0.04	2.52	0.01*
UK	1.30	0.94	0.94	2.55	0.09*
Brazil	0.99	0.42	0.23	1.33	0.07*

understanding of the dependence with reduced parameter space and simultaneously an improved estimation accuracy.

Denote the probability density function of each independent factor to be $f_j(z)$ for $j = 1, \ldots, p$. The log-likelihood is defined as:

Table 10.11 Summary statistics of the CVI data, Jul 2009–Dec 2013

	Mean(10^{-3})	SD(10^{-3})	Skewness	Kurtosis	JB-stats(10^3)
China	1.30	0.68	1.02	3.18	0.29*
Hong Kong	0.27	0.13	0.87	2.54	0.22*
India	0.18	0.08	−0.09	1.70	0.12*
Indonesia	0.25	0.10	2.96	12.52	8.62*
Japan	0.18	0.08	0.64	3.55	0.13*
US	0.70	0.43	1.38	4.93	0.78*
Germany	0.53	0.27	1.18	4.49	0.53*
Greece	0.83	0.58	1.12	3.70	0.38*
Ireland	0.70	0.96	2.43	8.19	3.46*
Italy	0.24	0.15	2.54	13.16	8.85*
Russian	0.31	0.25	1.47	6.31	1.34*
Spain	0.25	0.14	0.73	2.65	0.15*
UK	0.31	0.13	1.11	3.69	0.37*
Brazil	0.54	0.16	1.00	3.73	0.31*

$$l(B) = \sum_{i=1}^{n}\sum_{j=1}^{p} \log f_j\left(b_j^\top \mathbf{X}_i\right) + n \log |det(B)| \tag{10.2}$$

where $b_j^\top$ denotes the j-th row of B. To achieve the sparsity of the loading matrix B, a penalty function, denoted as ρ_λ is added to the log-likelihood, where λ is a tuning parameter. The penalized log-likelihood is defined as:

$$\mathbf{P}(B) = \sum_{i=1}^{n}\sum_{j=1}^{p} \log f_j(b_j^\top \mathbf{X}_i) + n \log |det(B)| - n \sum_{j=1}^{p}\sum_{k=1}^{p} \rho_\lambda(|b_{jk}|) \tag{10.3}$$

where b_{jk} denotes the (j, k)-th element of the loading matrix B. Take the gradient of the penalized likelihood function with respect to the loading matrix, we obtain:

$$\frac{\partial \mathbf{P}}{\partial B} = \sum_{i=1}^{n} \begin{bmatrix} \frac{f_1'(b_1^\top \mathbf{X}_i)}{f_1(b_1^\top \mathbf{X}_i)} \\ \frac{f_2'(b_2^\top \mathbf{X}_i)}{f_2(b_2^\top \mathbf{X}_i)} \\ \cdots \\ \frac{f_p'(b_p^\top \mathbf{X}_i)}{f_p(b_p^\top \mathbf{X}_i)} \end{bmatrix} \mathbf{X}_i^\top + n[B^\top]^{-1} - n\Omega$$

where $[B^\top]^{-1}$ is the inverse of transpose matrix of B, $\Omega_{jk} = sgn(b_{jk})\rho_\lambda'(|b_{jk}|)$ is the first derivative of the penalty function with respect to each element of the loading matrix, and $f_i'(s)/f_i(s)$ is the first derivative of log-density function of each

Table 10.12 Correlation matrix of the 14 economies from 1st April 1999 to 28th Feb 2001

	China	HK	India	Indo	Japan	US	DE	Greece	Ireland	Italy	Russian	Spain	UK	Brazil
China	1	−0.49	−0.45	−0.40	0.06	−0.28	−0.22	−0.42	−0.20	−0.37	−0.37	−0.22	−0.43	−0.23
HK		1	0.49	0.81	0.11	0.61	0.56	0.59	0.28	0.69	0.68	0.43	0.72	0.47
India			1	0.67	−0.03	0.07	0.02	0.19	0.08	0.25	0.63	0.50	0.30	0.35
Indo				1	−0.11	0.24	0.19	0.14	0.04	0.32	0.91	0.56	0.34	0.32
Japan					1	0.13	0.36	0.38	0.01	0.35	−0.22	−0.01	0.32	0.11
US						1	0.81	0.68	0.61	0.61	0.10	0.28	0.87	0.50
DE							1	0.69	0.32	0.79	0.07	0.43	0.78	0.39
Greece								1	0.48	0.77	−0.01	0.04	0.86	0.40
Ireland									1	0.29	−0.07	0.05	0.58	0.26
Italy										1	0.18	0.34	0.80	0.42
Russian											1	0.52	0.17	0.16
Spain												1	0.29	0.37
UK													1	0.53
Brazil														1

Table 10.13 Correlation matrix of the 14 economies from 1st Mar 2001 to 30th Nov 2001

	China	HK	India	Indo	Japan	US	DE	Greece	Ireland	Italy	Russian	Spain	UK	Brazil
China	1	0.35	−0.10	0.34	0.31	−0.11	−0.42	−0.28	−0.06	0.27	0.31	−0.14	−0.33	−0.32
HK		1	0.23	0.08	0.75	0.61	0.25	−0.13	0.31	0.36	0.36	0.57	0.32	0.01
India			1	0.28	−0.23	0.08	0.24	0.21	−0.51	0.14	0.66	0.16	−0.09	−0.37
Indo				1	−0.03	−0.22	−0.67	−0.22	−0.32	0.47	0.69	−0.45	−0.66	−0.54
Japan					1	0.60	0.14	0.01	0.62	0.58	−0.05	0.58	0.46	0.34
US						1	0.65	0.12	0.71	0.12	−0.18	0.64	0.77	0.42
DE							1	0.44	0.32	−0.29	−0.40	0.74	0.85	0.46
Greece								1	−0.04	0.27	−0.23	0.54	0.46	0.52
Ireland									1	0.06	−0.55	0.42	0.67	0.55
Italy										1	0.40	0.25	−0.06	0.09
Russian											1	−0.21	−0.58	−0.60
Spain												1	0.81	0.55
UK													1	0.73
Brazil														1

Table 10.14 Correlation matrix of the 14 economies from 1st Dec 2001 to 30th Nov 2007

	China	HK	India	Indo	Japan	US	DE	Greece	Ireland	Italy	Russian	Spain	UK	Brazil
China	1	−0.49	−0.54	−0.55	−0.58	−0.56	−0.58	−0.23	−0.52	−0.61	−0.41	−0.66	−0.53	−0.56
HK		1	0.81	0.80	0.83	0.72	0.92	0.66	0.76	0.89	0.50	0.88	0.88	0.90
India			1	0.87	0.84	0.72	0.80	0.53	0.65	0.88	0.38	0.85	0.79	0.83
Indo				1	0.93	0.89	0.83	0.49	0.82	0.89	0.41	0.85	0.85	0.91
Japan					1	0.88	0.85	0.46	0.83	0.87	0.46	0.89	0.87	0.89
US						1	0.83	0.44	0.91	0.81	0.41	0.83	0.91	0.85
DE							1	0.64	0.86	0.91	0.54	0.93	0.94	0.94
Greece								1	0.44	0.57	0.38	0.56	0.59	0.56
Ireland									1	0.81	0.45	0.82	0.89	0.86
Italy										1	0.49	0.92	0.85	0.91
Russian											1	0.56	0.49	0.51
Spain												1	0.91	0.92
UK													1	0.90
Brazil														1

Table 10.15 Correlation matrix of the 14 economies from 1st Dec 2007 to 30th Jun 2009

	China	HK	India	Indo	Japan	US	DE	Greece	Ireland	Italy	Russian	Spain	UK	Brazil
China	1	0.65	0.77	0.86	0.83	0.74	0.87	0.83	0.74	0.78	0.74	0.68	0.80	0.83
HK		1	0.89	0.78	0.75	0.76	0.74	0.86	0.51	0.79	0.58	0.66	0.85	0.82
India			1	0.93	0.91	0.89	0.87	0.95	0.74	0.94	0.79	0.77	0.91	0.88
Indo				1	0.95	0.88	0.94	0.97	0.83	0.91	0.90	0.82	0.91	0.92
Japan					1	0.91	0.91	0.93	0.87	0.93	0.92	0.80	0.91	0.86
US						1	0.81	0.87	0.85	0.94	0.85	0.78	0.93	0.76
DE							1	0.93	0.78	0.91	0.80	0.81	0.85	0.94
Greece								1	0.79	0.93	0.83	0.83	0.94	0.93
Ireland									1	0.83	0.84	0.77	0.81	0.70
Italy										1	0.81	0.83	0.90	0.87
Russian											1	0.68	0.84	0.71
Spain												1	0.79	0.82
UK													1	0.82
Brazil														1

Table 10.16 Correlation matrix of the 14 economies from 1st Jul 2009 to 31st December 2013

	China	HK	India	Indo	Japan	US	DE	Greece	Ireland	Italy	Russian	Spain	UK	Brazil
China	1	−0.25	−0.68	0.73	0.64	0.69	0.54	−0.45	−0.06	−0.19	0.02	−0.32	0.08	0.08
HK		1	0.45	0.09	0.17	0.18	0.51	0.79	−0.02	0.86	0.50	0.79	0.83	0.32
India			1	−0.24	−0.60	−0.51	−0.25	0.35	−0.20	0.34	0.57	0.39	0.21	0.58
Indo				1	0.61	0.63	0.66	−0.30	−0.06	0.07	0.58	−0.10	0.35	0.56
Japan					1	0.81	0.71	0.12	0.36	0.19	0.02	−0.05	0.44	−0.04
US						1	0.71	0.02	0.10	0.22	0.10	−0.02	0.47	0.06
DE							1	0.19	0.14	0.47	0.37	0.33	0.68	0.29
Greece								1	0.13	0.69	0.12	0.61	0.62	−0.07
Ireland									1	0.02	−0.16	−0.08	0.03	−0.21
Italy										1	0.44	0.75	0.75	0.25
Russian											1	0.31	0.52	0.87
Spain												1	0.66	0.15
UK													1	0.38
Brazil														1

independent factor. The sparse loading matrix is estimated using the gradient method. Given the loading matrix estimator, the independent factors are recovered in (10.1).

10.3.1 Independent Component's Density: NIG

The density of IC is unknown. Hyvärinen (1999b) developed the maximum likelihood estimation approach of independent factor extraction under a simple but unrealistic distribution with one distributional parameter, and proved consistency of the estimator, see also Pham and Garat (1997), Bell (1995). The log-likelihood function is defined under a simple but unrealistic distribution with one distributional parameter. Financial risk factors are however neither Gaussian distributed nor the special cases of the exponential power family. Instead, the factors are often asymmetric and with extreme values. This motivates the adoption of the normal inverse Gaussian (NIG) distribution for its desirable probabilistic features. With 4 distributional parameters, the NIG distribution is able to mark data characteristics from the central locations to the tails behaviours.

In our study, each factor is assumed to be normal inverse Gaussian (NIG) distributed with individual distributional parameters. The density is of the form:

$$f_{\mathrm{NIG}}(z_j) = \frac{\phi_j \delta_j}{\pi} \frac{K_1\left\{\phi_j \sqrt{\delta_j^2 + (z_j - \mu_j)^2}\right\}}{\sqrt{\delta_j^2 + (z_j - \mu_j)^2}} \exp\{\delta_j \sqrt{\phi_j^2 - \beta_j^2} + \beta_j (z_j - \mu_j)\},$$

where μ_j, δ_j, β_j and ϕ_j are NIG parameters for $j = 1, \cdots, p$. $K_1(\cdot)$ is the modified Bessel function of the third type. The distributional parameters fulfill the conditions $\mu_j \in \mathbb{R}$, $\delta_j > 0$, and $|\beta_j| \leq \phi_j$. The limiting distributions of NIG have been well developed in bn (1997); Blæsild (1999) including the Normal distribution, the Cauchy distribution and the Student-t distribution.

- For $\beta = 0$, $\phi \rightarrow \infty$ and $\delta/\phi = \sigma^2$, $NIG(\phi, \beta, \mu, \delta) \rightarrow N(\mu, \sigma^2)$
- For $\phi, \beta \rightarrow \infty$, $\mu = 0$ and $\delta = 1$, $NIG(\phi, \beta, \mu, \delta) \rightarrow Cauchy$
- For $\phi, \beta \rightarrow 0$, $\mu = 0$ and $\delta = 1$, $NIG(\phi, \beta, \mu, \delta) \rightarrow Student - t_1$

See bn (1997) for more details. Moreover, all independent factors are assumed to have unit variance to avoid identification ambiguity.

10.3.2 Penalty Function: SCAD

Question remains on the selection of penalty function in the estimation. Various penalty function has been proposed in literature, including the first order norm penalty of Lasso (Tibshirani 1996), the second order norm penalty of Ridge

(Frank and Friedman 1993) and the smoothly clipped absolute deviation (SCAD) penalty (Fan and Li 2001) and so on. Among them, the SCAD penalty is theoretically desirable with oracle property and has been widely used in quantile regression, logistic regression, high dimensional data analysis, large scale genomic data analysis and many others, see Gou et al. (2014), Xie and Huang (2009). In our study, we use the SCAD penalty, which is defined in the form of its first derivative:

$$\rho_{\lambda}^{'}(\theta) = \lambda\{I(\theta \leq \lambda) + \frac{(a\lambda - \theta)_{+}}{(a-1)\lambda} I(\theta > \lambda)\} \tag{10.4}$$

where $\theta > 0$ and $a = 3.7$ suggested in Fan and Li (2001).

10.3.3 Estimation

Substitute the NIG density and the SCAD penalty function into (10.3):

$$\mathbf{P}(B) = \sum_{i=1}^{n}\sum_{j=1}^{p} \log f_j(b_j^{\top}\mathbf{X}_i) + n\log|det(B)| - n\sum_{j=1}^{p}\sum_{k=1}^{p} p_{\lambda}(|b_{jk}|) \tag{10.5}$$

$$= \sum_{i=1}^{n}\sum_{j=1}^{p}\left\{\log \frac{\phi_j\delta_j}{\pi} \frac{K_1\left(\phi_j\sqrt{\delta_j^2 + (b_j^{\top}\mathbf{X}_i - \mu_j)^2}\right)}{\sqrt{\delta_j^2 + (b_j^{\top}\mathbf{X}_i - \mu_j)^2}} + \delta_j\sqrt{\phi_j^2 - \beta_j^2} + \beta_j(b_j^{\top}\mathbf{X}_i - \mu_j)\right\}$$

$$+ n\log|det(B)| - n\sum_{j=1}^{p}\sum_{k=1}^{p} p_{\lambda}(|b_{jk}|) \tag{10.6}$$

and the gradient of the log-likelihood function is:

$$\frac{\partial l}{\partial B} = \sum_{i=1}^{n} \begin{bmatrix} \frac{f_1^{'}(b_1^{\top}\mathbf{X}_i)}{f_1(b_1^{\top}\mathbf{X}_i)} \\ \frac{f_2^{'}(b_2^{\top}\mathbf{X}_i)}{f_2(b_2^{\top}\mathbf{X}_i)} \\ \cdots \\ \frac{f_p^{'}(b_p^{\top}\mathbf{X}_i)}{f_p(b_p^{\top}\mathbf{X}_i)} \end{bmatrix} \mathbf{X}_i^{\top} + n[B^{\top}]^{-1} - \Omega$$

where $\Omega_{jk} = sgn(b_{jk})\rho_{\lambda}^{'}(|b_{jk}|)$ and $\frac{f_j^{'}(s)}{f_j(s)} = \beta_j + \phi_j \frac{K_1^{'}(\phi_j\sqrt{\delta_j^2+(s-\mu_j)^2})}{K_1(\phi_j\sqrt{\delta_j^2+(s-\mu_j)^2})} \frac{s-\mu_j}{\sqrt{\delta_j^2+(s-\mu_j)^2}} - \frac{s-\mu_j}{\delta_j^2+(s-\mu_j)^2}$.

The optimization problem is solved in two steps, where maximum is achieved by changing the loading matrix B and the NIG parameters iteratively until the algorithm converges. The algorithm starts with an initial estimator of B_0, e.g. the estimation obtained by the conventional ICA:

1. Given the previous estimator of B, optimize the penalized log-likelihood function to obtain the NIG distributional parameters estimator. The EM algorithm is adopted for the estimation of NIG parameters, see Karlis (2002).

2. Based on the estimated NIG estimator, update the estimator of B by maximizing the penalized log-likelihood function.
3. Scale the estimator of B and the NIG parameters to have unit variance of each independent factor.
4. Repeat, until converge.

The penalized maximum likelihood estimation involves the choice of the tuning parameter λ. While too large tuning parameter leads to over sparse loading matrix, too small tuning parameter has over fitting effect to identify the true model. Cross validation (Kohavi 1995) and generalized cross validation (Li 1987) can be used. However the approaches are computational intensive. Even worse, there is a positive probability of model over-fitting by generalized cross validation (Wang et al. 2007). Alternatively, several information criteria have been proposed and widely used in time series analysis. In our study, we consider using the Schwarz–Bayesian information criterion (BIC) (Schwarz 1978) for its computation tractability and its consistency in model selection. The BIC is defined as:

$$BIC = -l(\hat{B}) + \log n \times \#\{\hat{B}_{ij} \neq 0\}$$

where $\hat{B}$ is the estimator of B. The penalty parameter with the lowest BIC is chosen to be optimal.

10.3.4 Property of Estimator

We prove the consistency of the PIF estimator under two conditions:

C1. The observations $(X_{i1}, \ldots, X_{ip})$ are IID with density $\big(g_1(X, B), \ldots, g_p(X, B)\big)$ with respect to some measure μ. The density has a common support and is identifiable. Furthermore, the first logarithmic derivatives of g_i satisfying the equation

$$E\frac{\partial \log g_a(X, B)}{\partial B_{jk}} = 0 \tag{10.7}$$

for all a, j and k.

C2. $E[-\Omega_a]$ is positive definite at point B with Ω_a defined as:

$$\Omega_a = \begin{vmatrix} \frac{\partial^2 g_a(B)}{\partial b_{11}\partial b_{11}} & \frac{\partial^2 g_a(B)}{\partial b_{11}\partial b_{12}} & \cdots & \frac{\partial^2 g_a(B)}{\partial b_{11}\partial b_{1p}} & \frac{\partial^2 g_a(B)}{\partial b_{11}\partial b_{21}} & \cdots & \frac{\partial^2 g_a(B)}{\partial b_{11}\partial b_{pp}} \\ \frac{\partial^2 g_a(B)}{\partial b_{12}\partial b_{11}} & \frac{\partial^2 g_a(B)}{\partial b_{12}\partial b_{12}} & \cdots & \frac{\partial^2 g_a(B)}{\partial b_{12}\partial b_{1p}} & \frac{\partial^2 g_a(B)}{\partial b_{12}\partial b_{21}} & \cdots & \frac{\partial^2 g_a(B)}{\partial b_{12}\partial b_{pp}} \\ \vdots & \vdots & \vdots & \vdots & \vdots & \vdots & \vdots \\ \frac{\partial^2 g_a(B)}{\partial b_{pp}\partial b_{11}} & \frac{\partial^2 g_a(B)}{\partial b_{pp}\partial b_{12}} & \cdots & \frac{\partial^2 g_a(B)}{\partial b_{pp}\partial b_{1p}} & \frac{\partial^2 g_a(B)}{\partial b_{pp}\partial b_{21}} & \cdots & \frac{\partial^2 g_a(B)}{\partial b_{pp}\partial b_{pp}} \end{vmatrix}$$

Theorem 10.1 *Let* $(X_{11}, X_{12}, \ldots, X_{1p}), \ldots, (X_{n1}, X_{n2}, \ldots, X_{np})$ *be IID measured vector, each with a density* $(g_1, g_2, \ldots, g_p)$ *that satisfies conditions (C1) and (C2).*

If $\max\{p''_{\lambda_n}(|B_{jk}|) : B_{jk} \neq 0\} \to 0$, *then there exists a local maximizer* $\hat{B}$ *of* $\mathbf{P}(B)$ *such that* $\|\hat{B} - B\| = \mathcal{O}_p(n^{-1/2} + a_n)$, *where* $a_n = \max\{p'_{\lambda_n}(|B_{jk}|) : B_{jk} \neq 0\}$

Note that, though the density of the observed variables g_a is unknown, Theorem 10.1 holds as long as the two conditions hold. Detailed proof can be found in Appendix.

10.4 Simulation

Before the implementation with real sovereign default probability data, we investigate the finite sample performance of the PIF method first by performing a number of simulation studies under the known data generating processes. Our interest is on the estimation accuracy of the proposed method and its robustness under various scenarios compared to the conventional ICA approach.

We design our simulation studies so that they properly reflect the real study at hand. All the parameters are obtained from analyzing the Corporate Vulnerability Index (CVI) data from April 1999 to February 2001, before the Dot Com bubble. In the first experiment, small dimensional data are generated based on the CVIs of India, Indonesia and Japan, 3 Asia countries of both emerging and advanced economies. We consider 3 scenarios with non-sparsity, medium sparsity and high sparsity in the loading matrix. In the second experiment, large dimensional data are produced, where the parameters are learned from the CVI data of the 14 economies from April 1999 to February 2001.

In the data generation process, we follow the model setting in (10.1) and generate dependent data with the loading matrix:

$$\mathbf{X}_i = B^{-1}\mathbf{Z}_i, \quad i = 1, \cdots, n.$$

The generated data are considered as the measured variables. Each experiment is repeated 100 times with $n = 200$ observations. Both the PIF and the conventional ICA methods are implemented. In addition to the two approaches, we also implement ICA with the NIG distributed source assumption, named as NIG-ICA in the following.

We evaluate the estimation accuracy of the PIF method, with focus on the factor loadings B and the identified factors $\mathbf{Z}_i$. We compare the estimation accuracy of the PIF method based on 3 measurements. For the loading matrix, our interests are the overall estimation accuracy and the elementary accuracy. While the Euclidean distance (ED) is used to measure the estimation error of the loading matrix estimator, the maximum norm (MN) reports the largest elementary bias of the matrix estimator. For the identified independent factors, we compute the root mean squared error (RMSE) to show the identification accuracy. The criteria are defined as follows:

$$ED = \sum_{jk} \left(b_{jk} - \hat{b}_{jk}\right)^2 \tag{10.8}$$

$$MN = \max\left(|b_{jk} - \hat{b}_{jk}|\right) \tag{10.9}$$

$$RMSE = \sqrt{\frac{1}{np}\sum_{ij}\left(Z_{ij} - \hat{Z}_{ij}\right)^2} \tag{10.10}$$

where b_{jk} refers to the (j, k)-th element of the matrix B, and $\hat{b}_{jk}$ represents the corresponding element estimators.

10.4.1 Experiment 1: 3 Dimensional Data

In the low dimensioned experiment, 3 scenarios are analyzed with 3 different loading matrices that are either non-sparse, sparse, or highly sparse:

Non-sparse loading matrix:

$$\begin{bmatrix} 52.7 & -10.7 & 14.4 \\ -32.3 & -17.3 & -5.2 \\ 18.1 & -6.3 & 12.8 \end{bmatrix};$$

Sparse loading matrix:

$$\begin{bmatrix} -3.2 & 31.2 & \mathbf{0} \\ 40.1 & -96.4 & -20.9 \\ -29.4 & 18.7 & \mathbf{0} \end{bmatrix};$$

Highly-sparse loading matrix:

$$\begin{bmatrix} -3.3 & 31.2 & \mathbf{0} \\ \mathbf{0} & 10.1 & \mathbf{0} \\ \mathbf{0} & 44.2 & -25.0 \end{bmatrix}.$$

Table 10.3 reports the simulation results based on the 100 replications. For all the 3 scenarios, the PIF is better than ICA in terms of estimation accuracy for both the loading matrix and the independent factors. In the sparsity scenario, the estimation accuracy of PIF is much better with lower ED of 6.67(SD: 3.98), MN of 5.54(SD: 3.65) and RMSE of 0.09(SD: 0.03) than that of ICA with ED of 27.19(SD: 17.47), MN of 20.40(SD: 13.61) and RMSE of 0.20(SD: 0.14). The improved accuracy is mostly contributed by the adoption of the NIG distributional assumption. In the highly-sparse scenario, the PIF is remarkably better than the conventional ICA. The improvement w.r.t to the NIG-ICA becomes larger.

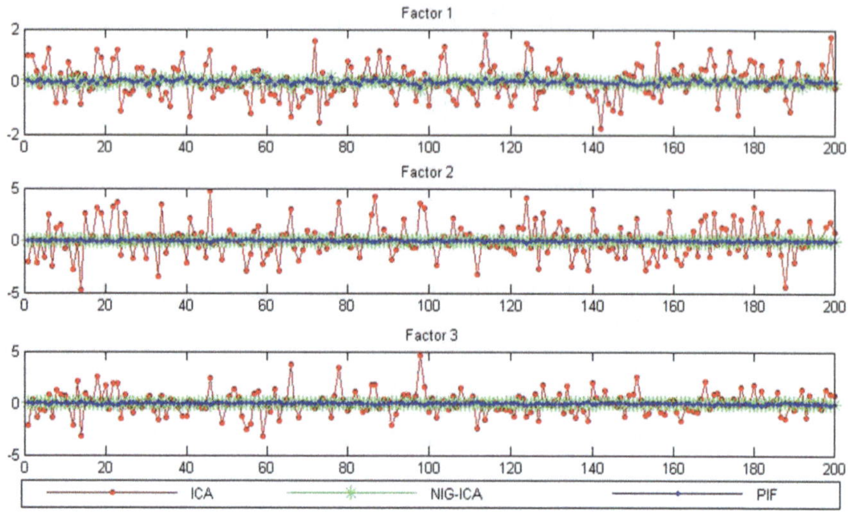

Fig. 10.2 Illustration of residuals of the factors in the sparsity setup. ICA is marked as *circle*, NIG-ICA is labeled with *star* and PIF with *dot*

Moreover, the tunning parameter λ is reasonably selected by using BIC. In the non-sparsity scenario, the optimal λ is 0, indicating non-necessity of penalty as the true loading matrix is not sparse. In the sparsity and high-sparsity scenarios, the optimal λ becomes 0.04 and 0.07 respectively, leading to a high detection rate of zero elements at 100 and 99% respectively. On the contrary, ICA and NIG-ICA are not able to detect any zero elements in the loading matrix. Furthermore, there is no mis-detection by PIF, meaning that no entries in the loading matrix are over pushed to zero.

Figure 10.2 illustrates one representation of the estimation error of the recovered independent factors by the PIF, NIG-ICA and ICA methods respectively in the high-sparsity scenario. While the ICA produces more variations with wider spread, the PIF and NIG-ICA recover the independent factor with smaller errors.

10.4.2 Experiment 2: Large Dimensional Data

In the second experiment with large dimensional data, we generate 14-dimensional dependent data with a sparse loading matrix learning from the CVI data, over a time span of April 1999 to February 2001. The loading matrix is shown in Table 10.4, where 35% of elements are zero.

The generation is repeated 100 times with $n = 200$ sample size. Table 10.5 reports the estimation result. The penalty parameter of PIF is chosen to be $\lambda = 0.08$ by minimizing BIC. The estimation accuracy of PIF is much better with ED of 88.60(SD: 26.11), MN of 60.00(SD: 24.63) and RMSE of 0.20(SD: 0.10) than that of ICA with ED of 419.24(SD: 56.11), MN of 204.00(SD: 36.54) and RMSE of 1.29(SD: 0.05) and slightly better than NIG-ICA with ED of 90.23(SD: 27.74), MN of 61.50(SD: 25.68) and RMSE of 0.22(SD: 0.10). In addition, PIF is able to detect 99.85% of zero entries in the loading matrix and without any miss-detection record of non-zeros.

The simulation study shows that the proposed PIF method has good performance compared to the alternative ICA and NIG-ICA methods with improved estimation accuracy. The good performance mostly attributes to the adoption of the NIG distribution and further by the sparsity of loading matrix. By adding the SCAD penalty function, the proposed PIF is able to identify zero entries in the sparse loading matrix and involves no miss-detection of non-zeros. Moreover, the penalty parameter can be reasonably chosen by using BIC. For example, in the non-sparse scenario, the penalty parameter is selected to be zero. The relative good performance of the PIF is stable with respect to the increase of sparsity and dimensionality.

10.5 Real Data Analysis

In this section, we analyze the sovereign default probabilities of 14 economies from April 1999 to December 2013. The sovereign default probabilities are quantified as daily equally-weighted CVI (Corporate Vulnerability Index) of each economy. The 14 economies are mixture of advanced and emerging economies including China, Hong Kong, India, Indonesia, Japan, US, Germany, Greece, Ireland, Italy, Russian, Spain, UK and Brazil. Data are obtained from the Risk Management Institute at National University of Singapore. We divide the time span into five subperiods based on the business cycles announced by the National Bureau of Economic Research among which two recessions happened: Dot Com bubble from March 2001 to November 2001 and the US sub prime crisis from December 2007 to June 2009. Our interest is to identify the statistical independent dominant factors and investigate the cross-dependence of the sovereign defaults among the economies.

We implement the proposed PIF method. Table 10.6 summarizes the sparse structure of the loading matrices over the 5 time periods. Each economy column reports the number of non-zero elements in the column of loading matrix, representing the number of factors participated in the economies. The total number of non-zero elements in the loading matrix is summarized in the column Total. Sparsity is reflected by the percentage of zero elements in the loading matrix. It shows that there is a V-shape sparsity in terms of US default probability over time, possibly driven by the cyclical pattern of the global economy. Five advanced economies Japan, Germany, Italy, Spain and UK display relatively stable low-sparse structure across the whole time. China and Hong Kong exhibit co-movement, indicating the connection between the two economies, though Hong Kong given its higher level of globalization appears

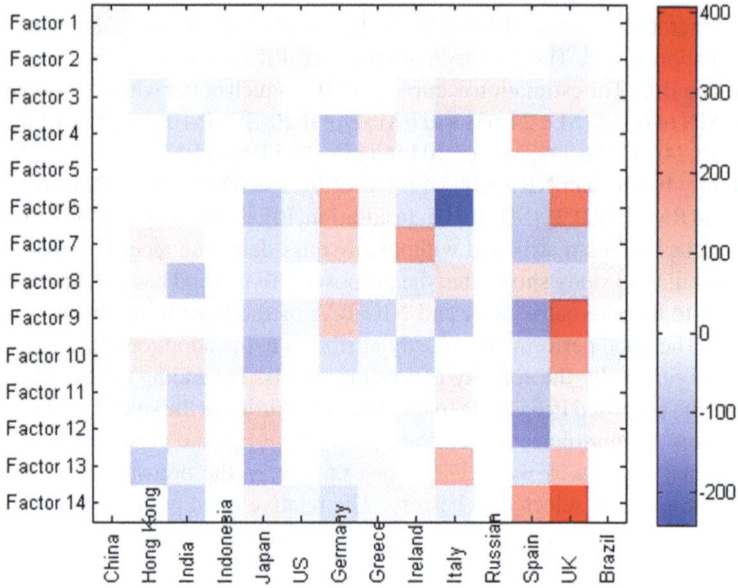

Fig. 10.3 Loading matrix: Apr 1999–Feb 2001

in more factors than China across all periods. The emerging economies of China, India and Indonesia show constant increasing in the number of participated factors along with their increased connection to the global economy especially in the fast growing export business.

Figures 10.3, 10.4, 10.5, 10.6 and 10.7 provides details of the estimated loading matrices over the five time periods. In each plot, we display the loadings of an independent factor with respect to the economies. Zero elements are colored in white. The loading matrix is interpretable. In the pre-Dot Com bubble period, the advanced economies including Japan, Germany, Ireland, Spain and UK participate the most number of factors, while the emerging economies such as China, Indonesia, Russia and Brazil are only related to a few factors. China, for example, only participates in one factor and moreover it is the only element of the factor, implying the closed market of China in the early time. During 1999 to 2001, most defaults in China happened due to the reforming of the state-owned enterprises, which were less likely affected or influenced by the global economy. On the contrary, Japan participates more than 10 factors implying its close connection to the global financial market. In the recent, the sparse inequality between the advanced and emerging economies decreases from period to period, see Fig. 10.8.

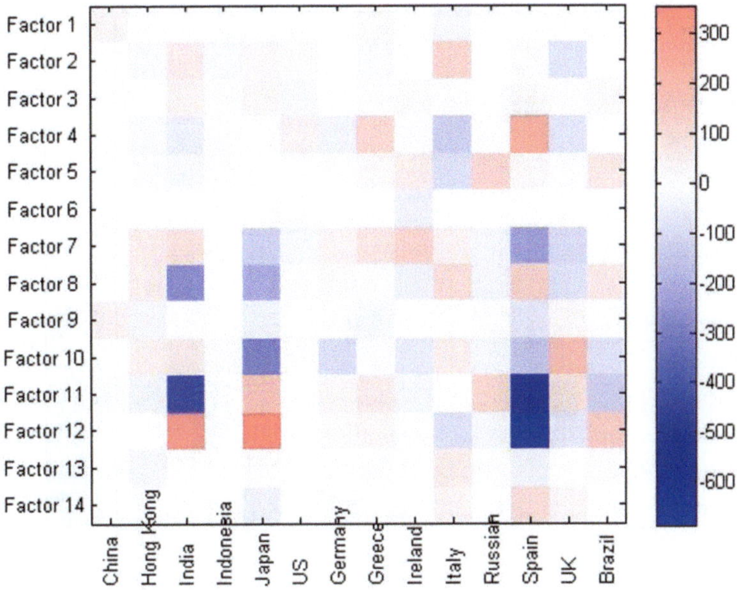

Fig. 10.4 Loading Matrix: Mar 2001–Nov 2001(Dot Com bubble)

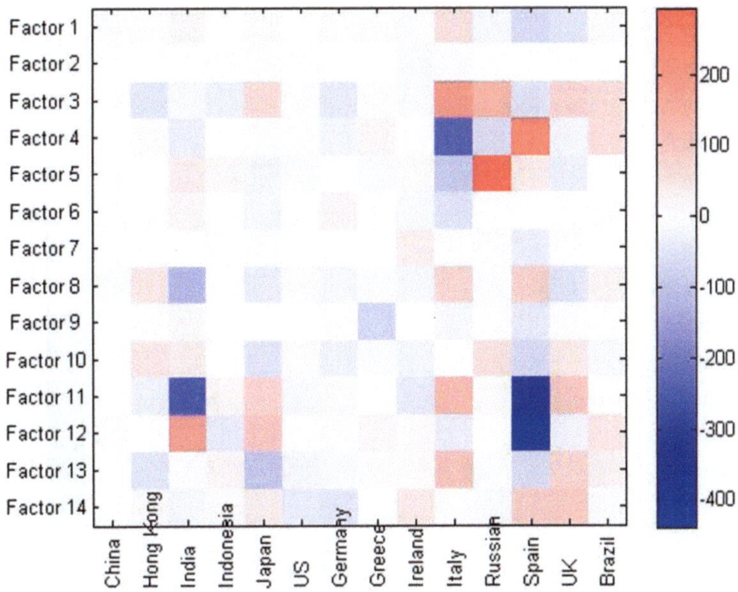

Fig. 10.5 Loading Matrix: Dec 2001–Nov 2007

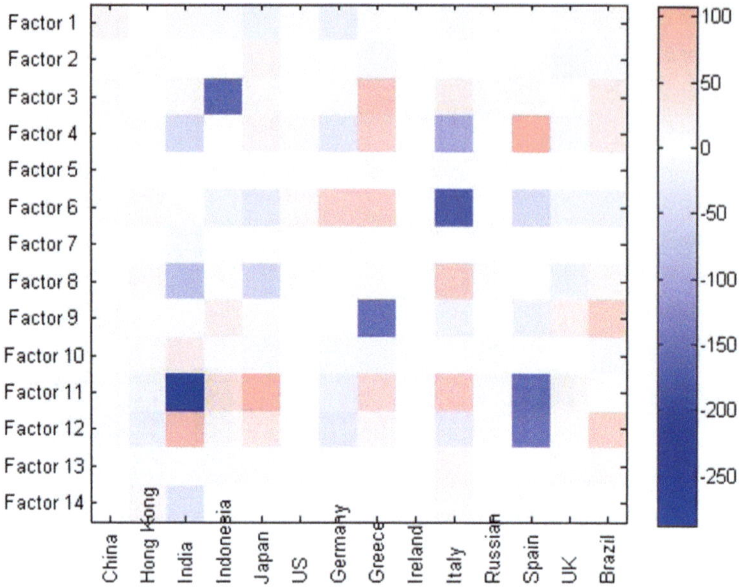

Fig. 10.6 Loading Matrix: Dec 2007–Jun 2009(Sub prime crisis)

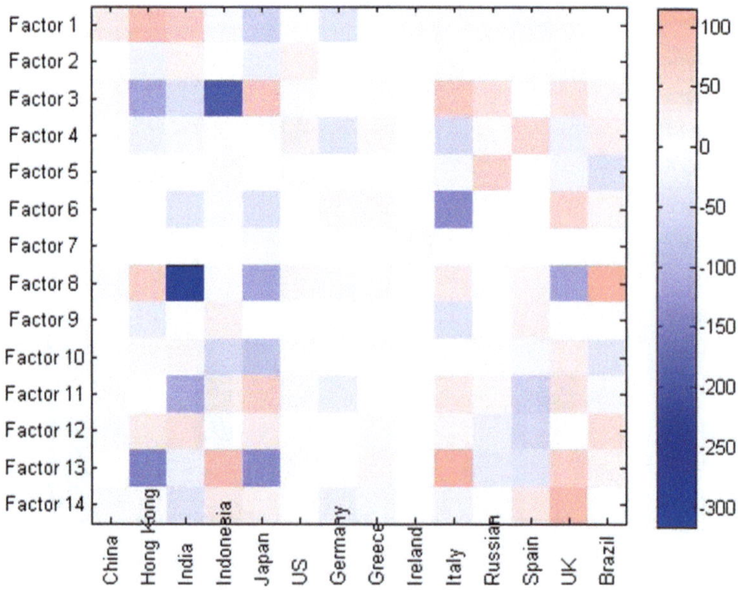

Fig. 10.7 Loading Matrix: Jul 2009–Dec 2013

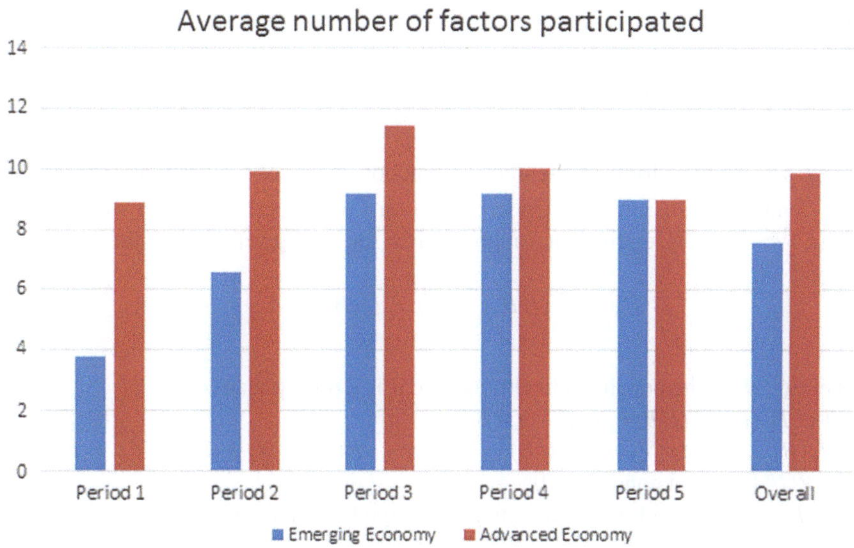

Fig. 10.8 Histogram of average number of factors participated by emerging economies and advanced economies across different period and overall

10.6 Conclusion

We propose the PIF method to transform the observed multivariate correlated variables into independent factors with a sparse loading matrix. We derive the consistency and convergence rate of the sparse loading matrix estimator. Based on the NIG distributional assumption, the estimation is done with a two step ML estimation algorithm by iterating NIG parameter updating and sparse loading matrix estimation. The optimal penalty parameter is chosen via minimizing BIC. We compare the performance of PIF with two alternatives, ICA and NIG-ICA in simulation. The results show the proposed PIF has good performance compared with the conventional ICA and NIG-ICA in both the loading matrix estimation and factor recovery. The estimation accuracy is much improved due to the imposing of NIG distribution. Furthermore, by adopting the SCAD penalty function in PIF, the estimation accuracy is further improved with sparse structure. Moreover, the optimal penalty parameter is reasonably selected by minimizing BIC. The performance of PIF is consistently better with respect to different level of sparse structure and dimensionality of the loading matrix. We implement the PIF to sovereign default probability using CVI data maintained at Credit Research Initiative, Risk Management Institute, National University of Singapore. The estimated loading matrix displays significant sparse structure. For example, China in the pre-Dot Com Bubble period only participates in one factor and is the only element, implying the independence of China's closed market and the global economy. The proposed model can be easily applied to other high-dimensional data.

Appendix

Proof of Theorem 1

Proof The explicit form of the density function g_j is not required, as long as the two conditions are fulfilled. Under condition C1 and C2, Equation $\|\hat{B} - B\| = O_P(n^{-1/2} + a_n)$ is equivalent to proof that for any given $\epsilon > 0$, there exist a large C s.t.

$$P\{\sup_{\|u\|=C} Q(B + \alpha_n u) < Q(B)\} \geq 1 - \epsilon \tag{10.11}$$

where $Q(B)$ is the penalized likelihood and u is a p-by-p matrix.

Let $D_n(u) = Q(B + \alpha_n u) - Q(B)$

$I_u(B) = -E(tr(\nabla_B tr(\nabla \frac{1}{n} l(B)^\top u)^\top u)) = -E(tr(\nabla_B d_u \frac{1}{n} l(B)^\top u)) > 0$ for any $y \in R^{p*p}$ based on condition (B)

If $D_n(u) < 0$ by choosing a sufficiently large C, then the proof is done.

$$\begin{aligned}
D(u) &= l(B + \alpha_n u) - l(B) - n \sum \{\rho_{\lambda_n}(|B_{jk} + \alpha_n u_{jk}|) - \rho_{\lambda_n}(|B_{jk}|)\} \\
&\leq l(B + \alpha_n u) - l(B) - n \sum_{B_{jk} \neq 0} \{\rho_{\lambda_n}(|B_{jk} + \alpha_n u_{jk}|) - \rho_{\lambda_n}(|B_{jk}|)\} \\
&\leq \alpha_n tr(\nabla l(B)^\top u) + \frac{1}{2}\alpha_n^2 tr(\nabla_B d_u l(B)^\top u)\{1 + o_P(1)\} \\
&- \sum_{B_{jk} \neq 0} [n\alpha_n \rho'_{\lambda_n}(|B_{jk}|) sgn(B_{jk}) u_{jk} + n\alpha_n^2 \rho''_{\lambda_n}(|B_{jk}|) u_{jk}^2 \{1 + o(1)\} \\
&\leq \alpha_n tr(\nabla l(B)^\top u) - \frac{1}{2} n\alpha_n^2 I_u(B)\{1 + o_P(1)\} \\
&- \sum_{B_{jk} \neq 0} [n\alpha_n \rho'_{\lambda_n}(|B_{jk}|) sgn(B_{jk}) u_{jk} + n\alpha_n^2 \rho''_{\lambda_n}(|B_{jk}|) u_{jk}^2 \{1 + o(1)\}
\end{aligned} \tag{10.12}$$

The first inequality is because $\rho_{\lambda_n}(0) = 0$ and $\rho_{\lambda_n}(\beta) \geq 0$. The next inequality is Taylor expansion. Then substitute $I_u(B)$ into the equation.

Base on condition (A), $n^{-1/2} tr(\nabla l(B)^\top u) = O_P(1)$, thus the first term of (8) is of order $O_P(n^{1/2}\alpha_n) = O_P(n\alpha_n^2)$. By choosing a sufficiently large C, the second term dominates the first term in $\|u\| = C$.

The last term in (8) is bounded by

$$\sqrt{s} n\alpha_n a_n \|u\| + n\alpha_n^2 max\{\rho''_{\lambda_n}(|B_{jk}|) : B_{jk} \neq 0\} \|u\|^2 \tag{10.13}$$

The first part of (9) is dominated by the second term in (8) when choosing a sufficiently large C. The second term in (9) is also dominated by the second term in (8) as $\max\{\rho''_{\lambda_n}(|B_{jk}|) : B_{jk} \neq 0\} \to 0$

Proof is completed.

References

Barndorff-Nielsen, O. (1997). Normal inverse gaussian distributions and stochastic volatility modelling. *Scandinavian Journal of Statistics, 24*, 1–13.

Bell, A. J., & Sejnowski, T. J. (1995). An information-maximization approach to blind separation and blind deconvolution. *Neural Computation, 7*, 1129–1159.

Blæsild, P. (1999). Generalized hyperbolic and generalized inverse Gaussian distributions, *Working Paper, University of Århus*.

Cardoso, J. F., & Souloumiac, A. (1993). Blind beamforming for non-Gaussian signals. *IEE Proceedings F Radar and Signal Processing, 140*, 362–370.

Chen, R. B., Chen, Y., & Härdle, W. K. (2014). TVICA Time varying independent component analysis and its application to financial data. *Computational Statistics & Data Analysis, 74*, 95–109.

Comon, P. (1994). Independent component analysis, A new concept? *Signal Processing, 36*, 287–287. Higher Order Statistics.

Duan, J. C., Sun, J., & Wang, T. (2012). Multiperiod corporate default prediction—A forward intensity approach. *Journal of Econometrics, 170*, 191–209.

Fan, J., & Li, R. (2001). Variable selection via nonconcave penalized likelihood and its oracle properties. *Journal of the American Statistical Association, 96*, 1348–1360.

Frank, L. E., & Friedman, J. H. (1993). A statistical view of some chemometrics regression tools. *Technometrics, 35*, 109–135.

Gou, J., Zhao, Y., Wei, Y., Wu, C., Zhang, R., Qiu, Y., et al. (2014). Stability SCAD: A powerful approach to detect interactions in large-scale genomic study. *BMC Bioinformatics, 133*, 140–159.

Hyvärinen, A. (1998). Analysis and projection pursuit. *Advances in Neural Information Processing Systems, 10*, 273.

Hyvärinen, A. (1999a). Fast and robust fixed-point algorithms for independent component analysis. *IEEE Transactions on Neural Networks, 10*, 626–634.

Hyvärinen, A. (1999b). The fixed-point algorithm and maximum likelihood estimation for independent component analysis. *Neural Processing Letterss, 10*, 1–5.

Hyvärinen, A., & Oja, E. (1997). A fast fixed-point algorithm for independent component analysis. *Neural Networks, 9*, 1483–1492.

Hyvärinen, A., & Raju, K. (2002). Imposing sparsity on the mixing matrix in independent component analysis. *Neurocomputing, 49*, 151–162.

Jones, M. C., & Sibson, R. (1987). What is projection pursuit? *Journal of the Royal Statistical Society. Series A (General), 24*, 1–10.

Karlis, D. (2002). An EM type algorithm for maximum likelihood estimation of the normal-inverse Gaussian distribution. *Statistics & Probability Letters, 57*, 43–52.

Kohavi, R., et al. (1995). A study of cross-validation and bootstrap for accuracy estimation and model selection. *Ijcai, 14*, 1137–1145.

Li, K. C. (1987). Asymptotic optimality for Cp, CL, cross-validation and generalized cross-validation: discrete index set. *The Annals of Statistics, 15*, 958–975.

Pham, D. T., & Garat, P. (1997). Blind separation of mixture of independent sources through a maximum likelihood approach. In *Proceedings of EUSIPCO*.

Schwarz, G., et al. (1978). Estimating the dimension of a model. *The Annals of Statistics, 6*, 461–464.

Tibshirani, R. (1996). Regression shrinkage and selection via the lasso. *Journal of the Royal Statistical Society. Series B (Methodological)*, *58*, 267–288.

Wang, H., Li, R., & Tsai, C. L. (2007). Tuning parameter selectors for the smoothly clipped absolute deviation method. *Biometrika*, *94*, 553–568.

Wu, E. H., Philip, L. H., & Li, W. K. (2006). An independent component ordering and selection procedure based on the MSE criterion. *Independent Component Analysis and Blind Signal Separation* (pp. 286–294).

Xie, H., & Huang, J. (2009). SCAD-penalized regression in high-dimensional partially linear models. *The Annals of Statistics*, *37*, 673–696.

Zhang, K., Peng, H., Chan, L., & Hyvärinen, A. (2009). ICA with sparse connections: revisited. *Independent component analysis and signal separation* (pp. 195–202).

Chapter 11
Term Structure of Loss Cascades in Portfolio Securitisation

L. Overbeck and C. Wagner

Abstract We report on the term structure of loss cascades generated through portfolio tranching. The results are based on the analytical form of the loss distribution for uniform loan portfolios and show that the expected loss of the first loss position increases roughly linear whereas the expected losses of the more senior tranches increase exponentially over time depending on the relation between mean default probability and tranching limits.

11.1 Introduction

Asset Backed Securities (ABS) and related portfolio dependent financial products like collateralised loan obligations (CLO) are used for several purposes, namely to transfer and manage credit risk, as a balance sheet management tool in order to obtain capital relief, and gain liquidity. From the methodological point of view these structures boil down to a repartition of interest earnings in exchange to loss burdens among possible investors whereas both are alloted according to the investors seniority. In the present note we treat only the second point, i.e. the allocation of

L. Overbeck (✉)
Mathematisches Institut Universität Giessen, Arndtstraße 2,
35392, Giessen, Germany
e-mail: Ludger.Overbeck@math.uni-giessen.de

C. Wagner
FMS Wertmanagement, Munich, Germany
e-mail: christoph.wagner@fms-wm.de

W.K. Härdle et al. (eds.), *Applied Quantitative Finance*, Statistics and Computing,
DOI 10.1007/978-3-662-54486-0_11

losses through the various tranches and their evolution in time. These structures have obtained a lot of attraction before and during the credit crisis 2007/08. Many banks built large trading desks for these structures, called correlation desks. In analogy to the volatility trading desks based on Black-Scholes model and its extension, the correlation desks were trading "implied correlation" which were based on the "base correlation approach" (cf. Li (2000), McGinty and Ahluwalia (2004) or Bluhm and Overbeck (2006)). This is a simplified default time model with a uniform Gaussian copula. For our focus, the timing of aggregate losses, we do not model the default time of the single entities in the portfolio, but model for each time step the aggregate loss in the portfolio. This is in some aspects a simple top-down approach, which is a more recent stream of modelling for structured products (cf. e.g. Sidenius et al. (2008), Bennani (2005), Schönbucher (2005) and Filipovic et al. (2011)). In the present chapter we assume a uniform portfolio and use loss distributions which are available in analytic form. The main result of the paper is that even is this simplified approach, the fact that losses are back loaded in senior tranche. This made it plausible that the down-rating in the credit crisis was especially severe on senior tranches. Also compared to migration behaviour of well rated counterparties, well rated tranches will migrate in a more non-linear way. Most of the downward migration will come and the end of the lifte-time of the transaction.

11.2 Loss Distribution of Uniform Portfolio

It is well known (cf. Vasicek (1987) or Bluhm et al. (2010)) that for a uniform portfolio of m loans, i.e. equal exposure $1/m$, equal default probability p and equal pairwise asset correlation ρ, the limiting distribution for $m \longrightarrow \infty$ is the so called normal inverse distribution $NID(p, \rho)$ (The underlying asset returns in this model are assumed to be normal distributed.). The distribution of the portfolio losses $0 \leq x \leq 1$ is given by the cumulative distribution function

$$NID(x, p, \rho) = N\left\{\frac{1}{\sqrt{\rho}}\left[\sqrt{1-\rho}N^{-1}(x) - N^{-1}(p)\right]\right\} \tag{11.1}$$

and its density

$$\phi(x, p, \rho) = \sqrt{\frac{1-\rho}{\rho}} \exp\left\{-\frac{1}{2\rho}\left\{(1-2\rho)\left[N^{-1}(x)\right]^2 - 2\sqrt{1-\rho}N^{-1}(x)N^{-1}(p) + \left[N^{-1}(p)\right]^2\right\}\right\} \tag{11.2}$$

with $0 < p, \rho < 1$, with mean p and variance $\sigma^2 = N_2\left(N^{-1}(p), N^{-1}(p); \rho\right) - p^2$, where N denotes the standard normal distribution function and $N_2(x, y; \rho)$ denotes the bivariate normal distribution function with zero expectation vector and covariance matrix showing units on the diagonal and ρ off the diagonal.

11.3 Time Slicing

Now, let us observe a portfolio and its losses on a discrete time grid $0 = t_0 < t_1 < t_2 \cdots < t_{n-1} < t_n$. Denote X_i the relative portfolio loss (relative to the remaining exposure) during time step i, then the absolute loss at i, assuming the balance at time 0 to be 1, is

$$Y_i = \prod_{j=1}^{i-1}(1 - X_j)\, X_i \qquad \text{for} \quad i = 1, \ldots n\,, \tag{11.3}$$

and accumulates over time to

$$\tilde{Y}_i = \sum_{j=1}^{i} Y_i\,. \tag{11.4}$$

Suppose further that our portfolio and the residues after losses can be considered as being uniform with possible changes only being reflected by time dependent portfolio parameters, p_i, ρ_i, $i = 1 \ldots n$. We can then draw the random variable X_i in step i from the normal inverse distribution

$$X_i \sim NID(x, p_i, \rho_i) \tag{11.5}$$

to obtain the absolut loss Y_i. The respective density function can in principle be calculated by product folding, but it does not seem to be possible to state the results in a closed form.

We therefore resort to Monte Carlo simulations of the loss distribution, whereby the random variables are generated according to Eq. (11.5). For this, we first take uniformly distributed random variables $Z \sim U(0, 1)$ and transform with

$$x = NID^{-1}(z, p_i, \rho_i) = N\left(\frac{1}{\sqrt{1-\rho}}\left(N^{-1}(p) - \sqrt{\rho}\, z\right)\right).$$

11.4 Loss Cascades

As already mentioned in the introduction during securitization transactions the portfolio losses L are allocated subsequently to various tranches according to their seniority, i.e. investor 1 holds for losses up to $\alpha_1\%$, investor 2 for remaining losses but smaller than $\alpha_2\%$, and so on. In mathematical notation this reads

$$L_i = (L - \alpha_{i-1})^+ \wedge (\alpha_i - \alpha_{i-1})$$

where $0 \le \alpha_0 < \alpha_1 < \cdots < \alpha_k$ are the boundaries of the tranches and L_i denotes the loss to be borne by tranche i. Thus, tranches are 'served' in cascades, if one tranche has overflowed further losses are allocated to the next senior tranche. The first tranche is usually kept by the issuer and is called first loss position (FLP). The mezzanine tranches are usually brought to the market as notes and the senior tranche is securitized by a credit default swap. For the rating and spreads of the various tranches an interesting quantity is the expected loss per tranche

$$EL_i = \int \frac{(x-\alpha_{i-1})^+ \wedge (\alpha_i - \alpha_{i-1})}{\alpha_i - \alpha_{i-1}} \, df(x) , \tag{11.6}$$

with $f(x)$ being the probability measure of some loss distribution.

11.5 Results

Table 11.1 shows the results of a Monte Carlo simulation with 10^6 simulations of a sequence of normal inverse distributed portfolio losses $X_i, i = 1, \ldots, 7$, Eq. (11.5), with constant portfolio parameters $p = 0.0026$ and $\rho = 0.17$. The first column denotes the year, the second the respective (forward) default rate p and the third and fourth column give the mean and the standard deviation of the accumulated loss, EL and UL. The remaining columns report on the accumulated expected loss per tranche, where some typical boundaries have been chosen. All quantities increase monotoneously, but the more interesting result can be seen in Fig. 11.1 (linear and logarithmic plot). Whereas the expected loss of the first tranche increases linearly (scaled on the right axis in the linear plot) the ELs of the other tranches increase exponentially over the years. In Fig. 11.2 we attempt a direct comparison of the default-probability term structure for the tranches $[2.4 - 3.9\%]$, $[3.9 - 6.5\%]$ and $[6.5 - 9\%]$ with respective corporate zero bonds (calibration based on rating reports of Standard & Poor's and Moody's Investors Services, Moody (2001). For this, we

Table 11.1 Vasicek (normal inverse) distribution

EL per tranche									
Year	p	EL	UL	0–2.4%	2.4–3.9%	3.9–6.5%	6.5–9%	9–11.5%	11.5–100%
1	0.0026	0.002593	0.004588	0.104352	0.004135	0.000830	0.000157	0.000041	0.000001
2	0.0026	0.005186	0.006491	0.206350	0.010987	0.002129	0.000389	0.000110	0.000001
3	0.0026	0.007771	0.007921	0.304995	0.021439	0.004068	0.000683	0.000177	0.000002
4	0.0026	0.010356	0.009126	0.399452	0.036819	0.006922	0.001108	0.000261	0.000003
5	0.0026	0.012934	0.010181	0.488493	0.058068	0.010884	0.001691	0.000392	0.000004
6	0.0026	0.015500	0.011127	0.570866	0.086047	0.016401	0.002502	0.000559	0.000006
7	0.0026	0.018059	0.011991	0.645940	0.121363	0.023847	0.003575	0.000777	0.000008

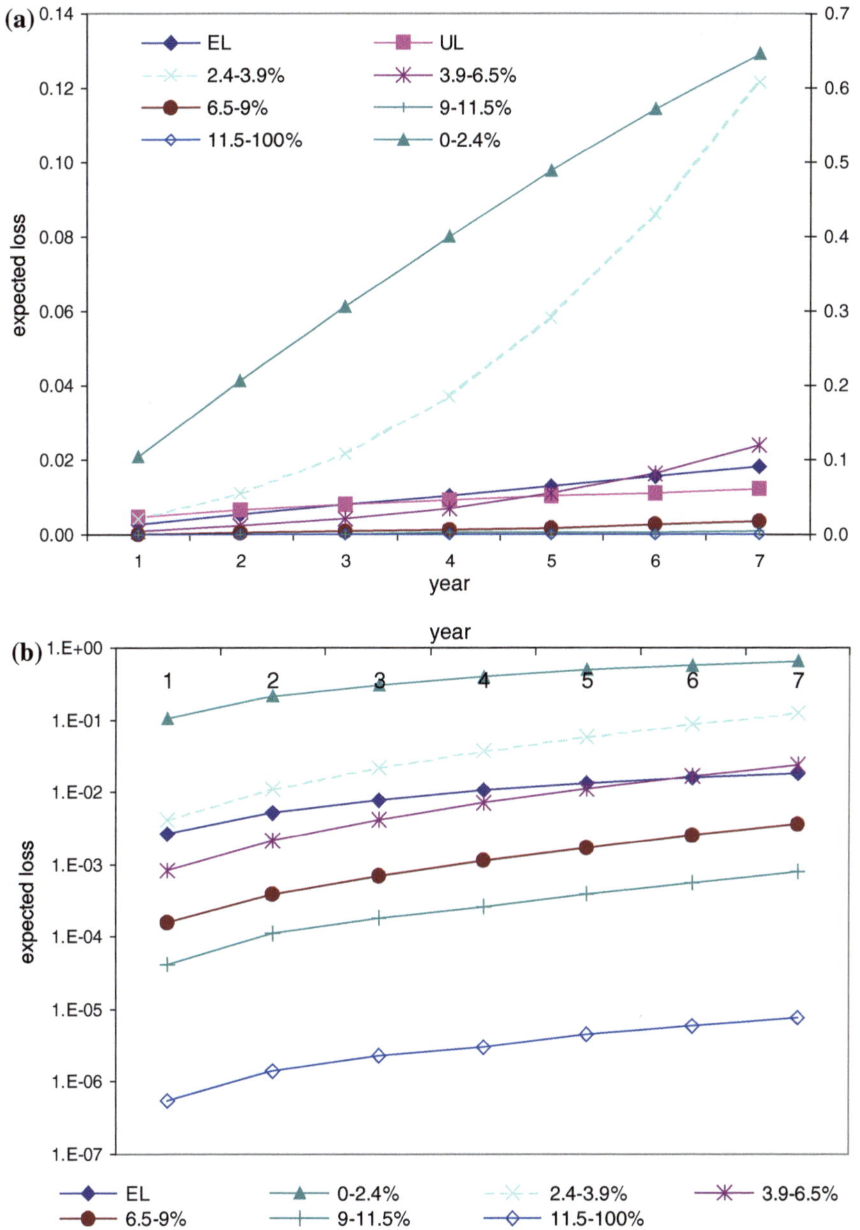

Fig. 11.1 Term structure of expected losses in tranches with $p = 0.0026$, $\rho = 0.17$, linear and logarithmic scale. Note that in the upper plot EL$[0 - 2.4\%]$ scales with the *right* axis

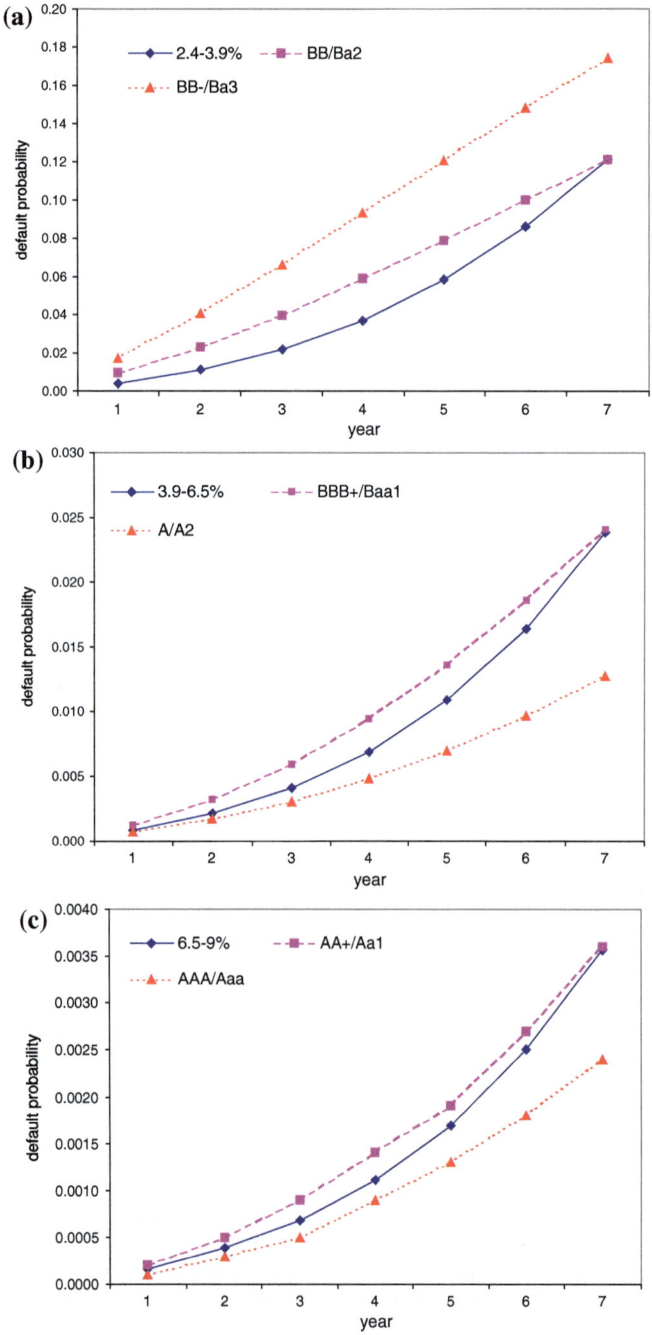

Fig. 11.2 Term structure of expected losses in tranches [2.4 – 3.9%], [3.9 – 6.5%], [6.5 – 9%], with $p = 0.0026$, $\rho = 0.17$, and corporate zero bonds for comparison

either try to match 'one-year' expected loss or the accumulated '7-years' expected loss per tranche to the respective default probabilities assigned by Moody's to a suitable corporate bond.

Since the first loss position is usually kept by the issuer he can expect linear increasing loss burdens over time whereas the investors buying the notes have to anticipate exponentially increasing loss burdens. This is different to the term structure of the expected loss for similar rated corporate bonds, these show a less convex increase in expected loss during time. The term structure of securitized tranches might therefore serve a non-linear risk appetite on the investors side. We can now estimate the respective rates r_i in tranche i given to the investors by calculating the net present value of the expected cash flows according to

$$\sum_{j=1}^{n-1} \frac{(1 - EL_i^j) r_i}{\prod_{l=1}^{j}(1 + z_l)} + \frac{(1 - EL_i^n)(1 + r_i)}{\prod_{l=1}^{n}(1 + z_l)} = 1 \,, \tag{11.7}$$

where EL_i^j denotes the accumulated expected loss in tranche i up to year j and z_l represents the risk free zero forward rate.

Using Eq. (11.7) and a constant risk free rate $z = z_l = 5.0\%$ we arrive at:

tranche	0–2.4%	2.4–3.9%	3.9–6.5%	6.5–9%	9–11.5%	11.5–100%
rate	20.560%	6.794%	5.339%	5.051%	5.011%	5.000%

In reality, the spreads given to investors are considerably higher. Clearly, this is again the discussion of real-world versus risk-neutral probabilities. But one justification for higher risk neutral spreads, besides liquidity or other additional risks, can be found in Fig. 11.3 where the ratios of unexpected to expected loss, UL/EL, for the whole portfolio (total) and all tranches are shown. All ratios decrease in time, but the more interesting result is that they differ considerably in orders of magnitude. Whereas the whole portfolio and the first loss piece $[0 - 2.4\%]$ yield a ratio of order one already the second tranche $[2.4 - 3.9\%]$ exhibits a ratio of order 10 and all more senior ratios increase roughly by a factor of two. This means that the variation of the losses around the expected value is much higher for the investors tranches than for the first loss position and requires an additional risk premium.

11.5.1 Other Loss Distributions

Since the tail behavior of loan loss distributions is a rather critical part in risk considerations we also experiment with other possible distributions. The following comparison is based on an EL/UL match, i.e. we choose the parameters such that the first two moments match to the ones obtained from the normal inverse distribution.

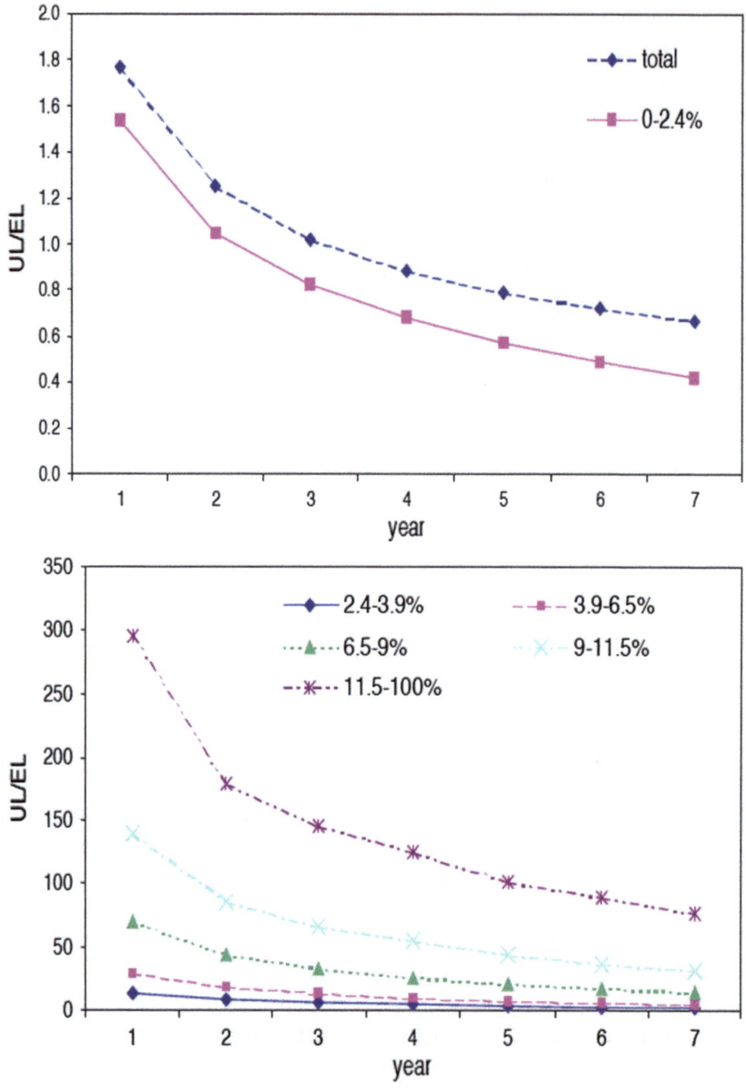

Fig. 11.3 Term structure of ratio UL/EL for the whole portfolio (total) and the tranches with $p = 0.0026$, $\rho = 0.17$

Beta Distribution

Choosing the parameters for the Beta-distribution with density

$$f_{\alpha,\beta}(x) = \frac{\Gamma(\alpha+\beta)}{\Gamma(\alpha)\Gamma(\beta)} x^{\alpha-1}(1-x)^{\beta-1}, \qquad 0 \le x \le 1, \tag{11.8}$$

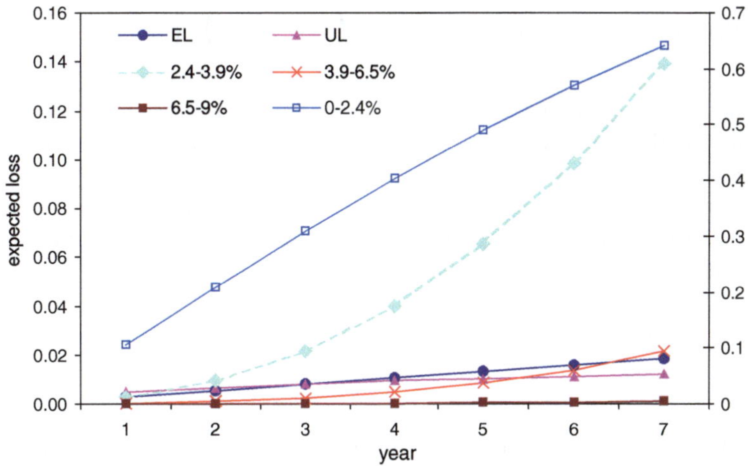

Fig. 11.4 Term structure of expected losses in tranches where the parameters of Beta-distribution ($\alpha = 0.315878$, $\beta = 121.176$) are chosen such that EL and UL match to the previous example (Vasicek-distribution with $p = 0.0026$, $\rho = 0.17$). Note that in the upper plot EL[$0 - 2.4\%$] scales with the *right* axis

Table 11.2 Beta distribution

EL per tranche								
Year	EL	UL	0–2.4%	2.4–3.9%	3.9–6.5%	6.5–9%	9–11.5%	11.5–100%
1	0.002569	0.004512	0.105027	0.002851	0.000209	4.48E-06	0	0
2	0.005176	0.006450	0.208788	0.009512	0.00085	2.14E-05	0	0
3	0.007789	0.007294	0.308685	0.021589	0.002135	6.06E-05	0	0
4	0.010391	0.009154	0.402956	0.039937	0.004462	0.000185	1.62E-06	0
5	0.012955	0.010195	0.489843	0.065128	0.008176	0.000377	2.56E-06	0
6	0.015510	0.011122	0.569401	0.098164	0.013608	0.000714	8.69E-06	0
7	0.018078	0.011989	0.641583	0.138913	0.021659	0.001298	2.98E-05	0

mean $\mu_B = \frac{\alpha}{\alpha+\beta}$ and variance $\sigma_B = \frac{\alpha\beta}{(\alpha+\beta+1)(\alpha+\beta)^2}$ as $\alpha = 0.315878$, $\beta = 121.176$ leads to a good EL/UL match with the Vasicek distribution under $p = 0.0026$, $\rho = 0.17$. Table 11.2 and Fig. 11.4 show the result where the yearly portfolio loss X_i is now drawn according to Eq. (11.8).

Negative Binomial Distribution

Another prominent loss distribution in extreme event statistics is the Negative-Binomial distribution with frequency function

$$P[Loss = n] = f_{\alpha,\beta}(n) = \frac{\Gamma(\alpha + n)}{n!\Gamma(\alpha)} \left(1 - \frac{\beta}{1+\beta}\right)^{\alpha} \left(\frac{\beta}{1+\beta}\right)^{n} \tag{11.9}$$

with mean $\mu_{NB} = \alpha\beta$ and variance $\sigma_{NB}^2 = \alpha\beta(1+\beta)$. Note that the Negative Binomial distribution can be constructed as a composition of a Poisson distribution conditional on Gamma distributed intensities Λ see e.g. Rice (1995), for a motivation in credit risk management see CreditRisk$^+$ (1997). The respective parameters α, β can then be found by matching first and second moment with the normal inverse distribution (Sect. 11.2) under the constraint

$$\frac{\mu_{NB}}{m} = p \quad \text{and} \quad \frac{\sigma_{NB}^2}{m^2} = \sigma^2 \qquad (m \text{ sufficiently large}).$$

Choosing $m = 10^6$ yields a fairly good approximation of the corresponding percentage loss, i.e. $Loss/m$, since the probabilities $P\big[Loss = k\big]$ are negligible for $k \geq 10^6$ and results in $\alpha = 0.3193$ and $\beta = 8.1416 \times 10^3$. Table 11.2 shows the results for the different tranches (Table 11.3). The expected (EL) loss, the unexpected loss (UL) and the expected loss in the first tranche $[0 - 2.4\%]$ match pretty well for all three distributions. As we move further into the tails to higher tranches, see also Fig. 11.5, we observe an increasing difference in the term structure between normal inverse respectively Beta-/Negative-Binomial distribution reflecting the different 'fatness' of tails, see especially tranche $[6 - 9\%]$. Due to the very asymmetric and 'extreme event'-like behavior of credit loss, we think that the normal inverse distribution more truthfully reflects the 'loss reality' (Bluhm et al. (2010)). Surprisingly, the term structure of beta-distribution and negative binomial distribution are fairly equal. For further investigation we generated a q-q-plot (Fig. 11.7) for both distributions with matched first two moments. Remember that in case of the Negativ-Binomial Distribution the discrete losses $n \in \mathbb{N}_0$ have to be divided by some large number m (For

Table 11.3 Negative Binomial distribution

EL per tranche								
Year	EL	UL	0–2.4%	2.4–3.9%	3.9–6.5%	6.5–9%	9–11.5%	11.5–100%
1	0.002587	0.004581	0.105659	0.003039	0.000205	0.000010	0.00E+00	0.00E+00
2	0.005197	0.006476	0.209482	0.009744	0.000864	0.000017	0.00E+00	0.00E+00
3	0.007777	0.007932	0.308187	0.021358	0.002205	0.000091	3.46E-06	0.00E+00
4	0.010356	0.009115	0.402161	0.038983	0.004387	0.000225	4.36E-06	0.00E+00
5	0.012926	0.010175	0.489310	0.063951	0.008158	0.000425	2.76E-05	0.00E+00
6	0.015511	0.011120	0.569918	0.096979	0.013770	0.000770	4.97E-05	0.00E+00
7	0.018083	0.011988	0.642122	0.138027	0.021825	0.001309	7.55E-05	0.00E+00

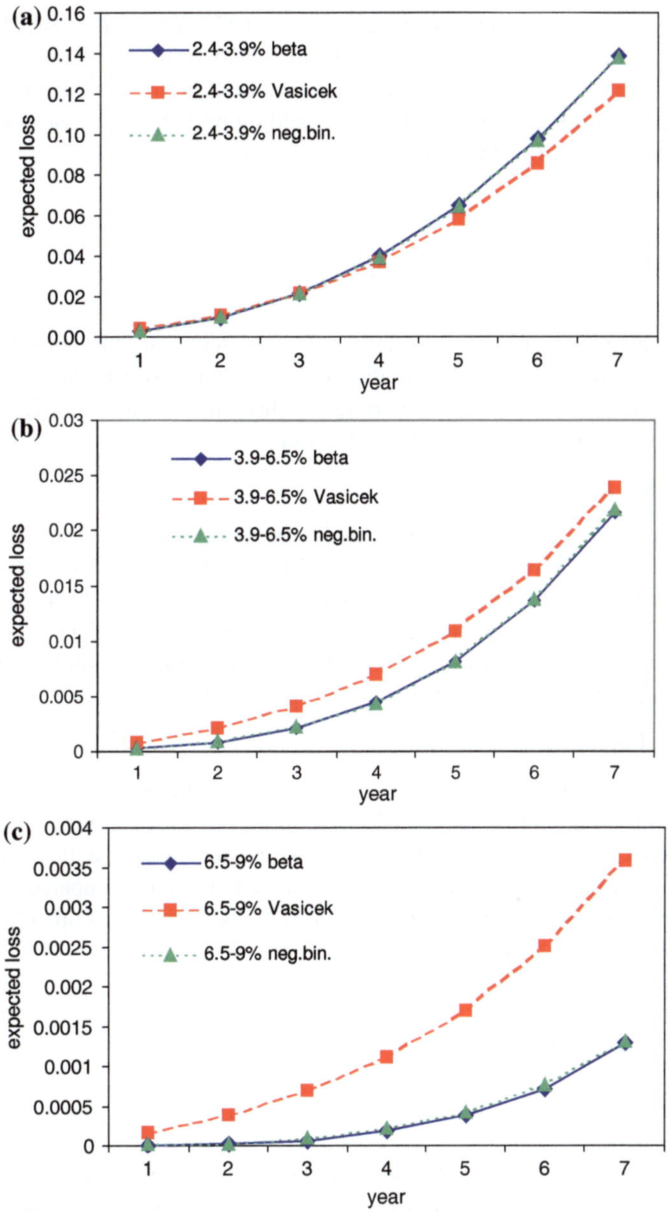

Fig. 11.5 Comparison of term structures of expected losses in tranches two, three and four with different underlying loss distributions under the constraint of matching the first two moments

the plot we chose $m = 1000$ whichresults in $\alpha = 0.323278$ and $\beta = 80.4258$.). The q-q-plot shows cummulative probabilities up to 99.995% and is well on the diagonal. Only in the right upper corner the points begin to fall below the diagonal, but this clearly depends on the cut-off m. These coinciding probability masses far out into the tails thus explain the identical tranching results.

11.5.2 Variable Portfolio Quality

Due to credit migrations we are often confronted with variable portfolio quality during the term of the transaction. In the following we investigate the consequences of two extreme cases, i.e. strictly deteriorating (back loaded) and strictly improving (front loaded) quality on our tranching structure.

Deteriorating Portfolio

A variable portfolio quality can represented through different one-year default probabilities, thus we simply choose a sequence of increasing 'forward' default probabilities p as in Table 11.4. The corresponding plots are shown in Fig. 11.6a. A comparison with the Table 11.1 resp. Figure 11.1 shows that the slopes of the EL-per-tranche curves over the years increases as expected. Only the first loss position flattens with year five since the first tranche begins to fill up (Fig. 11.7).

Improving Portfolio

Conversely, we can represent an improving portfolio quality by decreasing 'forward' yearly default probabilities, Table 11.5. The plot in Fig. 11.6b depicts again that the cumulative expected loss of the first loss position increases less than linear. The reason is that now the upper limit of the first tranche is small compared to the high default probabilities in the first years, i.e. again the first tranche fills up rather quickly and losses are passed to the next higher tranche.

Table 11.4 Increasing default probability

EL per tranche									
Year	p (forward)	EL	UL	0–2.4%	2.4–3.9%	3.9–6.5%	6.5–9%	9–11.5%	11.5–100%
1	0.0026	0.002593	0.004579	0.104369	0.004101	0.000831	0.000147	0.000039	0.000001
2	0.0036	0.006178	0.007552	0.241895	0.016947	0.003589	0.000685	0.000200	0.000004
3	0.0043	0.010451	0.010198	0.393470	0.045190	0.010055	0.001928	0.000516	0.000008
4	0.0048	0.015207	0.012650	0.542116	0.096050	0.023047	0.004522	0.001182	0.000016
5	0.0051	0.020238	0.014826	0.673930	0.172279	0.044946	0.009036	0.002325	0.000030
6	0.0053	0.025426	0.016816	0.780726	0.271508	0.078663	0.016656	0.004250	0.000054
7	0.0054	0.030681	0.018618	0.860469	0.386342	0.125166	0.028539	0.007366	0.000094

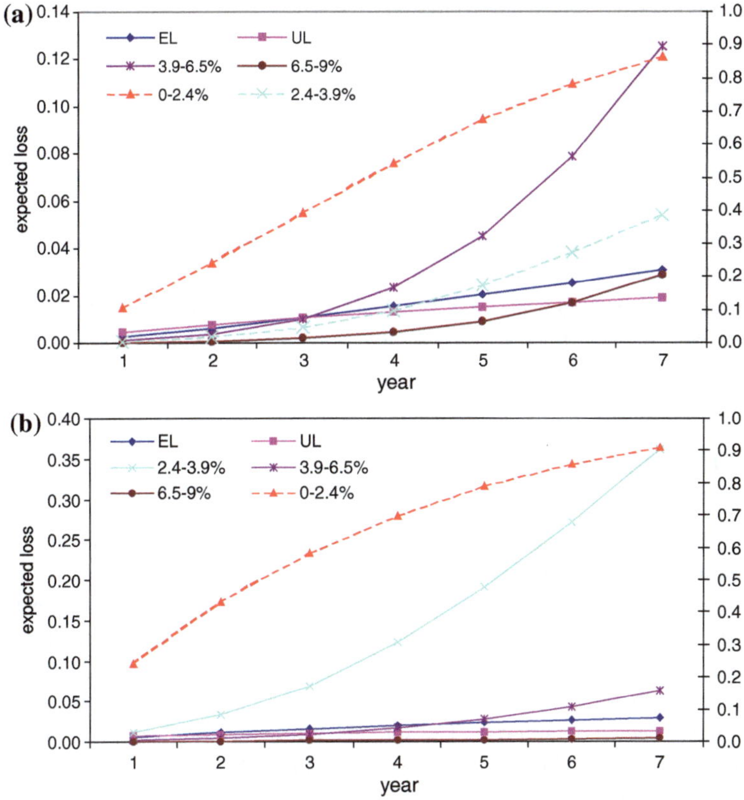

Fig. 11.6 Expected loss in tranches for a portfolio with deteriorating quality **a** and improving quality **b**. The *dashed lines* scale with the *right* axis, the *solid lines* scale with the *left* axis

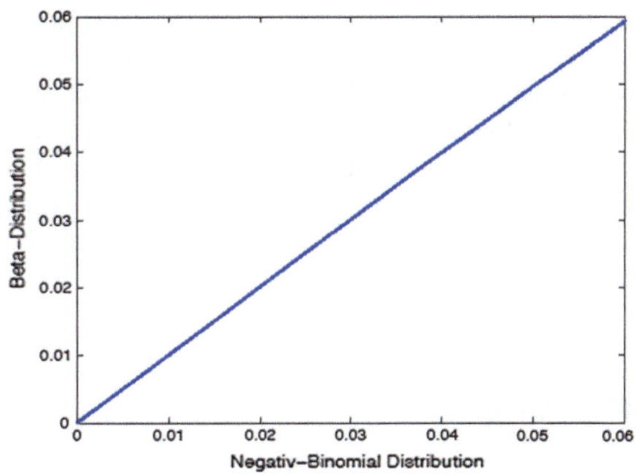

Fig. 11.7 q-q-plot for Beta- and Negativ-Binomial Distribution under the constraint of matched first two moments. The Negativ-Binomial Distribution is rescaled to $[0, 1]$ via a division by $m = 1000$. The cumulative probabilities are shown up to 99.995%

Table 11.5 Decreasing default probability

EL per tranche									
Year	p (forward)	EL	UL	0–2.4%	2.4–3.9%	3.9–6.5%	6.5–9%	9–11.5%	11.5–100%
1	0.0060	0.005992	0.006269	0.241622	0.010208	0.001379	0.000144	2.31E-05	1.27E-07
2	0.0050	0.010961	0.008220	0.431596	0.032250	0.004144	0.000389	6.05E-05	3.00E-07
3	0.0043	0.015214	0.009431	0.580569	0.068669	0.008830	0.000719	9.70E-05	5.02E-07
4	0.0039	0.019060	0.010333	0.698890	0.121829	0.016286	0.001245	1.47E-04	6.63E-07
5	0.0036	0.022595	0.011046	0.790304	0.190679	0.027340	0.002020	2.25E-04	9.83E-07
6	0.0033	0.025819	0.011616	0.857477	0.270135	0.042336	0.003105	3.23E-04	1.39E-06
7	0.0032	0.028936	0.012125	0.907148	0.359962	0.062859	0.004687	4.54E-04	1.86E-06

11.6 Conclusion

We have investigated the term structure of loss cascades in a tranched portfolio structure, commonly found in loan portfolio securitisation. The yearly loss is first simulated via a normal inverse distribution. The resulting expected losses per tranche increase roughly linear for the first loss position and exponential for the higher tranches. We show how to derive a corresponding rating for each tranche, first by comparison with corporate zero bonds, and second by calculating the spreads representing the expected default risk. The spreads of tranches implied by our analysis of a securitized portfolio show a more convex term structure than for comparable corporate bonds. Next, using other possible loss distribution (Beta-/Negative-Binomial distribution) we find that the expected losses for tranches higher than the first depend heavily on the chosen distribution. The respective parameters have been calibrated by matching the first two moments and reveal again the tail-'fatness' of the normal inverse distribution compared to the other two. It is well known (eg. Bluhm et al. (2010)) that the tail behavior of the normal inverse distribution captures the extreme type behavior of credit losses better than the other two distributions. Interestingly, Beta- and Negativ-Binomial distribution yield coinciding results. A brief look at the q-q-plot reveals that both distributions seem to coincide as used above. Eventually, we have a brief look on how variable portfolio quality can be treated in our context and how tranching limits and yearly default probabilities interact in the term structure of loss cascades.

References

Bennani, N. (2005). *The forward loss model: A dynamic term structure approach for the pricing of portfolio credit derivatives*, Working Paper (The Royal Bank of Scotland).

Bluhm, C., Overbeck, L., & Wagner, C. (2010). *Introdcution to credit risk modeling*. Boca Raton: Chapman & Hall/CRC Press.

Bluhm, C., & Overbeck, L. (2006). *Structured credit portfolio analysis, baskets and CDOs*. Boca Raton: Chapman & Hall/CRC Press.
CreditRisk^{+} - A Credit Risk Management Framework (1997), New York: Credit Suisse Financial Products.
Filipovic, D., Overbeck, L., & Schmidt, T. (2011). Dynamic CDO term structure modeling. *Mathematical Finance*, *21*(1), 53–71.
Li, D.X. (2000). On default correlation: A copula function approach.
McGinty, L., Ahluwalia, R. (2004). A model for base correlation calculation. Technical report (JP Morgan).
Moody's investor service. (2001). *Default and recovery rates of corporate bond issuers*. New York: Moody's.
Rice, J. A. (1995). *l mathematical statistics and data analysis* (2nd ed.). North Scituate: Duxbury Press.
Sidenius, J., Piterbarg, V., & Andersen, L. (2008). A new framework for dynamic credit portfolio loss modelling. *International Journal of Theoretical and Applied Finance*, *11*(2), 163–197.
Schönbucher, P. (2005). Portfolio losses and the term structure of loss transition rates: A new methodology for the pricing of Portfolio credit derivatives. Working Paper (ETH Zürich).
Vasicek, O. A. (1987). *Probability of loss on loan Portfolio*. San Francisco: KMV Corporation.

Chapter 12
Credit Rating Score Analysis

Wolfgang Karl Härdle, K.F. Phoon and D.K.C. Lee

Abstract We analyse a sample of funds and other securities each assigned a total rating score by an unknown *expert* entity. The scores are based on a number of risk and complexity factors, each assigned a category (factor score) of Low, Medium, or High by the expert entity. A principal component analysis of the data reveals that based on the chosen risk factors alone we cannot identify a single underlying latent source of risk in the data. Conversely, the chosen complexity factors are clearly related to one or two underlying sources of complexity. For the sample we find a clear positive relation between the first principal component and the total expert score. An attempt to match the securities' expert score by linear projection of their individual factor scores yields a best case correlation between expert score and projection of 0.9952. However, the sum of squared differences is, at 46.5552, still notable.

W.K Härdle (✉)
C.A.S.E.-Center for Applied Statistics and Economics, Humboldt-Universität zu Berlin, Unter den Linden 6, 10099 Berlin, Germany
e-mail: haerdle@hu-berlin.de

W.K. Härdle
School of Economics, 6th Level, Research fellow in Sim Kee Boon Institute for Financial Economics, Singapore Management University, 90 Stamford Road, Singapore 178903, Singapore

K.F. Phoon
School of Business, Singapore University of Social Sciences, Singapore, Singapore
e-mail: kfphoon@suss.edu.sg

D.K.C. Lee
Singapore University of Social Sciences, Singapore, Singapore
e-mail: Davidleekc@suss.edu.sg

W.K. Härdle et al. (eds.), *Applied Quantitative Finance*, Statistics and Computing,
DOI 10.1007/978-3-662-54486-0_12

12.1 Introduction

We are provided with a sample of $n = 100$ funds and other securities that have been assigned a rating score by an unknown *expert entity* – the expert (rating) score in the following. We assume the rating score to depend on a set of six risk factors and five complexity factors, each modelled as random variables on an ordinal scale of Low, Medium, High. The risk factors are volatility, liquidity, credit rating, duration/cash flow, leverage, and diversification degree. The complexity factors comprise of the number of structural layers, expansiveness of derivatives, availability and known pricing models, number of return outcome scenarios, and transparency/ease of understanding. In addition to the rating score, we know the category (i.e. Low, Medium, High) assigned to each factor for any given security included in the sample. Figures 12.1 and 12.2 show histograms for each of the risk and complexity factors, respectively.

To get a better impression regarding the relation between individual securities in the sample, we perform cluster analyses based on (i) only the risk factors, (ii) only the complexity factors, and (iii) both risk and complexity factors in the sample. In particular, we apply the Ward clustering algorithm using an Euclidean distance matrix. This algorithm is chosen to ensure that individual clusters are as homogenous as possible. However, other algorithms such as the single linkage or complete linkage

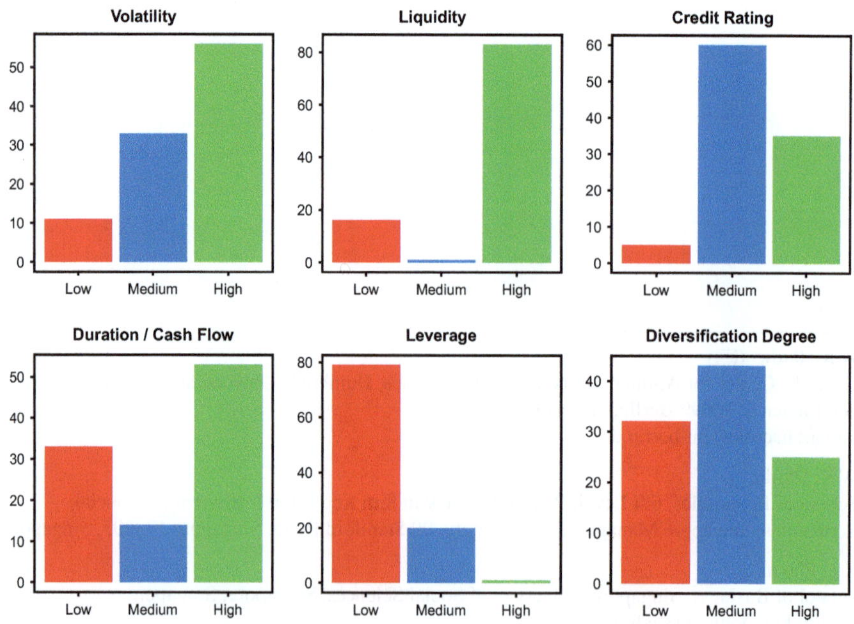

Fig. 12.1 Histograms of risk factor scores

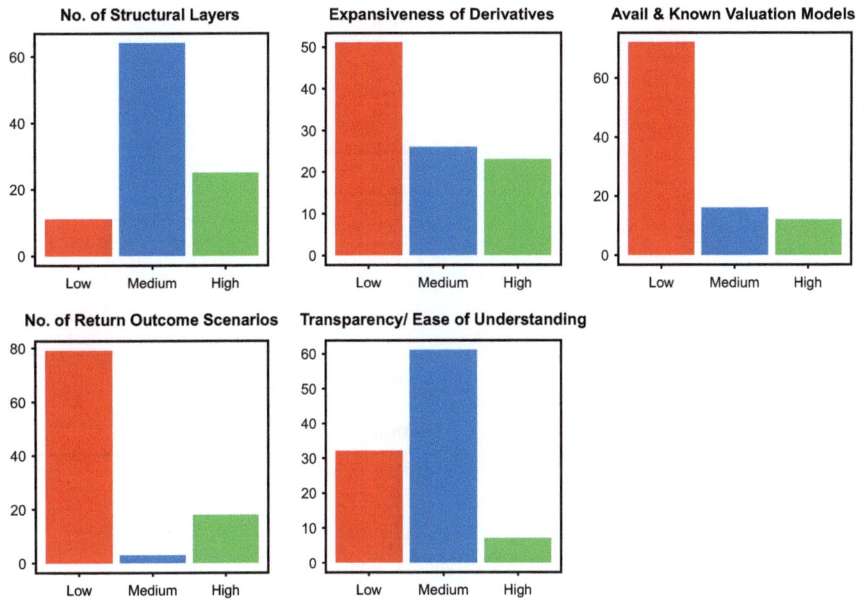

Fig. 12.2 Histograms of complexity factor scores

algorithms can be applied as well Härdle and Simar (2015). The results are depicted in Fig. 12.3.

12.2 Principal Components Analysis of Factor Scores

Principal components analysis (PCA) allows for the identification of uncorrelated latent factors that drive the variation in a sample of multivariate random variables. We consider a random variable $Y = (Y_1, \ldots, Y_j, \ldots, Y_k)^\top$ with $Y_j \in \{Low, Medium, High\}, 1 \leq j \leq k$. Y represents a vector of the risk and complexity categories assigned to a security i by the expert entity. To later be able to perform PCA on our sample we assign a discrete scale $\{1, 2, 3\}$ to each Y_j yielding a random variable $X = (X_1, \ldots, X_j, \ldots, X_k)^\top$ with $X_j \in \{1, 2, 3\}, 1 \leq j \leq k$ (i.e. $Y_j = High$ is equivalent to $X_j = 3$). For easier reference let us refer to each of the X_j as a factor score.

Our sample is now represented by a discrete matrix $X \in \{1, 2, 3\}^{n \times k}$, with each row i representing a security and each column j representing a factor. The element $x_{i,j}$ is therefore security i's score for the j-th factor. We still cannot apply PCA to X directly, however, without violating the basic assumption of normally distributed continuous random variables made in PCA. To circumvent this issue, we apply a discrete PCA using the polychoric correlation matrix of the factor scores.

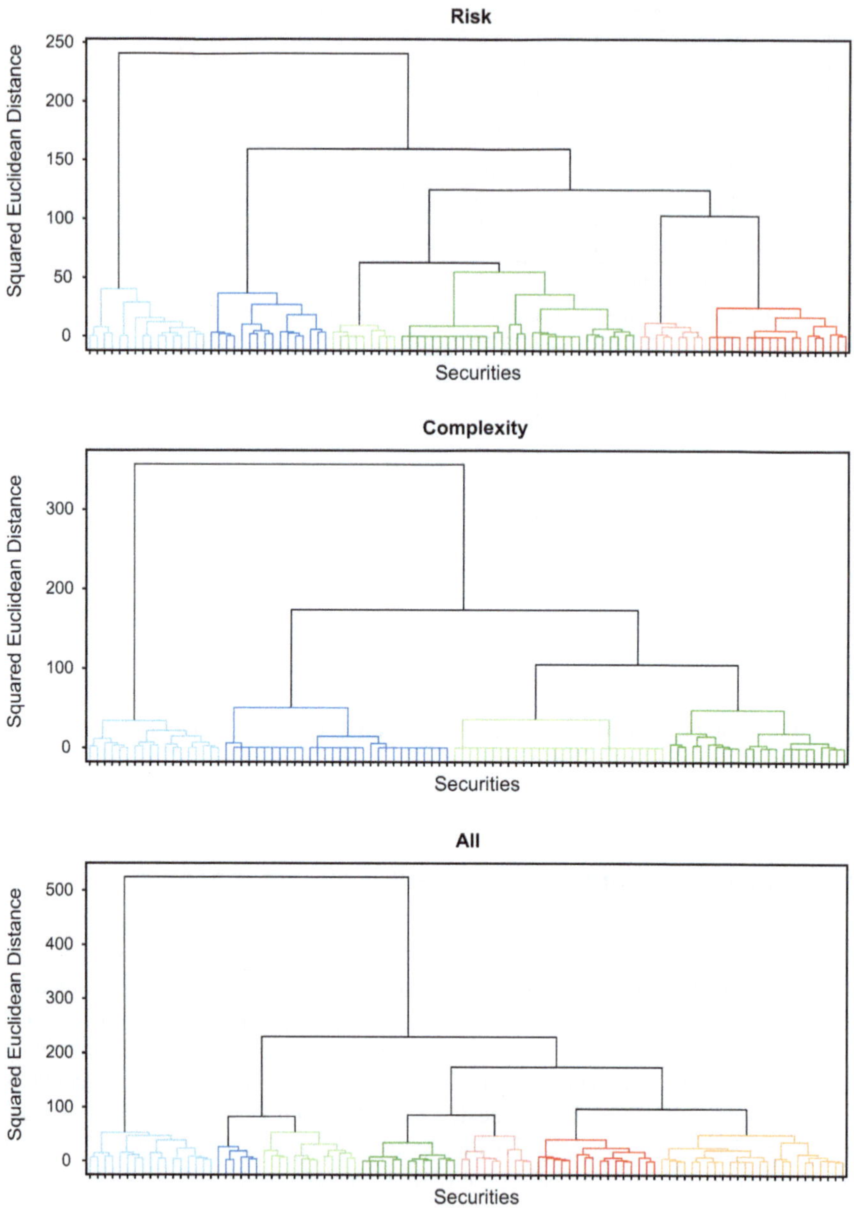

Fig. 12.3 Dendrograms of cluster analysis. Ward algorithm using Euclidean distances. Clusters formed below a threshold of 60 are coloured

Table 12.1 Projection Vector of PC_1 Projection vectors for PC_1 obtained from the eigendecompositions of the polychoric correlation matrices of X^{Risk}, X^{Comp}, and X^{All}

	X^{Risk}	X^{Comp}	X^{All}
w_1	−0.2141	0.3279	−0.1594
w_2	0.6013	0.4030	0.4275
w_3	0.0905	0.5185	0.1237
w_4	0.5106	0.4896	0.2687
w_5	0.1308	0.4707	−0.1087
w_6	0.5537		0.1166
w_7			−0.1929
w_8			−0.3142
w_9			−0.4440
w_{10}			−0.4553
w_{11}			−0.3722

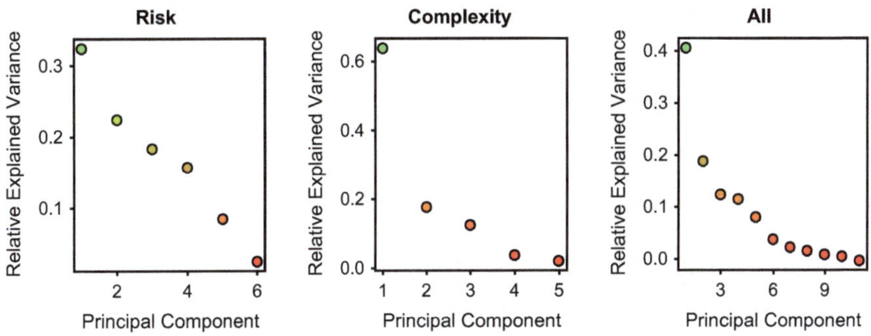

Fig. 12.4 Fraction of variance explained by each of the principal components

Just as the cluster analysis, PCA is performed on three sub-samples of X; X^{Risk}, X^{Comp}, and X^{All}. The number of columns of X therefore depends on the sub-sample (i.e. X^{Risk} is 100×6, X^{Comp} is 100×5, and X^{All} is 100×11). Table 12.1 shows the resulting projection vectors for the first principal component (PC), PC_1.

One method of analysing the relation between PCs and the underlying sample is to look at fractions of sample variance explained by each PC. This is possible, because the sum of PC variances matches the sum of variances of the underlying random variables in a sample (i.e. $\sum_{j=1}^{k} Var[PC_j] = \sum_{j=1}^{k} s_{x_j,x_j}$). The fraction of variance explained by each PC can therefore be measured as $\frac{Var[PC_j]}{\sum_{j=1}^{k} Var[PC_j]}$. If the fraction of explained variance for the first one or two PCs is very high, we know that the underlying random variables are in fact mainly driven by some latent factors represented by those two PCs. Figure 12.4 depicts the fractions of sample variance explained by each of the principal components (PCs).

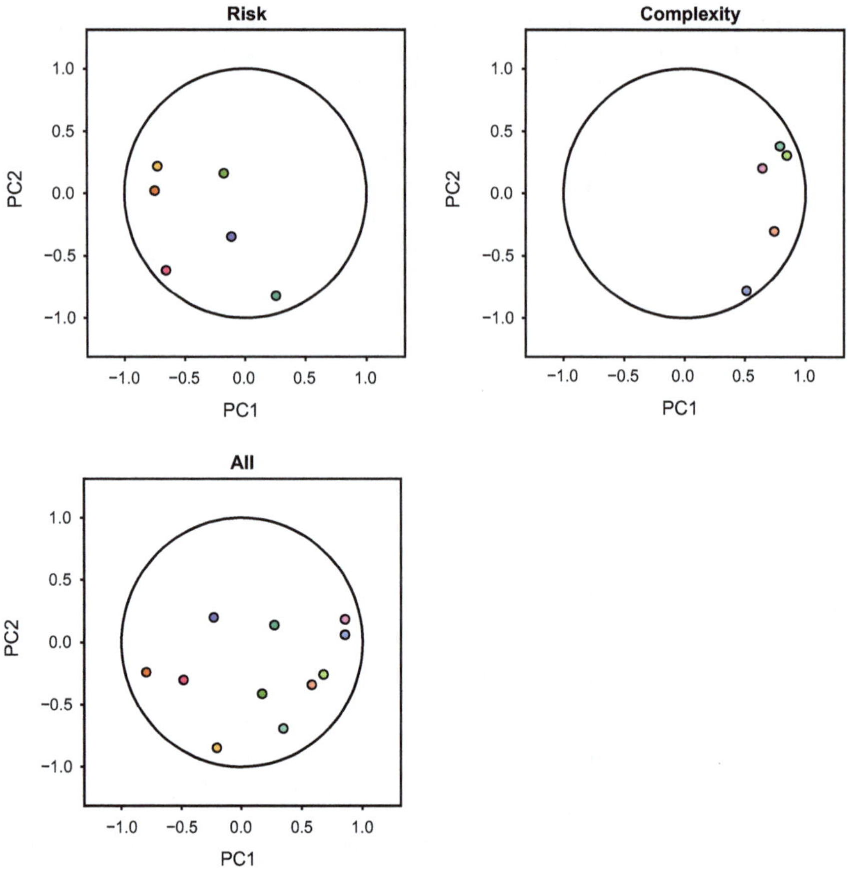

Fig. 12.5 Correlations of the factors with the first two PCs, based on the PCA of only risk, only complexity, and both risk and complexity factors. The risk factors are volatility, liquidity, credit rating, duration/cash flow, leverage, and diversification degree. The complexity factors comprise of the number of structural layers, expansiveness of derivatives, availability & known pricing models, number of return outcome scenarios, and transparency/ease of understanding

When only considering risk factors, the sample variance appears to be distributed fairly evenly among PCs. If we assume risk to be some latent variable that we expect the risk factors to be proxies of, the finding contradicts this assumption. Instead, the chosen risk factors appear to proxy for various independent latent factors. The opposite is true for the group of complexity factors, where the first PC explains more than 60 percent of the sample variation. All remaining PCs each explain less than 20 percent at the most. This reveals that the chosen complexity factors – at least in large parts – track the same underlying latent complexity factor. When including both risk and complexity factors in the PCA, the first PC explains around 40 percent of the sample variation and the next three or four PCs add another 10 to 20 percent each.

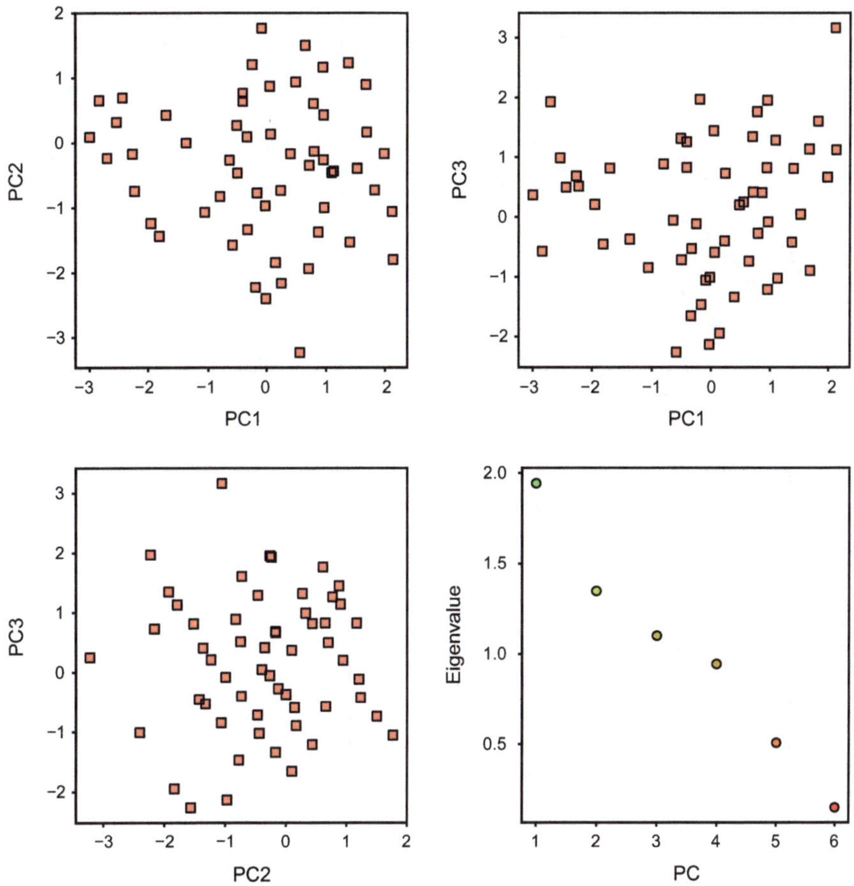

Fig. 12.6 The first three PCs derived from the PCA of the risk factors plotted against each other (*top left*, *top right*, and *bottom left*) and the eigenvalues of the polychoric correlation matrix of risk factors (*bottom right*)

In Fig. 12.5 we plot the correlation of each of the risk and complexity factors with the first two PCs for each of the factor sample subsets. Note that only the absolute correlation value is relevant when interpreting these correlations because PCs are not determined in their sign. Our results support the previous discussion regarding the explained sample variance. While the absolute correlation for risk factors with both PC_1 and PC_2 range from zero to 1.0 (top left panel), absolute correlations for complexity factors lie clearly within a range from 0.5 to 1.0 with a strong tendency towards higher values (top right panel). In the bottom left panel we note the absence of a clear correlation pattern between factors and the first two PCs. With the exception of the "number of structural" layers factor all complexity factors maintain a strong correlation with PC_1. Risk factors deviate very clearly from their correlations with both PCs in the top left panel. Figures 12.6, 12.7, and 12.8 plot the first three PCs

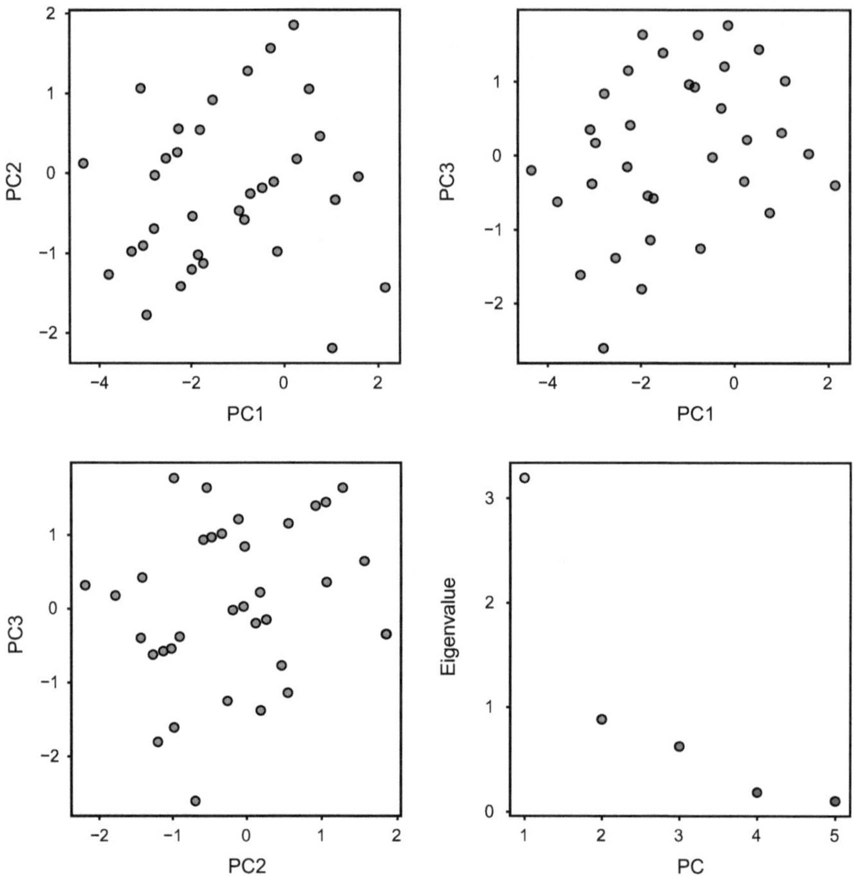

Fig. 12.7 The first three PCs derived from the PCA of the complexity factors plotted against each other (*top left*, *top right*, and *bottom left*) and the eigenvalues of the polychoric correlation matrix of complexity factors (*bottom right*)

against each other and show the correlation matrix eigenvalues associated with each principle component.

Finally, we plot the expert score of each security in the sample against its first PC in Fig. 12.9. As can be seen there is a clear relation between the total score and the first PC for risk, complexity, and both risk and complexity factors. This relation is most evident for the latter two groups.

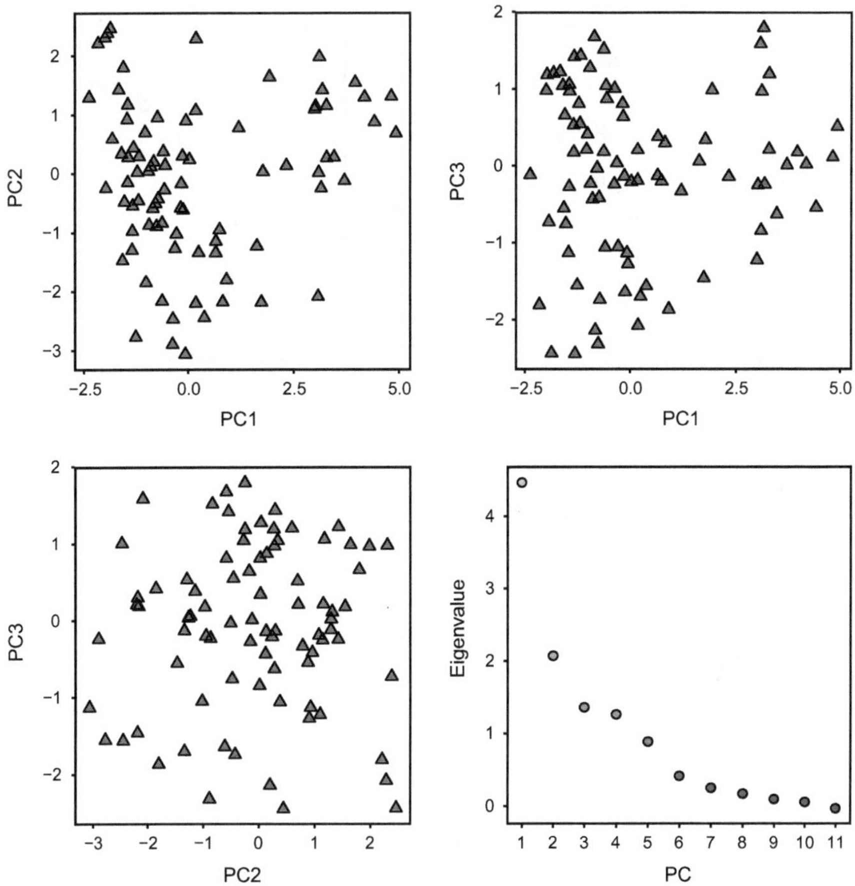

Fig. 12.8 The first three PCs derived from the PCA of the risk and complexity factors plotted against each other (*top left*, *top right*, and *bottom left*) and the eigenvalues of the polychoric correlation matrix of risk and complexity factors (*bottom right*)

12.2.1 Cross Validation via Leave-One-Out

The PCA results are cross validated by employing a leave-one-out (LOO) procedure. We compute the first PC for a security i based on weights obtained from a PCA of the sample excluding security i. In Fig. 12.10 we plot the LOO PCs against their regular counterparts. Additionally, we define a function

$$R_1 = \sum_{i=1}^{n} \left\{ f_1(x_i) - \hat{f}_1(x_i) \right\}^2, \tag{12.1}$$

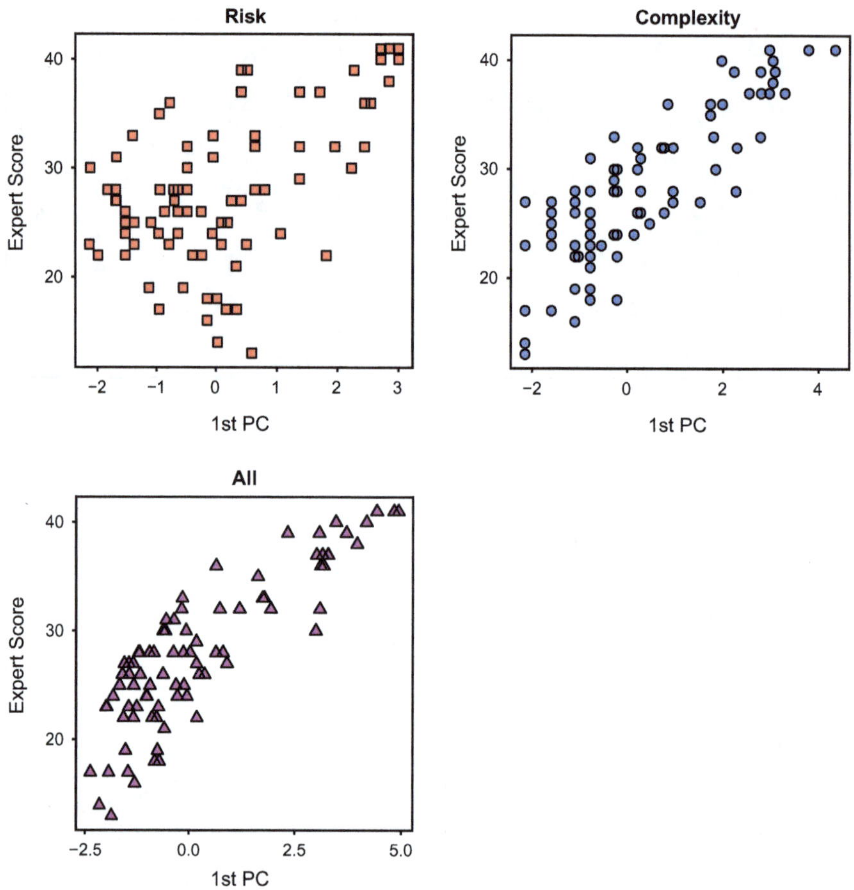

Fig. 12.9 The first PC from the PCA of risk (*top left*), complexity (*top right*), and risk and complexity (*bottom left*) factor scores plotted against the expert score of the corresponding securities

where $f_1(x_i)$ is the first PC for security i resulting from a PCA of the whole sample and $\hat{f}_1(x_i)$ is the first PC for security i computed from the weights of a PCA of the sample of $n-1$ securities (i.e. excluding security i). The values of R_1 for the three samples X^{Risk}, X^{Comp}, and X^{All} are 10.0655, 0.0219, and 0.2899, respectively. From these results we take that the PCA has some stability issues when only considering risk factors. Otherwise results are stable.

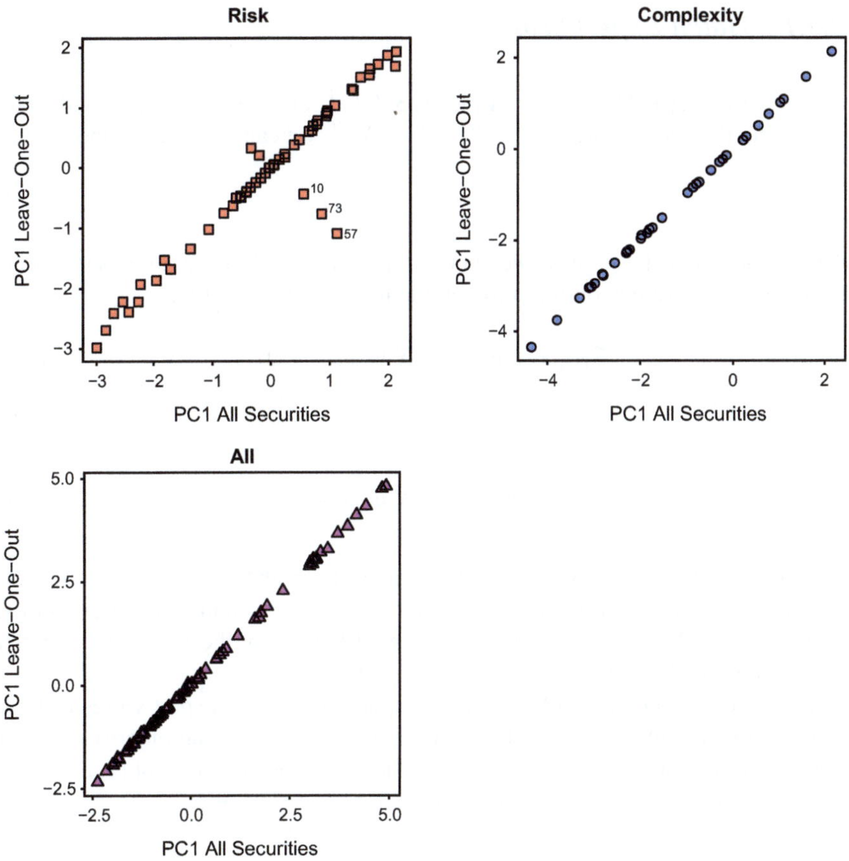

Fig. 12.10 $\hat{f}_1(x_i)$ plotted against $f_1(x_i)$ for risk factors (*top left*), for complexity factors (*top right*), for risk and complexity factors (*bottom*). 10.0655, 0.0219, and 0.2899 in each setup respectively Outliers are labeled with their security index in the sample

12.3 Adjusted Weighting of Factor Scores

In the following we consider two different applications of adjusting the weights applied to X. First, we try to find a weighting vector $w \in \mathbb{R}^k$ such that the projection $x_i\, w$ for each security i is as close as possible to its known expert score. Second, we evaluate the maximum distance between the projections of X through randomly chosen random vectors w.

12.3.1 Match Expert Score

Given a matrix $X_1 \in \{1, 2, 3\}^{n \times k}$, $X_2 \in \{1, 3, 5\}^{n \times k}$, or $X_3 \in \{1, 4, 9\}^{n \times k}$ and again considering the sub-samples X^{Risk}, X^{Comp}, or X^{All}, we can compute a function

$$R_2(X, w) = X\, w - f, \tag{12.2}$$

where w is an $k \times 1$ vector of weights and f is an $n \times 1$ vector of expert scores. From this we derive two optimisation problems (OPs) OP_1 and OP_2,

$$\widehat{w}_{OP_1} = \arg\min_{w_{OP_1}} \parallel X\, w_{OP_1} - f \parallel_1, \tag{12.3}$$

and

$$\widehat{w}_{OP_2} = \arg\min_{w_{OP_2}} \parallel X\, w_{OP_2} - f \parallel_2^2, \tag{12.4}$$

respectively. Table 12.2 shows the optimal weights for both OPs using one of X_1, X_2, or X_3 and either risk factors, complexity factors, or both risk and complexity factors. Figures 12.11, 12.12, 12.13, 12.14, 12.15 and 12.16 show the resulting weighted scores $X\widehat{w}$ plotted against the known expert scores.

As can be seen in our results, the linear approximation of expert scores is hard, even when using all 11 factors. The sum of squared approximation errors, R_2^*, in Table 12.2 is lowest for X_1 and the use of all factors. A discrete scale of $\{1, 2, 3\}$ thus appears better suited than the alternatives $\{1, 3, 5\}$ and $\{1, 4, 9\}$.

12.3.2 Cross Validation via Leave-One-Out

As with the PCA, we perform a LOO analysis to see how strongly the optimisation results for (12.4) depend on individual securities (Table 12.3).

We only consider OP_2 for X_1 because the overall results are best in this specification. The results, depicted in Fig. 12.17, are fairly robust against sample modifications. This is particularly true for X_1^{All}.

12.3.3 Widest Projection Spread

Given some random $k \times 1$ weighting vector we can compute the maximum spread between each projection in $X\, w$ and its nearest neighbour. We define $z = X\, w$ and then consider the order statistics of the elements z_i of z (i.e. $\forall i = 1, \dots, n-1 :$ $z_{(i)} \leq z_{(i+1)}$). The maximum spread between all $z_{(i)}$ and their respective nearest

Table 12.2 Match Expert Score Weights. Optimal (normalised) weights $\widehat{w} \in \mathbb{R}^k$, correlations between $A\,\widehat{w}$ and f, as well as the optimal target function value R_2^* (this is the actual target function and not R_2 itself) for OP_1 and OP_2 using matrices $X_1 \in \{1, 2, 3\}^{k \times n}$, $X_2 \in \{1, 3, 5\}^{k \times n}$, and $X_3 \in \{1, 4, 9\}^{k \times n}$. The weights have been normalised to unit vectors to facilitate a comparison with PCA weights and simulation weights

Panel A: Risk factors

	X_1		X_2		X_3	
	OP_1	OP_2	OP_1	OP_2	OP_1	OP_2
$\widehat{w}_1$	0.7773	0.7882	0.8020	0.8123	0.6574	0.7511
$\widehat{w}_2$	−0.2064	−0.1632	−0.1433	−0.0702	0.0750	0.0529
$\widehat{w}_3$	0.0346	0.0289	0.1299	0.1182	0.1069	0.0946
$\widehat{w}_4$	−0.1650	−0.1170	−0.1433	−0.1328	−0.1737	−0.1412
$\widehat{w}_5$	0.4952	0.5499	0.4596	0.5077	0.6352	0.5829
$\widehat{w}_6$	0.2821	0.1876	0.2962	0.2142	0.3424	0.2536
$\rho_{A\,\widehat{w},f}$	0.7729	0.8296	0.7562	0.7920	0.6380	0.7664
R_2^*	334.8355	1845.7312	408.8966	2698.5647	487.7317	3924.8506

Panel B: Complexity factors

	X_1		X_2		X_3	
	OP_1	OP_2	OP_1	OP_2	OP_1	OP_2
$\widehat{w}_1$	0.5636	0.5290	0.6732	0.6777	0.5420	0.6216
$\widehat{w}_2$	0.2873	0.3255	0.2891	0.2960	0.3391	0.2848
$\widehat{w}_3$	−0.0136	0.1554	−0.0008	0.0916	0.0947	0.0498
$\widehat{w}_4$	0.5849	0.5154	0.4811	0.3873	0.1521	0.2456
$\widehat{w}_5$	0.5075	0.5695	0.4814	0.5429	0.7477	0.6854
$\rho_{A\,\widehat{w},f}$	0.9825	0.9924	0.9745	0.9755	0.9535	0.9515
R_2^*	339.0912	1755.0434	453.0134	3363.2401	677.2657	6798.8500

Panel C: Risk and complexity factors

	X_1		X_2		X_3	
	OP_1	OP_2	OP_1	OP_2	OP_1	OP_2
$\widehat{w}_1$	0.6622	0.6279	0.5914	0.6130	0.3821	0.5261
$\widehat{w}_2$	−0.0003	−0.0284	0.2187	0.1362	0.4382	0.2817
$\widehat{w}_3$	−0.2645	−0.2019	−0.1590	−0.0705	−0.1454	−0.0126
$\widehat{w}_4$	0.1326	0.1535	0.1182	0.1511	0.0754	0.1186
$\widehat{w}_5$	0.2651	0.3006	0.2368	0.2868	0.2879	0.3930
$\widehat{w}_6$	−0.1326	−0.1486	−0.1183	−0.1187	−0.0669	−0.0568
$\widehat{w}_7$	0.2652	0.2960	0.2368	0.2921	0.1633	0.2432
$\widehat{w}_8$	0.2650	0.2776	0.2365	0.2566	0.1853	0.2278
$\widehat{w}_9$	0.1329	0.1695	0.1950	0.2079	0.1391	0.2617
$\widehat{w}_{10}$	0.3969	0.3551	0.4945	0.3880	0.6406	0.3581
$\widehat{w}_{11}$	0.2653	0.3296	0.3139	0.3695	0.2390	0.4052
$\rho_{A\,\widehat{w},f}$	0.9942	0.9952	0.9768	0.9792	0.9117	0.9513
R_2^*	38.0562	46.5552	91.9700	225.6082	147.5615	589.0314

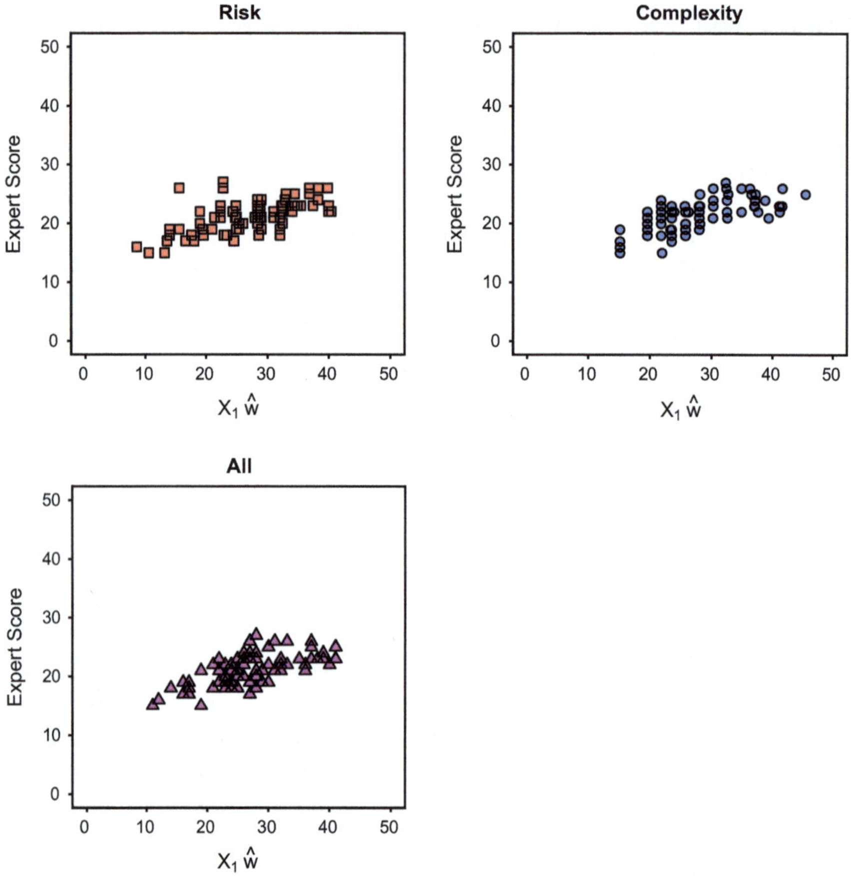

Fig. 12.11 The expert score (f) plotted against $X_1\,\widehat{w}$ for OP_1. We distinguish between results for risk factors (*top left*), complexity factors (*top right*), and risk and complexity factors (*bottom left*)

neighbour is then given by

$$R_3(z) = \max_{i}^{n-1} \left(z_{(i+1)} - z_{(i)} \right). \tag{12.5}$$

To examine the influence of the weighting vector w on the maximum projection spread we generate 1000 $k \times 1$ uniform random vectors ($w \sim \mathcal{U}(-1, 1)^k$). These vectors are then scaled to unit vectors.

Figure 12.18 shows the resulting 1000 simulated maximum spreads. The mean maximum spreads for the risk, complexity, and both risk and complexity cases are $\bar{s}^{Risk} = 0.6807$, $\bar{s}^{Compl} = 0.74725$, $\bar{s}^{All} = 0.6904$. A box plot of the results is shown in Fig. 12.19.

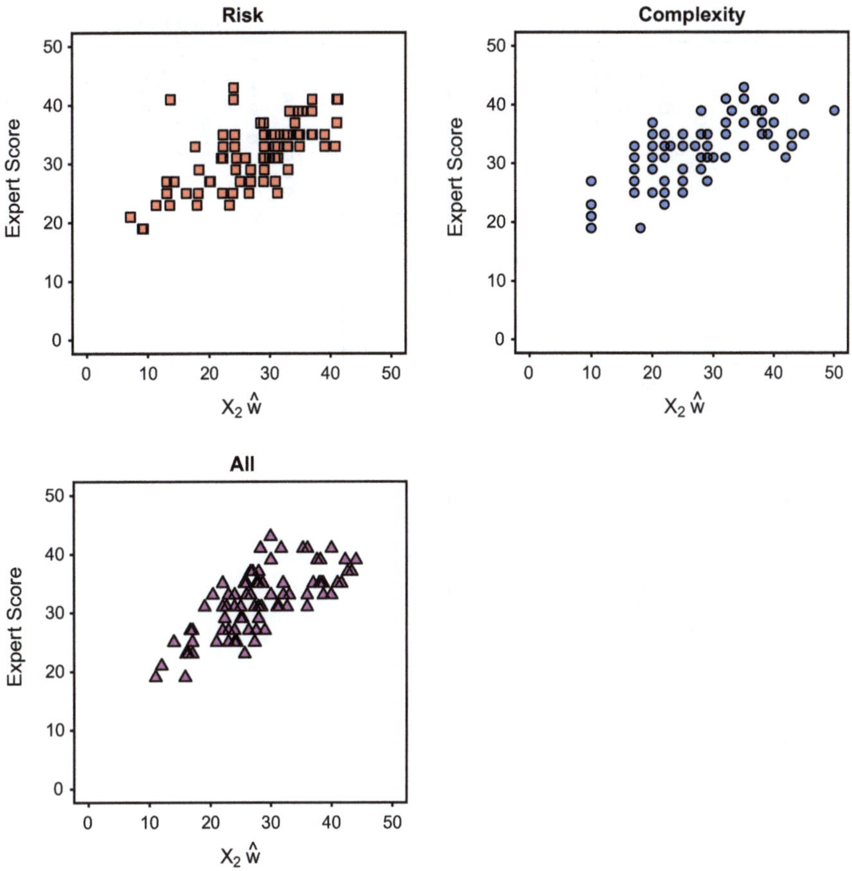

Fig. 12.12 The expert score (f) plotted against $X_2\,\widehat{w}$ for OP_1. We distinguish between results for risk factors (*top left*), complexity factors (*top right*), and risk and complexity factors (*bottom left*)

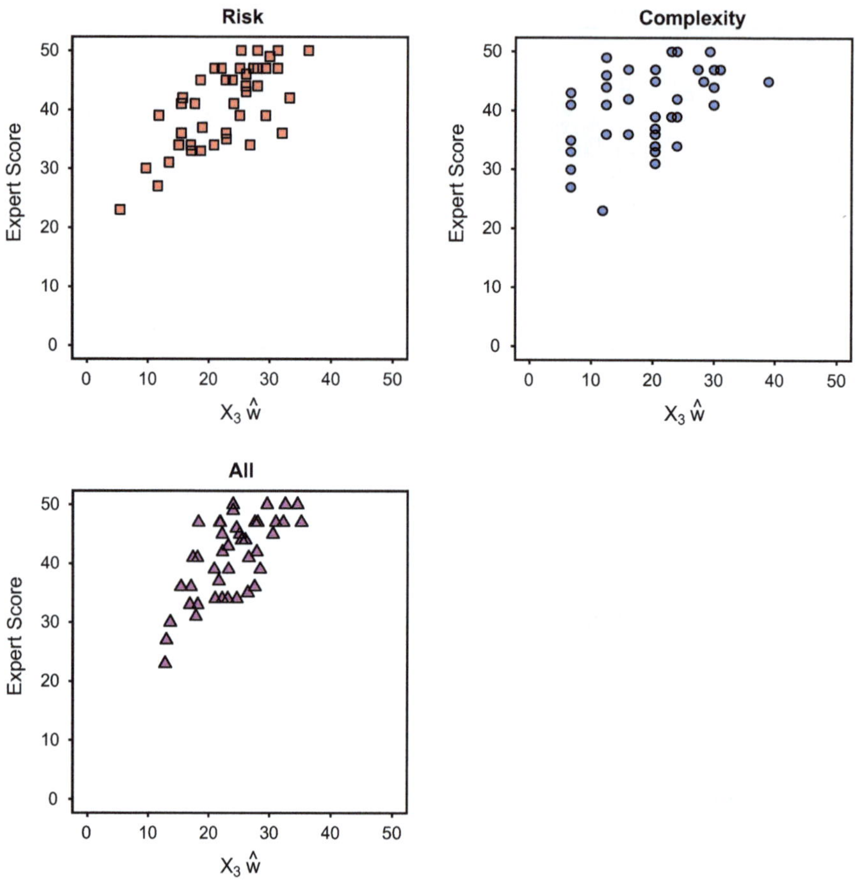

Fig. 12.13 The expert score (f) plotted against $X_3\,\widehat{w}$ for OP_1. We distinguish between results for risk factors (*top left*), complexity factors (*top right*), and risk and complexity factors (*bottom left*)

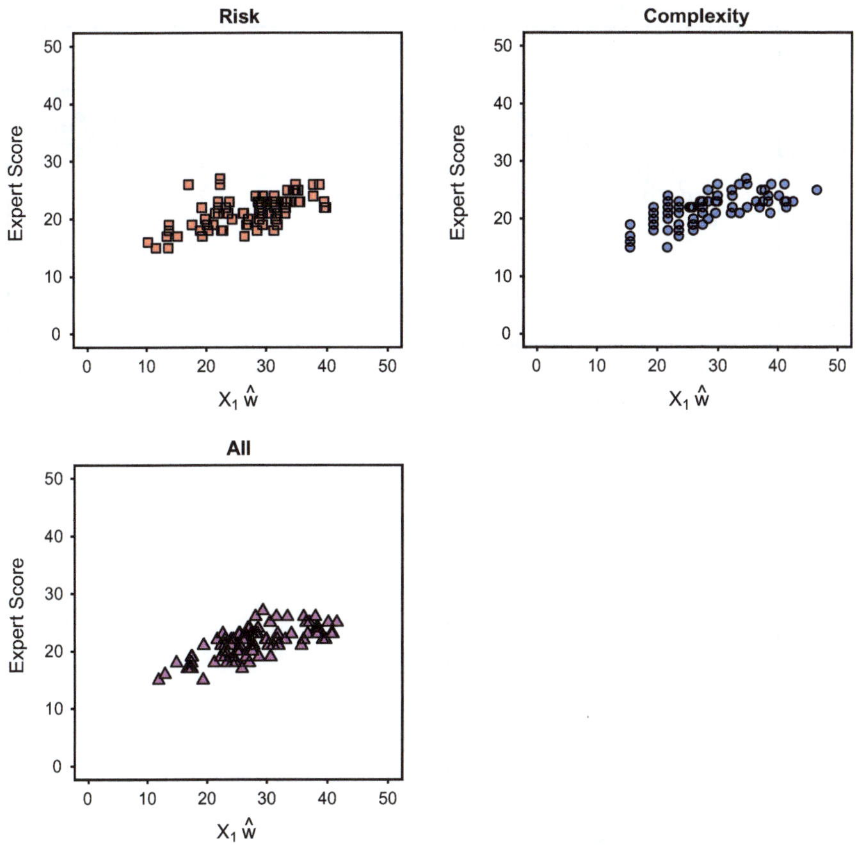

Fig. 12.14 The expert score (f) plotted against $X_1\,\widehat{w}$ for OP_2. We distinguish between results for risk factors (*top left*), complexity factors (*top right*), and risk and complexity factors (*bottom left*)

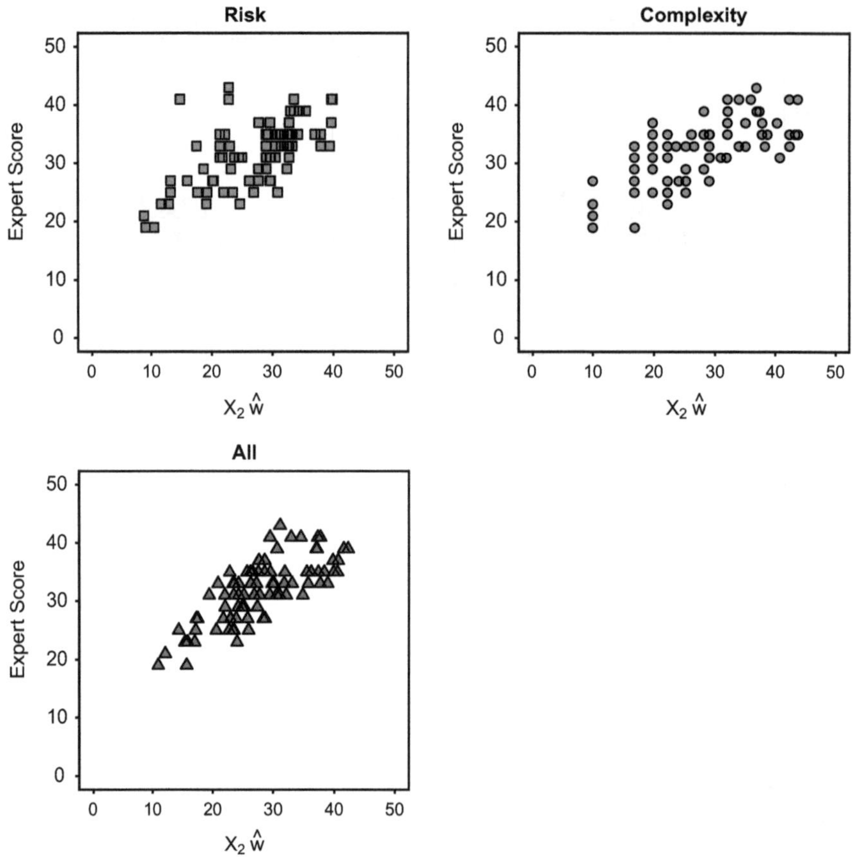

Fig. 12.15 The expert score (f) plotted against $X_2\,\widehat{w}$ for OP_2. We distinguish between results for risk factors (*top left*), complexity factors (*top right*), and risk and complexity factors (*bottom left*)

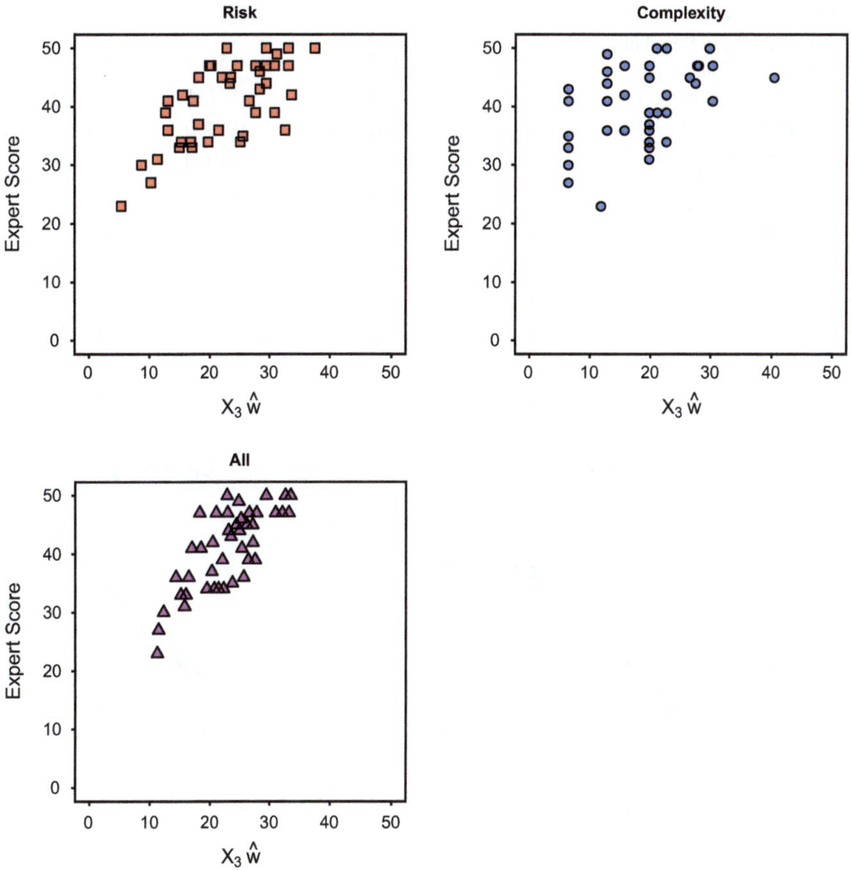

Fig. 12.16 The expert score (f) plotted against $X_3\,\widehat{w}$ for OP_2. We distinguish between results for risk factors (*top left*), complexity factors (*top right*), and risk and complexity factors (*bottom left*)

Table 12.3 Top Ten Mean Maximum Spread Simulation Weights The mean of the weighting vectors projecting the ten largest spreads from the original score matrix. The mean vectors for X^{Risk}, X^{Comp}, and X^{All} are normalised to unit vectors

	X^{Risk}	X^{Comp}	X^{All}
w_1	−0.9275	0.5161	0.4319
w_2	0.1824	0.4141	0.0833
w_3	−0.2483	0.3300	0.0257
w_4	0.1164	−0.3030	−0.2453
w_5	0.0130	0.6013	−0.3244
w_6	0.1764		−0.2807
w_7			−0.1371
w_8			−0.3083
w_9			0.1572
w_{10}			−0.5835
w_{11}			−0.2874

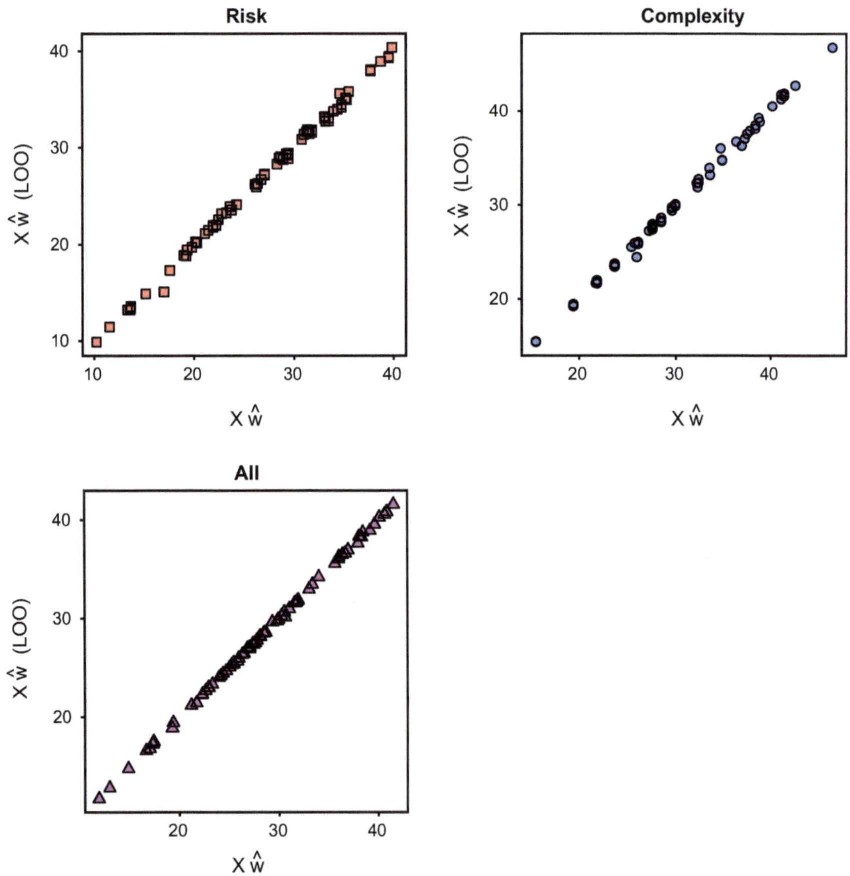

Fig. 12.17 $X_1\,\widehat{w}_{LOO}$ plotted against $X_1\,\widehat{w}$ for OP_2. We distinguish between results for risk factors (*top left*), complexity factors (*top right*), and risk and complexity factors (*bottom left*)

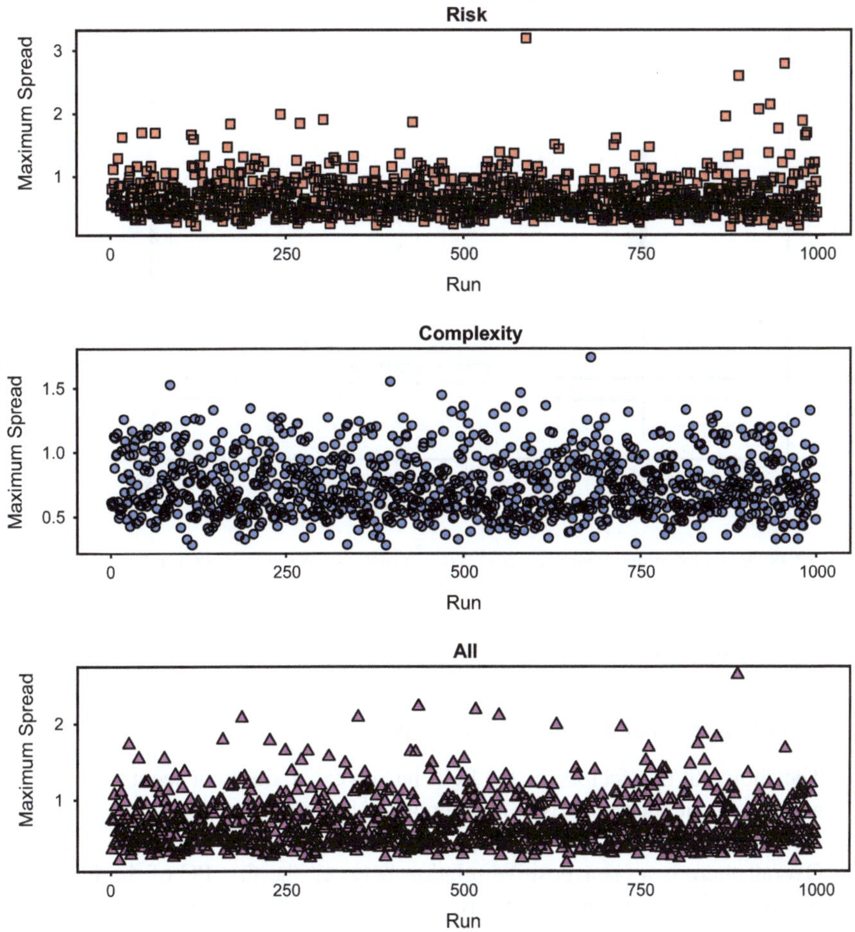

Fig. 12.18 Maximum spread among projections $\mathbf{X}_1\,\mathbf{w}$ for 1000 randomly chosen $\mathbf{w}$

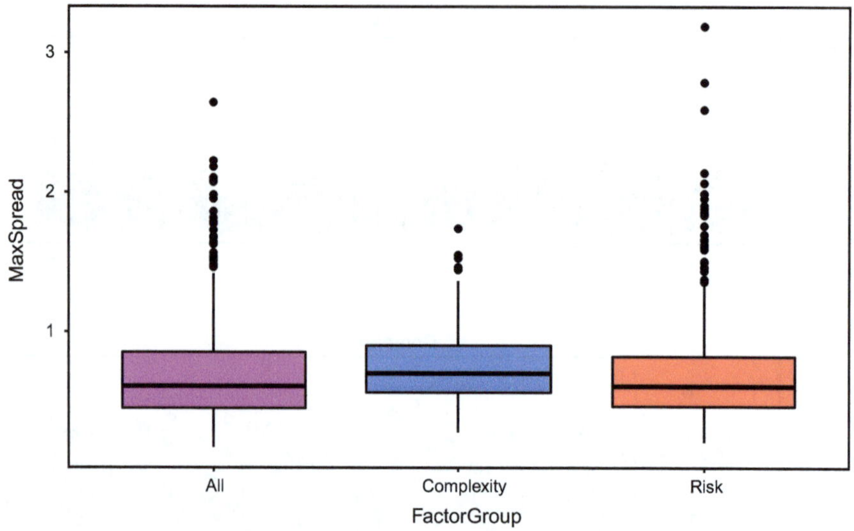

Fig. 12.19 Box plot of maximum spreads among projections $\mathbf{X}_1\,\mathbf{w}$ for 1000 randomly chosen $\mathbf{w}$

12.4 Conclusion

We can summarise our results in a few key points:

1. The choice of risk factors, as the PCA has revealed, does not seem to proxy for a single latent source of risk. The opposite is true for the choice of complexity factors.
2. Overall there is a clear positive relation between the first PC of the full PCA, involving all factors, and the expert score of a security as shown in Fig. 12.9.
3. Approximation of the total expert scores through linear projection of the score matrix is possible, but not perfect. We obtain best results by using a score scale of $\{1, 2, 3\}$ and applying the L^2 norm during optimisation.

Reference

Härdle, W. K., & Simar, L. (2015). *Applied multivariate statistical analysis* (4th ed.). Berlin: Springer.

Part III
Dynamics Risk Measurement

Chapter 13
Copulae in High Dimensions: An Introduction

Ostap Okhrin, Alexander Ristig and Ya-Fei Xu

Abstract This paper reviews the latest proceeding of research in high dimensional copulas. At the beginning the bivariate copulas are given as a fundamental followed with the multivariate copulas which are the concentration of the paper. In multivariate copula sections, the hierarchical Archimedean copula, the factor copula and vine copula are introduced. In the following section the estimation methods for multivariate copulas including parametric and nonparametric routines, are presented. Also the introduction of the goodness of fit tests in copula context is given. An empirical study of multivariate copulas in risk management is performed thereafter.

13.1 Introduction

Researches of dependence modeling were burgeoning during the last decade. The traditional approaches that concentrate on the elliptical distributions such as Gaussian models are giving way to copula-based models. Albeit these Gaussian models sometimes own the convenience in model construction and computation, yet an abundant amount of empirical evidences do not support the underlying assumptions. De facto, shortcomings in the elliptical and especially Gaussian family are mainly in lack of asymmetrical and tail dependence which have been deeply discussed in numer-

The original version of this chapter was revised: For detailed information please see Erratum.
The erratum to this chapter is available at https://doi.org/10.1007/978-3-662-54486-0_19

O. Okhrin
Chair of Econometrics and Statistics esp. in Transport, Technische Universität Dresden, Dresden, Germany

A. Ristig
Department of Statistics and Operations Research, Universität Wien, Vienna, Austria

Y.-F. Xu (✉)
School of Business and Economics, Humboldt-Universität zu Berlin, Berlin, Germany
e-mail: yafei.xu@hu-berlin.de

W.K. Härdle et al. (eds.), *Applied Quantitative Finance*, Statistics and Computing,
DOI 10.1007/978-3-662-54486-0_13

ous papers. Furthermore and of great importance, margins of elliptical distributions belong to the same elliptical family.

The seminal result of Sklar (1959) provides a partial solution to these problems. It allows to separate the marginal distributions from the dependency structure between the random variables. Since the theory on modeling and estimation of univariate distributions is well established compared to the multivariate case, the initial problem reduces to modeling the dependency by copulas. In particular, this approach dramatically widens the class of candidate distributions and allows a simple construction of distributions with less parameters than imposed by elliptical models.

In the beginning of the copula study, researches were mainly focused on the bivariate dependence but as time passes problems raised by the financial, technological, biological industries dictated the rules of further developments, namely moves to higher dimensions. Nonetheless, it has been realized as clearly stated in Mai and Scherer (2013), that "the step from one-dimensional modeling is clearly large. But, unfortunately, the step from two to three (or even more) dimensions is not a bit smaller".

Numerous steps are accomplished in order to contribute to research on high-dimensional modeling approaches and these main branches have been established: pair copula construction, see Joe (1996), Bedford and Cooke (2001), Bedford and Cooke (2002) and Kurowicka and Cooke (2006), hierarchical Archimedean copula, see Savu and Trede (2010), Hofert (2011) and Okhrin et al. (2013a), and factor copula, see Krupskii and Joe (2013) and Oh and Patton (2015).

This chapter attempts at discussing such non-standard multivariate copula models and the subsequent sections are organized as follows. We introduce bivariate copulae and review modern multivariate copula families. Then, corresponding estimation methods and goodness of fit tests are presented. Last but not least, we study a risk management topic empirically.

13.2 Bivariate Copula

Modeling the dependence between only two random variables using copulae is the subject of this section. There are several equivalent definitions of the copula function. We define it as a bivariate distribution function and the simplest one is as follows:

Definition 13.1 The copula $C(u, v)$ is a bivariate distribution with margins being $U[0, 1]$.

Term copula was mentioned for the first time in the seminal result of Sklar (1959). The separation of the bivariate distribution function into the copula function and margins is formally stated in the subsequent theorem. One possible proof is presented in Nelsen (2006), for others we refer to Durante et al. (2012), Durante et al. (2013) and Durante and Sempi (2005)

Theorem 13.1 *Let F be a bivariate distribution function with margins F_1 and F_2, then there exists a copula C such that*

$$F(x_1, x_2) = C\{F_1(x_1), F_2(x_2)\}, \quad x_1, x_2 \in \overline{\mathbb{R}} = \mathbb{R} \cup \{\infty, -\infty\}. \tag{13.1}$$

If F_1 and F_2 are continuous then C is unique. Otherwise C is uniquely determined on $F_1(\overline{\mathbb{R}}) \times F_2(\overline{\mathbb{R}})$.

Conversely, if C is a copula and F_1 and F_2 are univariate distribution functions, then function F in (13.1) is a bivariate distribution function with margins F_1 and F_2.

As indicated above, the theorem allows decomposing any continuous bivariate distribution into its marginal distributions and the dependency structure. Since by definition, the latter is the copula function with uniform margins, it follows that the copula density can be determined in the usual way

$$c(u_1, u_2) = \frac{\partial^2 C(u_1, u_2)}{\partial u_1 \partial u_2}, \quad u_1, u_2 \in [0, 1]. \tag{13.2}$$

Being armed with the Theorem 13.1 and (13.2), the density function $f(\cdot)$ of the bivariate distribution F can be rewritten in terms of copula

$$f(x_1, x_2) = c\{F_1(x_1), F_2(x_2)\} f_1(x_1) f_2(x_2), \quad x_1, x_2 \in \overline{\mathbb{R}}.$$

A very important property of copulae is given in Nelsen (2006) stating that copulae are invariant under strictly monotone transformations of margins. Seen from this angle, copulae capture only those features of the dependency which are invariant under increasing transformations.

13.2.1 Copula Families

Naturally, there is an infinite number of different copula functions satisfying the properties of Definition 13.1 and the number of them being deeply studied is expanding. In this section, we discuss three copula classes namely simple, elliptical and Archimedean copulae.

Simplest Copulae

To form basic intuition for copula functions, we first study some extreme special cases, like stochastically independent, perfect positive or negative dependent random variables. According to Theorem 13.1, the copula of two stochastically independent random variables X_1 and X_2 is given by the product (independence) copula defined as

$$\Pi(u_1, u_2) = u_1 u_2, \quad u_1, u_2 \in [0, 1].$$

The contour diagrams of the bivariate density function with product copula and either Gaussian or t_3-distributed margins are given in Fig. 13.1. Two additional extremes are the lower and upper Fréchet–Hoeffding bounds. They represent the perfect negative and positive dependence of two random variables respectively

$$W(u_1, u_2) = \max(0, u_1 + u_2 - 1) \quad \text{and} \quad M(u_1, u_2) = \min(u_1, u_2), \quad u_1, u_2 \in [0, 1].$$

If $C = W$ and $(X_1, X_2) \sim C(F_1, F_2)$ then X_2 is a decreasing function of X_1. Similarly, if $C = M$, then X_2 is an increasing function of X_1. In general, we can argue that an arbitrary copula which represents some dependency structure lies between these two bounds, i.e.

$$W(u_1, u_2) \leq C(u_1, u_2) \leq M(u_1, u_2), \quad u_1, u_2 \in [0, 1].$$

The bounds serve as benchmarks for the evaluation of the dependency magnitude. There are numerous techniques for building new copulae by mixing at least two of the presented simplest copula. For example, copula families B11 and B12, see Joe (1997), arise as a combination of the upper Fréchet–Hoeffding bound and the product copula

$$C_{B11}(u_1, u_2, \theta) = \theta M(u_1, u_2) + (1-\theta)\Pi(u_1, u_2) = \theta \min\{u_1, u_2\} + (1-\theta)u_1 u_2,$$
$$C_{B12}(u_1, u_2, \theta) = M(u_1, u_2)^{\theta}\Pi(u_1, u_2)^{1-\theta} = (\min\{u_1, u_2\})^{\theta}(u_1 u_2)^{1-\theta}, \quad u_1, u_2, \theta \in [0, 1].$$

Family B11 builds on the fact that every convex combination of copulas is a copula as well. Family B12 is also known as Spearman or Cuadras–Augé copula, which is a weighted geometric mean of the upper Fréchet–Hoeffding bound and the product copula. Further generalization is done by using power mean over the upper Fréchet–Hoeffding bound and the product copula

$$\begin{aligned} C_p(u_1, u_2, \theta_1, \theta_2) &= \{\theta_1 M^{\theta_2}(u_1, u_2) + (1-\theta_1)\Pi^{\theta_2}(u_1, u_2)\}^{1/\theta_2} \\ &= \{\theta_1 \min(u_1, u_2)^{\theta_2} + (1-\theta_1)(u_1 u_2)^{\theta_2}\}^{1/\theta_2}, \end{aligned}$$

with $\theta_1 \in [0, 1]$, $\theta_2 \in \mathbb{R}$. Last but not least, a convex combination of the Fréchet–Hoeffding lower bound, upper bound and product copula forms the Fréchet copula

$$C_F(u_1, u_2, \theta_1, \theta_2) = \theta_1 W(u_1, u_2) + (1 - \theta_1 - \theta_2)\Pi(u_1, u_2) + \theta_2 M(u_1, u_2),$$

subject to $0 \leq \theta_1 + \theta_2 \leq 1$. Note that any bivariate copula can be approximated by the Fréchet family and a bound of the resulting approximation error can be estimated. Nelsen (2006) provides further methods for constructing multivariate copulas and discusses convex combination in more detail.

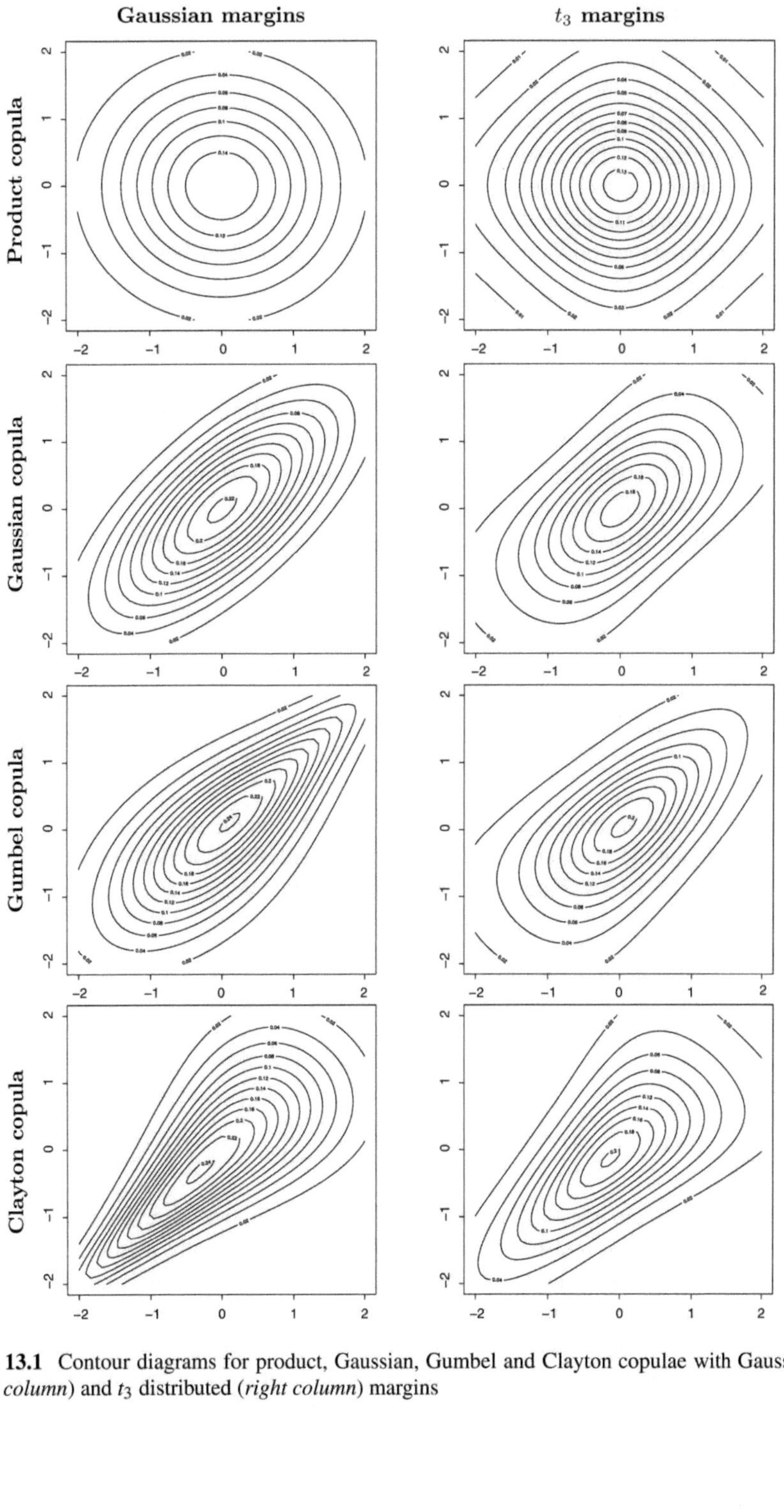

Fig. 13.1 Contour diagrams for product, Gaussian, Gumbel and Clayton copulae with Gaussian (*left column*) and t_3 distributed (*right column*) margins

Elliptical Family

Due to the popularity of the Gaussian and t-distribution in several applications, elliptical copulae play an important role as well. The construction of this type of copulae is directly based on Sklar's Theorem showing how new bivariate distributions can be constructed. The copula-based modeling approach substantially widens the family of elliptical distributions by keeping the same elliptical copula function and varying the marginal distributions or vice versa.

To determine the copula function of a given bivariate distribution, we employ the transformation

$$C(u_1, u_2) = F\{F_1^{-1}(u_1), F_2^{-1}(u_2)\}, \quad u_1, u_2 \in [0, 1], \tag{13.3}$$

where F_i^{-1}, $i = 1, 2$, are (generalized) inverses of the marginal distribution functions. Based on (13.3), arbitrary elliptical distributions can be derived. The problem, however, is that such copulae depend on the inverse distribution functions of the marginals which are rarely available in an explicit form.

For instance, from Formula 13.3 follows that the Gaussian copula and its density are given by

$$\begin{aligned} C_N(u_1, u_2, \delta) &= \Phi_\delta(\Phi^{-1}(u_1), \Phi^{-1}(u_2)), \\ c_N(u_1, u_2, \delta) &= (1-\delta^2)^{-\frac{1}{2}} \exp\left\{ -\frac{1}{2}(1-\delta^2)^{-1}(u_1^2 + u_2^2 - 2\delta u_1 u_2)\right\} \\ &\quad \times \exp\left\{\frac{1}{2}(u_1^2 + u_2^2)\right\}, \quad \text{for all } u_1, u_2 \in [0, 1],\ \delta \in [-1, 1], \end{aligned}$$

where Φ is the distribution function of N(0, 1), Φ^{-1} is the functional inverse of Φ and Φ_δ denotes the bivariate standard normal distribution function with correlation coefficient δ. In the bivariate case, the t-copula and its density are given by

$$\begin{aligned} C_t(u_1, u_2, \nu, \delta) &= \int_{-\infty}^{t_\nu^{-1}(u_1)} \int_{-\infty}^{t_\nu^{-1}(u_2)} \frac{\Gamma\left(\frac{\nu+2}{2}\right)}{\Gamma\left(\frac{\nu}{2}\right)\pi\nu\sqrt{(1-\delta^2)}} \\ &\quad \times \left\{1 + \frac{x_1^2 - 2\delta x_1 x_2 + x_2^2}{(1-\delta^2)\nu}\right\}^{-\frac{\nu}{2}-1} dx_1 dx_2, \\ c_t(u_1, u_2, \nu, \delta) &= \frac{f_{\nu,\delta}\{t_\nu^{-1}(u_1), t_\nu^{-1}(u_2)\}}{f_\nu\{t^{-1}(u_1)\} f_\nu\{t^{-1}(u_2)\}}, \quad u_1, u_2, \delta \in [0, 1], \end{aligned}$$

where δ denotes the correlation coefficient, ν is the number of degrees of freedom. $f_{\nu,\delta}$ and f_ν are joint and marginal t-distributions respectively, while t_ν^{-1} denotes the quantile function of the t_ν distribution. In-depth analysis of the t-copula is done in Rachev et al. (2008) and Luo and Shevchenko (2010). Long-tailed distributed margins lead to more mass and variability in the tail areas of the corresponding bivariate distribution. However, the contour-curves of the t-copula are symmetric, which reflects the ellipticity of the underlying copula. This property is theoretically supported by Nelsen (2006), stating that a bivariate copula is elliptical and thus, has reflection symmetry, if and only if

$$C(u_1, u_2, \theta) = u_1 + u_2 - 1 + C(1 - u_1, 1 - u_2, \theta), \quad u_1, u_2 \in [0, 1].$$

The next class of copulae and their generalizations provide an important flexible and rich family of alternatives to elliptical copulae.

Archimedean Family

In contrast to elliptical copulae, Archimedean copulae are not constructed via (13.3), but are related to Laplace transforms of bivariate distribution functions. The function $C : [0, 1]^2 \to [0, 1]$ defined as

$$C(u_1, u_2) = \phi\{\phi^{-1}(u_1) + \phi^{-1}(u_2)\}, \quad u_1, u_2 \in [0, 1],$$

is a 2-dimensional Archimedean copula, where $\phi \in \mathcal{L} = \{\phi : [0; \infty) \to [0, 1] \mid \phi(0) = 1,\ \phi(\infty) = 0;\ (-1)^j \phi^{(j)} \geq 0;\ j = 1, \dots, \infty\}$ is referred to as the generator of the copula. The generator usually depends on some parameters, however, mostly generators with a single parameter θ are considered. Nelsen (2006) and Joe (2014) provide a thoroughly classified list of popular generators for Archimedean copulae and discuss their properties.

The useful applications in finance, see Patton (2012), appearing to be the Gumbel copula with the generator function $\phi(x, \theta) = \exp\{-x^{1/\theta}\}$, $1 \leq \theta < \infty$, $x \in [0, 1]$, leading to the copula function

$$C(u_1, u_2, \theta) = \exp\left\{-\left[(-\log u_1)^\theta + (-\log u_2)^\theta\right]^{1/\theta}\right\}, \quad u_1, u_2 \in [0, 1].$$

Genest and Rivest (1989) showed that a bivariate distribution based on the Gumbel copula with extreme valued marginal distributions is the only bivariate extreme value distribution belonging to the Archimedean family. Moreover, all distributions based on Archimedean copulae belong to its domain of attraction under common regularity conditions. In contrary to elliptical copulae, the Gumbel copula leads to asymmetric contour diagrams in Fig. 13.1. It exhibits a stronger linkage between positive values, however, more variability and more mass in the negative tail area. Opposite is observed for the Clayton copula with the generator $\phi(x, \theta) = (\theta x + 1)^{-\frac{1}{\theta}}$ with $-1 < \theta < \infty$, $\theta \neq 0$, $x \in [0, 1]$, and copula function

$$C(u_1, u_2, \theta) = (u_1^{-\theta} + u_2^{-\theta} - 1)^{-\frac{1}{\theta}}, \quad u_1, u_2 \in [0, 1].$$

Also, the Frank generator $\phi(x, \theta) = \theta^{-1} \log\{1 - (1 - e^{-\theta})e^{-x}\}$ with $0 \leq \theta < \infty$, $x \in [0, 1]$, enjoys increased popularity and induces the copula function

$$C(u_1, u_2, \theta) = -\theta^{-1} \log\left\{\frac{1 - e^{-\theta} - (1 - e^{-\theta u_1})(1 - e^{-\theta u_2})}{1 - e^{-\theta}}\right\}, \quad u_1, u_2 \in [0, 1].$$

The respective Frank copula is the only elliptical Archimedean copula.

13.2.2 Bivariate Copula and Dependence Measures

Since copulae define the dependence structure between random variables, there is a relationship between copulae and different dependency measures. The classical measures for continuous random variables are Kendall's τ and Spearman's ρ. Similarly as copula functions, these measures are invariant under strictly increasing transformations. They are equal to 1 or -1 under perfect positive or negative dependence respectively. In contrast to τ and ρ, the Pearson correlation coefficient measures the linear dependence and, therefore, is not suitable for measuring non-linear relationships. Next, we discuss the relationship between τ, ρ and the underlying copula function.

Definition 13.2 Let F be a continuous bivariate cumulative distribution function with the copula C. Moreover, let $(X_1, X_2) \sim F$ and $(X_1', X_2') \sim F$ be independent random pairs. Then Kendall's τ is given by

$$\begin{aligned}\tau_2 &= \mathrm{P}\{(X_1 - X_1')(X_2 - X_2') > 0\} - \mathrm{P}\{(X_1 - X_1')(X_2 - X_2') < 0\} \\ &= 2\,\mathrm{P}\{(X_1 - X_1')(X_2 - X_2') > 0\} - 1 = 4 \int\limits_{[0,1]^2} C(u_1, u_2)\, dC(u_1, u_2) - 1.\end{aligned}$$

Kendall's τ represents the difference between the probability of two random concordant pairs and the probability of two random discordant pairs. For most copula functions with a single parameter θ there is a one-to-one relationship between θ and the Kendall's τ_2. For example, it holds that

$$\tau_2(\text{Gaussian and } t) = \frac{2}{\pi} \arcsin \delta, \quad \tau_2(\text{Archimedean}) = 4 \int_0^1 \frac{\phi^{-1}(t)}{(\phi^{-1})'}\, dt + 1,$$
$$\tau_2(\Pi) = 0, \qquad \tau_2(W) = 1, \qquad \tau_2(M) = -1.$$

For instance, this implies that an unknown copula parameter θ of the Gaussian, t and an arbitrary Archimedean copulae can be estimated using a type of method of moments procedure with a single moment condition. This requires, however, an estimator of τ_2, c.f. Kendall (1970). Naturally, it is computed by

$$\tau_{2n} = \frac{4}{n(n-1)} P_n - 1,$$

where n stands for the sample size and P_n denotes the number of concordant pairs, e.g. such pairs (X_1, X_2) and (X_1', X_2') that $(X_1 - X_1')(X_2 - X_2') > 0$. Next we provide the definition and similar results for the Spearman's ρ.

Definition 13.3 Let F be a continuous bivariate distribution function with the copula C and the univariate margins F_1 and F_2 respectively. Assume that $(X_1, X_2) \sim F$. Then the Spearman's ρ is given by

$$\rho_2 = 12 \int_{\overline{\mathbb{R}}^2} F_1(x_1) F_2(x_2)\, dF(x_1, x_2) = 12 \int_{[0,1]^2} u_1 u_2\, dC(u_1, u_2) - 3.$$

Similarly as for Kendall's τ, the relationship between Spearman's ρ and specific copulae is given through

$$\rho_2(\text{Gaussian and } t) = \frac{6}{\pi} \arcsin \frac{\delta}{2},$$
$$\rho_2(\Pi) = 0, \qquad \rho_2(W) = 1, \qquad \rho_2(M) = -1.$$

Unfortunately, there is no explicit representation of Spearman's ρ_2 for Archimedean in terms of generator functions as by Kendall's τ. The estimator of ρ is easily computed using

$$\rho_{2n} = \frac{12}{n(n+1)(n-1)} \sum_{i=1}^{n} R_i S_i - 3\frac{n+1}{n-1},$$

where R_i and S_i denote the ranks of two samples. The exact regions determined by Kendall's τ and Spearman's ρ have been recently given by Schreyer et al. (2017).

13.3 Multivariate Copula: Primer and State-of-Art

As mentioned in the introduction, step from bivariate copulas to multivariate is large. Nevertheless, many works have been written properly different high-dimensional copulas. This section introduces simple multivariate models and most prominent families like hierarchical Archimedean copula (HAC), pair-copula constructions and factor copula.

A d-dimensional copula is also the distribution function on $[0, 1]^d$ having all marginal distributions uniform on $[0, 1]$. In Sklar's Theorem, the importance of copulas in the area of multivariate distributions is re-stated in an exquisite way.

Theorem 13.2 *Let F be a multivariate distribution function with margins $F_1, \dots,$ F_d, then there exists the copula C such that*

$$F(x_1, \dots, x_d) = C\{F_1(x_1), \dots, F_k(x_d)\}, \quad x_1, \dots, x_d \in \overline{\mathbb{R}}.$$

If F_i are continuous for $i = 1, \dots, d$ then C is unique. Otherwise C is uniquely determined on $F_1(\overline{\mathbb{R}}) \times \cdots \times F_d(\overline{\mathbb{R}})$.

Conversely, if C is a copula and $F_1, \dots, F_d$ are univariate distribution functions, then function F defined above is a multivariate distribution function with margins $F_1, \dots, F_d$.

As in the bivariate case, the representation in Sklar's Theorem can be used for constructing new multivariate distributions by changing either the copula function of marginal distributions. For an arbitrary continuous multivariate distribution we can determine its copula from the transformation

$$C(u_1,\dots,u_d) = F\{F_1^{-1}(u_1),\dots,F_d^{-1}(u_d)\},\quad u_1,\dots,u_d \in [0,1], \tag{13.4}$$

where F_i^{-1} are inverse marginal distribution functions. Copula density and density of the multivariate distribution with respect to copula are

$$c(u_1,\dots,u_d) = \frac{\partial^k C(u_1,\dots,u_d)}{\partial u_1 \dots \partial u_d},\quad u_1,\dots,u_d \in [0,1],$$

$$f(x_1,\dots,x_d) = c\{F_1(x_1),\dots,F_d(x_d)\}\prod_{i=1}^{d} f_i(x_i),\quad x_1,\dots,x_d \in \overline{\mathbb{R}}.$$

For the multivariate case as well as for the bivariate case copula functions are invariant under monotone transformations.

13.3.1 Extensions of Simple and Elliptical Bivariate Copulae

The independence copula and the upper and lower Fréchet–Hoeffding bounds can be straightforwardly generalized to the multivariate case. The independence copula is defined by the product $\Pi(u_1,\dots,u_d) = \prod_{i=1}^d u_i$ and the bounds are given by

$$W(u_1,\dots,u_d) = \max\Big(0, \sum_{i=1}^{d} u_i + 1 - d\Big),$$

$$M(u_1,\dots,u_d) = \min(u_1,\dots,u_d),\quad u_1,\dots,u_d \in [0,1].$$

An arbitrary copula $C(u_1,\dots,u_d)$ lies between the Fréchet–Hoeffdings bounds

$$W(u_1,\dots,u_d) \le C(u_1,\dots,u_d) \le M(u_1,\dots,u_d),$$

where the Fréchet–Hoeffding lower bound is not a copula function for $d > 2$ though. The generalization of elliptical copulas to $d > 2$ is straightforward as well. For example, the Gaussian case yields

$$C_N(u_1,\dots,u_d,\Sigma) = \Phi_\Sigma\{\Phi^{-1}(u_1),\dots,\Phi^{-1}(u_d)\},$$

$$c_N(u_1,\dots,u_d,\Sigma) = |\Sigma|^{-1/2} \exp\Big[-\frac{1}{2}\{\Phi^{-1}(u_1),\dots,\Phi^{-1}(u_d)\}^\top(\Sigma^{-1} - I)\{\Phi^{-1}(u_1),\dots,\Phi^{-1}(u_k)\}\Big]$$

for all $u_1,\dots,u_d \in [0,1]$, where Φ_Σ is a d-dimensional Gaussian distribution with zero mean and correlation matrix Σ. Individual dispersion is imposed via the marginal distributions. Note that in the multivariate case the implementation of elliptical copulas can be involved due to technical difficulties with multivariate cdf's.

13.3.2 Hierarchical Archimedean Copula

A simple multivariate generalization of the Archimedean copulas is defined as

$$C(u_1, \dots, u_d) = \phi\{\phi^{-1}(u_1) + \dots + \phi^{-1}(u_d)\}, \quad u_1, \dots, u_d \in [0, 1], \tag{13.5}$$

where $\phi \in \mathcal{L}$. This definition provides a simple, but rather limited technique for the construction of multivariate copulas, since a possibly complicated multivariate dependence structure is determined by a single copula parameter. Furthermore, multivariate Archimedean copulas imply that the variables are exchangeable. This means, that the distribution of $(u_1, \dots, u_d)$ is the same as of $(u_{j_1}, \dots, u_{j_d})$ for all $j_\ell \neq j_v$. This is certainly not an acceptable assumption in practical applications.

A more flexible method is provided by hierarchical Archimedean copula (HAC) sometimes also called the nested Archimedean copula which replaces a uniform margin of a simple Archimedean copula by an additional Archimedean copula. The iterative substitution of margins by copulas widens the spectrum of attainable dependence structures. For example, the copula function for fully nested HAC is given by

$$\begin{aligned} C(u_1, \dots, u_d) &= \phi_{d-1}\big\{\phi_{d-1}^{-1} \circ \phi_{d-2}\big(\dots [\phi_2^{-1} \circ \phi_1\{\phi_1^{-1}(u_1) + \phi_1^{-1}(u_2)\} \\ &\quad + \phi_2^{-1}(u_3)] + \dots + \phi_{d-2}^{-1}(u_{d-1})\big) + \phi_{d-1}^{-1}(u_d)\big\} \\ &= \phi_{d-1}[\phi_{d-1}^{-1} \circ C(\{\phi_1, \dots, \phi_{d-2}\})(u_1, \dots, u_{d-1}) + \phi_{d-1}^{-1}(u_d)] \end{aligned} \tag{13.6}$$

for $\phi_{d-i}^{-1} \circ \phi_{d-j} \in \mathcal{L}^*$, $i < j$, where

$$\mathcal{L}^* = \{\omega : [0; \infty) \to [0, \infty) \,|\, \omega(0) = 0,\ \omega(\infty) = \infty;\ (-1)^{j-1}\omega^{(j)} \geq 0;\ j = 1, \dots, \infty\},$$

As indicated above, contrarily to the usual Archimedean copula (13.5), HAC defines the dependency structure in a recursive way. At the lowest level of the so called HAC-tree, the dependency between the two variables is modeled by a copula function with the generator ϕ_1, i.e. $z_1 = C(u_1, u_2) = \phi_1\{\phi_1^{-1}(u_1) + \phi_1^{-1}(u_2)\}$. At the second level, an another copula function is used to model the dependency between z_1 and u_3, etc. The generators ϕ_i can come from the same family and differ only through the parameter or, to introduce more flexibility, come from different generator families, c.f. Hofert (2011). As an alternative to the fully nested model, so-called partially nested copulas combine arbitrarily many copula functions at each copula level. For example the following 4-dimensional copula, where the first and the last two variables are joined by individual copulas with generators ϕ_{12} and ϕ_{34}. Further, the resulted copulas are combined by a copula with the generator ϕ.

$$C(u_1, u_2, u_3, u_4) = \phi(\phi^{-1}[\phi_{12}\{\phi_{12}^{-1}(u_1) + \phi_{12}^{-1}(u_2)\}] + \phi^{-1}[\phi_{34}\{\phi_{34}^{-1}(u_3) + \phi_{34}^{-1}(u_4)\}]).$$

The estimation of HAC is a challenging task, since both the copula structure and parameters of the generator functions have to be estimated. The variety of possible structures does not permit the enumeration of all possible structures and selecting that structure-parameter combination with the largest log-likelihood value.

Okhrin et al. (2013a) first propose methods for determining the optimal structure of HAC with (non-)parametrically estimated margins and provide asymptotic theory for the estimated parameters. The basic idea of the estimation procedure uses the fact that HAC are recursively defined and that dependencies decrease from the lowest to the highest hierarchical level for common parametric families. To sketch the procedure suppose margins are known: Parameters related to strongly dependent random variables are estimated first and the variables grouped at the bottom of the HAC-tree. The determined HAC-tree is spanned by at least two random variables and the tree itself determines a univariate random variable. After removing all random variables spanning the tree from the set of variables and adding the univariate random variable determined by the tree, the parameter of the subsequent level is determined by the selecting that pair of variables with the strongest dependency again. An additional level is added to the tree referring to the pair of variables with the strongest dependence and the set of variables is modified as explained above. The sketched steps are iteratively repeated until the HAC-tree is spanned by all random variables. This method is implemented in the `HAC` package for `R`, see Okhrin and Ristig (2014).

Segers and Uyttendaele (2014) introduce an algorithm for non-parametric structure determination by firstly decomposing the HAC's tree structure into four variants of trivariate structures. Then, the whole tree structure is subsequently determined based on testing the distance between trivariate copulas and Kendall's distribution function. Górecki et al. (2016) generalize the approach of Okhrin et al. (2013a) and propose an algorithm for simultaneous estimation of the structure and parameters based on the inversion of Kendall's τ_2, i.e. based on the link between Kendall's τ_2 and Archimedean generators.

Properties and simulation procedures are comprehensively studied in Joe (1997), Whelan (2004), Savu and Trede (2010), Hofert (2011), Okhrin et al. (2013b), Rezapour (2015) and Górecki et al. (2016). Note that HAC became a standard tool for pricing credit derivatives in academia such as collateralized debt obligations, see Hering et al. (2010), Hofert and Scherer (2011) and Choroś-Tomczyk et al. (2013).

Brechmann (2014) proposed hierarchical Kendall copula, which does not suffer from parameter restriction, but are slightly more complicated in estimation. Similar approach to avoid parameter restrictions and family limitations are proposed by using Lévy subordinated HAC, see Hering et al. (2010) and the corresponding application see Zhu et al. (2016).

13.3.3 Factor Copula

In classical factor analysis, a function links the observed and latent variables under the assumption that the latent variables explain the observed variables, e.g., see Johnson and Wichern (2013) and Härdle and Simar (2015). For example, a random variable $X_i, i = 1, \ldots, d$, is generated by an additive factor model, if

$$X_i = \sum_{j=1}^{m} \alpha_{ij} W_j + \varepsilon_i, \tag{13.7}$$

where W_j, $j = 1, \ldots, m$, are latent common factors and ε_i, $i = 1, \ldots, d$, are mutually independent idiosyncratic disturbances. The basic idea of factor models and their natural interpretation can be exported to the copula world in order to induce dependencies between independent idiosyncratic disturbances via common factors. Factor copula models, however, can be split into two complementary groups both having strengths and weaknesses. On the one hand, there are (implicit) factor copula models inducing dependencies among random variables via a functional which links latent factors and idiosyncratic disturbances. Such models are a straightforward extension of factor models from multivariate analysis. On the other hand, factor copulas and dependencies also arise from integrating the product of conditionally independent distributions –given a latent factor– with respect to this factor. This approach benefits from the fact, that the copula collapses to the product copula in case of known factors.

Oh and Patton (2015) concentrate on (implicit) factor copulas for $X = (X_1, \ldots, X_d)^\top$ arising from a functional relation between the factor(s) and mutual independent idiosyncratic errors. In this sense, the dependence component of the joint distribution of X is implied from the factors' distribution, the distribution of the idiosyncratic disturbances and the link function. In particular, X follows a multivariate distribution specified via a copula, i.e. $X \sim F(x_1, \ldots, x_d) = C\{F_1(x_1), \ldots, F_d(x_d)\}$. For instance, the additive single factor copula model is represented as

$$\begin{aligned} &X_i = W + \varepsilon_i,\ i = 1, \ldots, d, \\ &W \sim F_W(\theta_W),\ \varepsilon_i \overset{i.i.d.}{\sim} F_\varepsilon(\theta_\varepsilon),\ W \perp \varepsilon_i, \quad \text{for all} \quad i = 1, \ldots, d,, \end{aligned} \tag{13.8}$$

where W is the single common factor following the distribution of $F_W(\theta_W)$ and $\varepsilon_1, \ldots, \varepsilon_d$ are mutually independent shocks with distribution function $F_\varepsilon(\theta_\varepsilon)$. This model is extended to the non-linear factor copula based on the following representation,

$$\begin{aligned} &Z_i = h(W, \varepsilon_i),\ i = 1, 2, \ldots, d, \\ &W \sim F_W(\theta_W),\ \varepsilon_i \overset{i.i.d.}{\sim} F_\varepsilon(\theta_\varepsilon),\ W \perp \varepsilon_i, \quad \text{for all} \quad i = 1, \ldots, d, \end{aligned} \tag{13.9}$$

where h is a non necessarily linear link function. Thus, the dependence structure can be built in a more flexible way compared to the linear additive version. Model (13.8) implies a joint Gaussian random vector $X = (X_1, \ldots, X_d)^\top$, if the common factor and the idiosyncratic factor are both Gaussian. Therefore, a joint density function is available as well.

Nonetheless, a nice analytical expression of the joint density function for a factor copula with non-Gaussian margins and non-Gaussian factor is rarely available which makes parameter estimation demanding. Oh and Patton (2013) propose an estimation method for copula models without analytical form of the density function. This relies on a simulated method of moments approach building on the simplicity to draw random samples from a factor model. The proposed estimator for $(\theta_W^\top, \theta_\varepsilon^\top)^\top$ is found numerically by minimizing the distance between scale free empirical dependence measures between X_k and X_ℓ, such as $\tau_{2n}^{k\ell}$, $k = 1, \ldots, d; \ell = k+1, \ldots, d$, and those obtained from a drawn sample. Oh and Patton (2013) prove under weak regularity conditions that the simulated method of moment estimator is consistent and asymptotically normal. However, as argued by Genest et al. (1995), method of moment estimators of copula parameters can be highly inefficient.

Another form of factor copulae relies on the assumption that the observed variables $U_1, \ldots, U_d$ are conditionally independent given latent factors $V_1, \ldots, V_m$. Note that all random variables U_i, $i = 1, \ldots, d$, and V_j, $j = 1, \ldots, m$, are assumed to be uniformly distributed. Then, the conditional distribution of U_i given m factors $V_1, \ldots, V_m$ is given by $C_{U_i|V_1,\ldots,V_m}$. By using $C_{U_i|V_1,\ldots,V_m}$, the dependence structure of the observed variables $U_1, \ldots, U_d$ can be specified by the following copula function, such that

$$C(u_1, \ldots, u_d) = \int_{[0,1]^m} \prod_{i=1}^{d} C_{U_i|V_1,\ldots,V_m}(u_i|v_1, \ldots, v_m) dv_1 \cdots dv_m \quad \text{with} \quad u_i \in (0, 1), \tag{13.10}$$

where the factors are out integrated. For the special case $m = 1$, the copula function (13.10) can be simplified to the form

$$C(u_1, \ldots, u_d) = \int_{[0,1]} \prod_{i=1}^{d} C_{U_i|V_1}(u_i|v_1) dv_1 \quad \text{with} \quad u_i \in (0, 1). \tag{13.11}$$

Let C_{U_i,V_1} and c_{U_i,V_1} be the joint cdf and density of the pairs of random variables (U_i, V_1), $i = 1, \ldots, d$. Moreover, let the conditional distribution of U_i given V_1 be denoted by $C_{U_i|V_1}(u_i|v_1) = \partial C_{U_i,V_1}(u_i, v)/\partial v|_{v=v_1}$. Then, the copula density of $C(u_1, \ldots, u_d)$ can be represented by

$$c(u_1, \ldots, u_d) = \frac{\partial^d C(u_1, \ldots, u_d)}{\partial u_1 \cdots \partial u_d} = \int_{[0,1]} \prod_{i=1}^{d} c_{U_i,V_1}(u_i, v_1) dv_1 \quad \text{with} \quad u_i \in (0, 1), \tag{13.12}$$

where $c_{U_i,V_1}(u_i, v_1) = \partial C(u_i|v_1)/\partial u_i$. Seen from this angle, the dependencies between d observed variables is determined by d bivariate copulas $C_{U_i,V_1}(u_i, v)$. Based on a parametric copula density $c(\cdot;\theta)$, Krupskii and Joe (2013) separate the parameter estimation into two steps. In the first step, the margins are estimated parametrically or non-parametrically. In the second step, the maximum likelihood (ML) method is employed to estimate the parameter θ.

Numerous literature about the factor copula's theory and applications can be referred to. Andersen et al. (2003), Hull and White (2004) and Laurent and Gregory (2005) have contributed works on generalization of one factor copula models. A comprehensive review of the factor copula theory is given in Joe (2014). Some applications by using factor copula models can be referred to Li (2000) for credit derivative pricing, Krupskii and Joe (2013) for fitting stock returns and Oh and Patton (2015) for measuring systemic risk.

13.3.4 Vine Copula

Vine copula or pair-copula constructions are originally proposed in Joe (1996) and developed in depth by Bedford and Cooke (2001), Bedford and Cooke (2002), Kurowicka and Cooke (2006) and Aas et al. (2009). The catchy name is due to similarities of the graphical representation of vine copulae and botanical vines. The fundamental idea of the vine copula is to construct a d-dimensional copula by decomposing the dependence structure into $d(d-1)/2$ bivariate copulas.

Let S be the index subset of $D = \{1, \dots, d\}$ referring to the index set of conditioning variables and T be the index set of conditioned variables with $T \cup S = D$. Let $\sharp M$ denote the cardinality of set M. The cdf of variables with index in S is denoted by F_S, so that $F(x) = F_D(x)$. The conditional cdf of variables with index in T conditional on S is denoted $F_{T|S}$. A similar notation is used for the corresponding copulas. To derive a vine copula for a given $x = (x_1, \dots, x_d)^\top$ in the spirit of Joe (2014), we start from a d-dimensional distribution function, i.e.

$$F(x) = \int_{(-\infty,x_S]} F_{T|S}(x_T|y_S)dF_S(y_S), \tag{13.13}$$

and replace the conditional distribution $F_{T|S}(x_T|x_S)$ by the corresponding $\sharp T$-dimensional copula $F_{T|S}(x_T|x_S) = C_{T;S}\{F_{j|S}(x_j|x_S) : j \in T\}$. The copula $C_{T;S}\{F_{j|S}(x_j|x_S) : j \in T\}$ is implied by Sklar's Theorem with margins $F_{j|S}(x_j|x_S),\ j \in T$. It is not a conditional distribution although with conditional distribution as margins. This yields a copula-based representation of the joint d-dimensional distribution function from (13.13), which is given by

$$F(x) = \int_{(-\infty,x_S]} C_{T;S}\{F_{j|S}(x_j|y_S) : j \in T\}dF_S(y_S). \tag{13.14}$$

Note that the support of the integral in (13.13) and (13.14) is a cube $(-\infty, x_S] \in \mathbb{R}^{\sharp S}$. Converting all univariate margins to uniformly distributed random variables allows rewriting $F(x)$ as a d-dimensional copula

$$C(u) = \int_{[0,u_S]} C_{T;S}\{G_{j|S}(u_j|v_S) : j \in T\} dC_S(v_S), \tag{13.15}$$

where $G_{j|S}(u_j|v_S)$ is a conditional distribution from copula $C_{S\cup\{j\}}$. If $T = \{i_1, i_2\}$, then

$$C_{S\cup\{i_1,i_2\}}(u_{S\cup\{i_1,i_2\}}) = \int_{[0,u_S]} C_{i_1,i_2;S}\{G_{i_1|S}(u_{i_1}|v_S), G_{i_2|S}(u_{i_2}|v_S)\} dC_S(v_S). \tag{13.16}$$

Since the essential idea of vine copula is based on building a joint dependence structure by $d(d-1)/2$ bivariate copulae, (13.16) is an important building block in the construction of vines referring to a ($\sharp S + 2$)-dimensional copula built from a bivariate copula $C_{i_1,i_2;S}$.

In case of continuous random variables, the d-dimensional distribution function from (13.13) admits a density function $f(x_1, \ldots, x_d)$, which can be decomposed and represented by bivariate copula densities in an analogue manner. Examples of density decompositions for the 6-dimensional case related to so called C-vine (canonical vine), D-vine (drawable vine) and R-vine (regular vine) copulas are given as follows.

The C-vine structure is illustrated in the left column of Fig. 13.2 and its density decomposition is

$$\begin{aligned} c\{F_1(x_1), \ldots, F_6(x_6)\} &= c_{12}\{F_1(x_1), F_2(x_2)\} \cdot c_{13}\{F_1(x_1), F_3(x_3)\} \\ &\cdot c_{14}\{F_1(x_1), F_4(x_4)\} \cdot c_{15}\{F_1(x_1), F_5(x_5)\} \cdot c_{16}\{F_1(x_1), F_6(x_6)\} \\ &\cdot c_{23;1}\{F(x_2|x_1), F(x_3|x_1)\} \cdot c_{24;1}\{F(x_2|x_1), F(x_4|x_1)\} \\ &\cdot c_{25;1}\{F(x_2|x_1), F(x_5|x_1)\} \cdot c_{26;1}\{F(x_2|x_1), F(x_6|x_1)\} \\ &\cdot c_{34;12}\{F(x_3|x_{12}), F(x_4|x_{12})\} \cdot c_{35;12}\{F(x_3|x_{12}), F(x_5|x_{12})\} \\ &\cdot c_{36;12}\{F(x_3|x_{12}), F(x_6|x_{12})\} \cdot c_{45;123}\{F(x_4|x_{123}), F(x_5|x_{123})\} \\ &\cdot c_{46;123}\{F(x_4|x_{123}), F(x_6|x_{123})\} \cdot c_{56;1234}\{F(x_5|x_{1234}), F(x_6|x_{1234})\}. \end{aligned} \tag{13.17}$$

The density of the D-vine structure –given in the centred column of Fig. 13.2— is

$$\begin{aligned} c\{F_1(x_1), \ldots, F_6(x_6)\} &= c_{12}\{F_1(x_1), F_2(x_2)\} \cdot c_{23}\{F_2(x_2), F_3(x_3)\} \\ &\cdot c_{34}\{F_3(x_3), F_4(x_4)\} \cdot c_{45}\{F_4(x_4), F_5(x_5)\} \cdot c_{56}\{F_5(x_5), F_6(x_6)\} \\ &\cdot c_{13;2}\{F(x_1|x_3), F(x_2|x_3)\} \cdot c_{24;3}\{F(x_2|x_3), F(x_4|x_3)\} \\ &\cdot c_{35;4}\{F(x_3|x_4), F(x_5|x_4)\} \cdot c_{46;5}\{F(x_4|x_5), F(x_6|x_5)\} \\ &\cdot c_{14;23}\{F(x_2|x_{23}), F(x_4|x_{23})\} \cdot c_{25;34}\{F(x_2|x_{34}), F(x_5|x_{34})\} \\ &\cdot c_{36;45}\{F(x_3|x_{45}), F(x_6|x_{45})\} \cdot c_{15;234}\{F(x_1|x_{234}), F(x_5|x_{234})\} \\ &\cdot c_{26;345}\{F(x_2|x_{345}), F(x_6|x_{345})\} \cdot c_{16;2345}\{F(x_1|x_{2345}), F(x_6|x_{2345})\}. \end{aligned} \tag{13.18}$$

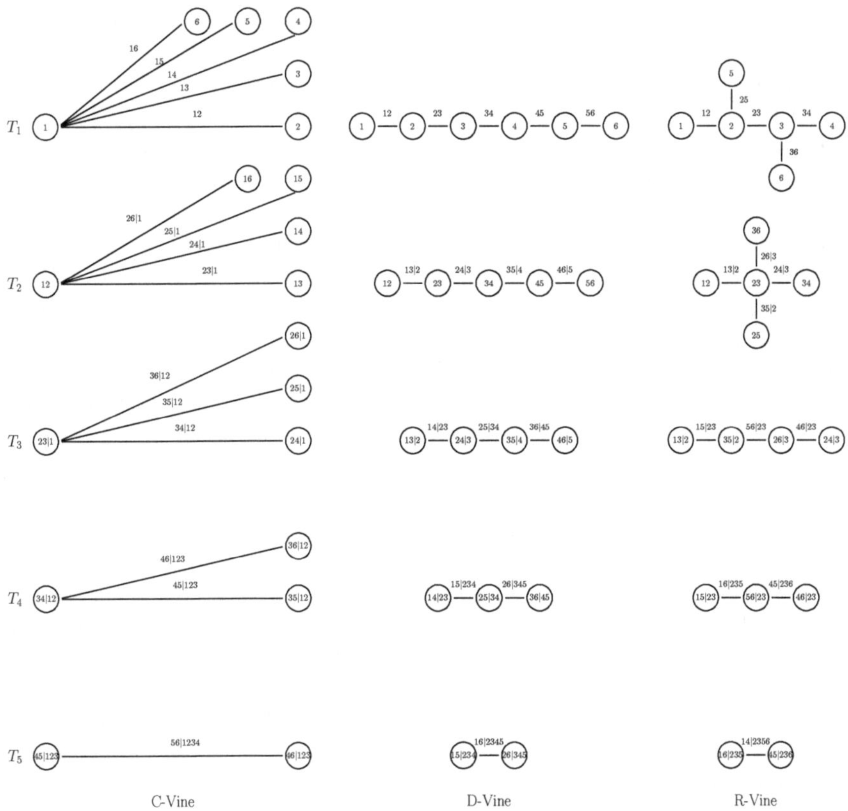

Fig. 13.2 Vine tree structures of C-vine, D-vine and R-vine

The density of the R-vine structure illustrated in the right column of Fig. 13.2 is

$$\begin{aligned} c\{F_1(x_1), \ldots, F_6(x_6)\} &= c_{12}\{F_1(x_1), F_2(x_2)\} \cdot c_{23}\{F_2(x_2), F_3(x_3)\} \\ &\cdot c_{34}\{F_3(x_3), F_4(x_4)\} \cdot c_{25}\{F_2(x_2), F_5(x_5)\} \cdot c_{36}\{F_3(x_3), F_6(x_6)\} \\ &\cdot c_{13;2}\{F(x_1|x_2), F(x_3|x_2)\} \cdot c_{24;3}\{F(x_2|x_3), F(x_4|x_3)\} \\ &\cdot c_{26;3}\{F(x_2|x_3), F(x_6|x_3)\} \cdot c_{35;2}\{F(x_3|x_2), F(x_5|x_2)\} \\ &\cdot c_{15;23}\{F(x_1|x_{23}), F(x_5|x_{23})\} \cdot c_{56;23}\{F(x_5|x_{23}), F(x_6|x_{23})\} \\ &\cdot c_{46;23}\{F(x_4|x_{23}), F(x_6|x_{23})\} \cdot c_{16;235}\{F(x_1|x_{235}), F(x_6|x_{235})\} \\ &\cdot c_{45;236}\{F(x_4|x_{236}), F(x_5|x_{236})\} \cdot c_{14;2356}\{F(x_1|x_{2356}), F(x_4|x_{2356})\}. \end{aligned} \quad (13.19)$$

In particular, the C-vine and D-vine have an intuitive graphical representation which can be immediately related to the decomposition of the copula density function into the product of bivariate copula densities. For example, the product of bivariate copula densities from the first two lines of the right hand side of Eq. 13.17 refers to a C-vine represented in the upper left graphic of Fig. 13.2. The formula and the corresponding graphic illustrate that the first variable X_1 is pairwise coupled with the second, third ... and sixth random variable. The subsequent two lines (3–4) of Eq. 13.17 are related to the second graphic of the left column of Fig. 13.2. Conditional on X_1, random variable X_2 is pairwise coupled with X_3, X_4, X_5 and X_6. Connecting the remaining graphics with formulas is left to the reader. While the "formula-graphic" matching follows a similar scheme in case of the D-vine, the R-vine belongs to a more general vine copula class and contains the C-vine and D-vine as special cases. A rigorous definition of an R-vine copula can be found in Joe (2014).

In fact, vines can be estimated by either full or stage-wise ML such as the inference function for margins (IFM) method discussed below in Sect. 13.4. Nonetheless, the inference approach derived in Haff (2013) namely the stepwise semi-parametric estimator deserves to be mentioned in more detail. Here, the marginal distributions are non-parametrically estimated by the empirical distribution function such as for factor copulae or HAC. In order to obtain a consistent and asymptotically Gaussian distributed estimator of a parametric vine copula, a so called simplifying assumption is required. The latter permits replacing "conditional" bivariate copula densities with unconditional densities. Then, it can be straightforwardly shown, that the log-likelihood can be maximized in a stage-wise manner. This is due to the decomposition of the density into the product of bivariate copula densities, so that the log-likelihood function is a sum of logarithmized copula densities. Coming back to the C-vine example from Fig. 13.2. At the first stage, all parameters of bivariate copulas represented in the upper left graphic of Fig. 13.2 are estimated, i.e. the parameters of the copulae for $(X_1, X_2), \ldots, (X_1, X_6)$. Keeping the corresponding parameters fixed at estimated values, the four parameters of copulae referring to the pairs from the second graphic of the left column of Fig. 13.2 are estimated. Holding these parameters fixed at estimated values again, all vine parameters of the remaining bivariate densities can be estimated iteratively. Literature on pair-copula construction is spreading steadily, and most recent information about it can be found on vine copula homepage `http://www.statistics.ma.tum.de/en/research/vine-copula-models/`.

13.4 Estimation Methods

The estimation of a copula-based multivariate distribution involves both the estimation of the copula parameters $\boldsymbol{\theta}$ and the estimation of the margins F_j, $j = 1, \ldots, d$. The properties and goodness of the estimator of $\boldsymbol{\theta}$ heavily depend on the estimators of F_j, $j = 1, \ldots, d$. We distinguish between a parametric and a non-parametric specification of the margins. If we are interested only in the dependency structure, the estimator of $\boldsymbol{\theta}$ should be independent of any parametric models for the margins. However, Joe (1997) argues that complete distribution models and, therefore, parametric models for margins are actually more appropriate for applications.

In the bivariate case, a standard method of estimating the univariate parameter θ is based on Kendall's τ_2 statistic by Genest and Rivest (1993). The estimator of τ_2 complemented by the method of moments allows to estimate the parameters. However, as shown in Genest et al. (1995), the ML method leads to substantially more efficient estimators. For non-parametrically estimated margins, Genest et al. (1995) show the consistency and asymptotic normality of ML estimators and derive the moments of the asymptotic distribution. The ML procedure can be performed simultaneously for the parameters of the margins and of the copula function. Alternatively, a two-stage procedure can be applied, where the parameters of margins are estimated at the first stage and the copula parameters at the second stage, see Joe (1997) and Joe (2005). Chen and Fan (2006) and Chen et al. (2006) analyze the case of non-parametrically estimated margins. Fermanian and Scaillet (2003) and Chen and Huang (2007) consider a fully non-parametric estimation of the copula. Next we provide details on both approaches. Note that estimation procedures for HAC, conditional-independence-based factor copulas and vines are in fact generalizations of the subsequent approaches taking specific needs of the copula into account, e.g., parameter restrictions.

13.4.1 Parametric Margins

Let $\boldsymbol{\alpha} = (\boldsymbol{\alpha}_1^\top, \ldots, \boldsymbol{\alpha}_d^\top)^\top$ denote the vector of parameters of marginal distributions and $\boldsymbol{\theta}$ parameters of the copula. The classical full ML estimator $\hat{\boldsymbol{\eta}}$ of $\boldsymbol{\eta} = (\boldsymbol{\alpha}^\top, \boldsymbol{\theta}^\top)^\top$ solves the system of equations

$$
\begin{aligned}
\frac{\partial \mathcal{L}(\boldsymbol{\eta}, \mathbf{X})}{\partial \boldsymbol{\eta}} &= \mathbf{0}, \\
\text{where}\quad \mathcal{L}(\boldsymbol{\eta}, \mathbf{X}) &= \sum_{i=1}^{n} \log \left\{ c(F_1(x_{1i}, \boldsymbol{\alpha}_1), \ldots, F_d(x_{di}, \boldsymbol{\alpha}_d), \boldsymbol{\theta}) \prod_{j=1}^{d} f_j(x_{ji}, \boldsymbol{\alpha}_j) \right\} \\
&= \sum_{i=1}^{n} \Big\{ \log c(F_1(x_{1i}, \boldsymbol{\alpha}_1), \ldots, F_d(x_{di}, \boldsymbol{\alpha}_k), \boldsymbol{\theta}) + \sum_{j=1}^{d} \log f_j(x_{ji}, \boldsymbol{\alpha}_j) \Big\}.
\end{aligned}
$$

Following the standard theory on ML estimation, the estimator $\hat{\boldsymbol{\eta}}$ is efficient and asymptotically normal. However, it is often computationally demanding to solve the system simultaneously. Alternatively the multistage optimization proposed in Joe (1997), also known as inference functions for margins, can be applied: Firstly, the parameters of the margins are separately estimated under the assumption that the copula is the product copula. Secondly, the parameters of the copula are estimated replacing the parameters of margins by estimates from the first step and treating them as known quantities. The above optimization problem is then replaced by

$$\left(\frac{\partial \mathcal{L}_1}{\partial \boldsymbol{\alpha}_1^\top}, \dots, \frac{\partial \mathcal{L}_d}{\partial \boldsymbol{\alpha}_d^\top}, \frac{\partial \mathcal{L}_{d+1}}{\partial \boldsymbol{\theta}^\top}\right)^\top = \mathbf{0}, \tag{13.20}$$

$$\begin{aligned} \text{where} \quad & \mathcal{L}_j = \sum_{i=1}^{n} l_j(\mathbf{X}_i), \quad \text{for} \quad j = 1, \dots, d+1, \\ & l_j(\mathbf{X}_i) = \log\ f_j(x_{ji}, \boldsymbol{\alpha}_j), \quad \text{for} \quad j = 1, \dots, d, i = 1, \dots, n, \\ \text{and} \quad & l_{d+1}(\mathbf{X}_i) = \log c\big\{F_1(x_{1i}, \boldsymbol{\alpha}_1), \dots, F_d(x_{di}, \boldsymbol{\alpha}_d), \boldsymbol{\theta}\big\}, \quad \text{for} \quad i = 1, \dots, n. \end{aligned}$$

The first d components in (13.20) correspond to the usual ML estimation of the parameters of the marginal distributions. The last component reflects the estimation of the copula parameters. Detailed discussion on this method can be found in Joe (1997). Note, that this procedure does not lead to efficient estimators, however, as argued by Joe (1997) the loss in the efficiency is modest and mainly depends on the strength of dependencies. This method is a special case of the generalized method of moments with an identity weighting matrix, see Cherubini et al. (2004). The advantage of the two-stage procedure lies in the dramatic reduction of the numerical complexity.

13.4.2 Non-parametric Margins

In this section, we consider a non-parametric estimation of the marginal distributions also referred to as canonical ML. The asymptotic properties of the multistage estimator for $\boldsymbol{\theta}$ do not depend explicitly on the type of the non-parametric estimator, but on its convergence properties. Here, we use the rectangular kernel (histogram) resulting in the estimator

$$\widehat{F}_j(x) = (n+1)^{-1} \sum_{i=1}^{n} \mathbf{1}(x_{ji} \leq x), \quad j = 1, \dots, d.$$

The factor $n/(n+1)$ is used to restrict fitted values to the open unit interval. This is necessary as several copula densities are not bounded at zero and/or one. Let $\widehat{F}_1, \dots, \widehat{F}_d$ denote the non-parametric estimators of $F_1, \dots, F_d$. The canonical ML estimator $\hat{\boldsymbol{\theta}}$ of $\boldsymbol{\theta}$ solves the system $\partial \mathcal{L}/\partial \boldsymbol{\theta}^\top = \mathbf{0}$ by maximizing the pseudo log-likelihood with estimated margins $\widehat{F}_1, \dots, \widehat{F}_d$, i.e.

$$\begin{aligned} \mathcal{L} &= \sum_{i=1}^{n} l(\mathbf{X}_i) \quad \text{for} \quad j = 1, \dots, p, \\ l(\mathbf{X}_i) &= \log c\big\{\widehat{F}_1(x_{1i}), \dots, \widehat{F}_d(x_{di}), \boldsymbol{\theta}\big\}, \quad \text{for} \quad i = 1, \dots, n. \end{aligned}$$

As in the parametric case, the semi-parametric estimator $\hat{\boldsymbol{\theta}}$ is asymptotically normal under suitable regularity conditions. This method was first used in Oakes (1994) and then investigated by Genest et al. (1995) and Shih and Louis (1995). Additional properties of the estimator, such as the covariance matrix, are stated in these papers.

13.5 Goodness-of-Fit Tests for Copulae

Having a dataset and an estimated copula at hand, it arises the natural question whether the selected copula describes the data properly. For this purpose, a series of different goodness-of-fit tests has been developed in the last decade. Under the H_0-hypothesis one assumes that the true copula belongs to some parametric family $H_0 : C \in \mathcal{C}_0$.

The most natural test approach is to measure the deviation of the parametric copula from the empirical one given through

$$C_n(u_1, \ldots, u_d) = n^{-1} \sum_{i=1}^{n} \prod_{j=1}^{d} I\{\widehat{F}_j(x_{ij}) \leq u_j\}.$$

Gaensler and Stute (1987) and Radulovic and Wegkamp (2004) show that C_n is a consistent estimation of the true underlying copula. Several tests are based on the empirical copula process, which is defined as follows

$$\mathbb{C}_n(u_1, \ldots, u_d) = \sqrt{n}\{C_n(u_1, \ldots, u_d) - C_{\hat{\theta}}(u_1, \ldots, u_d)\}.$$

Fermanian (2005) and Genest and Rèmillard (2008) propose to compute different measures to quantify the deviation of the assumed parametric copula from the empirical copula, one of those is Cramér–von Mises distance

$$S_n^E = \int_{[0,1]^d} \mathbb{C}_n(u_1, \ldots, u_d)^2 dC_n(u_1, \ldots, u_d)$$

or the weighted Cramér–von Mises distance, with tuning parameters $m \geq 0$ and $\zeta_m \geq 0$ given as

$$R_n^E = \int_{[0,1]^d} \left\{ \frac{\mathbb{C}_n(u_1, \ldots, u_d)}{[C_{\hat{\theta}}(u_1, \ldots, u_d)\{1 - C_{\hat{\theta}}(u_1, \ldots, u_d)\} + \zeta_m]^m} \right\}^2 dC_n(u_1, \ldots, u_d).$$

The usual Kolmogorov–Smirnov distance as for classical univariate tests is also applicable here

$$T_n^E = \sup_{\{u_1, \ldots, u_d\} \in [0,1]^d} |\mathbb{C}_n(u_1, \ldots, u_d)|.$$

The other group of tests developed and investigated by Genest and Rivest (1993), Wang and Wells (2000), Genest et al. (2006) are based on the probability integral transform and in particular on so called Kendall's transform. Having

$$(X_1, \dots, X_d) \sim F(x_1, \dots, x_d) = C_\theta\{F_1(x_1), \dots, F_d(x_d)\},$$

one concludes similar to $F_i(X_i) \sim U(0, 1)$ that the copula-based random variable is

$$C_\theta\{F_1(X_1), \dots, F_d(X_d)\} \sim K_\theta(v)$$

where $K_\theta(v)$ is the univariate Kendall's distribution (not necessarily uniform), see Barbe et al. (1996), Jouini and Clemen (1996). Empirically, the distribution function K can be estimated as

$$K_n(v) = n^{-1} \sum_{i=1}^{n} I\left[C_n\{\widehat{F}_1(x_{i1}), \dots, \widehat{F}_d(x_{id})\} \le v\right], \quad v \in [0, 1].$$

Further usual test statistics for the univariate distributions like Cramér–von Mises or Kolmogorov–Smirnov, see Genest et al. (2006), can be applied

$$S_n^{(K)} = \int_0^1 \mathbb{K}_n(v)^2 dK_{\hat{\theta}}(v), \qquad T_n^{(K)} = \sup_{v \in [0,1]} |\mathbb{K}_n(v)|,$$

where $\mathbb{K}_n = \sqrt{n}(K_n - K_{\hat{\theta}})$ is the Kendall's process. Here is, however, a little challenge in using this tests: as in testing for Kendall's distribution one tests in null hypothesis has $H_0^{''} : K \in \mathcal{K}_0 = \{K_\theta : \theta \in \Theta\}$, and as $H_0 \subset H_0^{''}$, the non-rejection of $H_0^{''}$ does not imply non rejection of H_0. For the bivariate Archimedean copulas $H_0^{''}$ and H_0 are equivalent.

Another series of goodness-of-fit tests, is constructed via the other important integral transform, that dates back to Rosenblatt (1952). Based on the conditional distribution of U_i by

$$\begin{aligned} C_d(u_i|u_1, \dots, u_{i-1}) &= \mathrm{P}\{U_i \le u_i | U_1 = u_1 \dots U_{i-1} = u_{i-1}\} \\ &= \frac{\partial^{i-1} C(u_1, \dots, u_i, 1, \dots, 1)/\partial u_1 \dots \partial u_{i-1}}{\partial^{i-1} C(u_1, \dots, u_{i-1}, 1, \dots, 1)/\partial u_1 \dots \partial u_{i-1}}, \end{aligned}$$

the Rosenblatt transform is defined as follows.

Definition 13.4 Rosenblatt's probability integral transform of a copula C is the mapping $\mathfrak{R} : (0, 1)^d \to (0, 1)^d$, $\mathfrak{R}(u_1, \dots, u_d) = (e_1, \dots, e_d)$ with $e_1 = u_1$ and $e_i = C_d(u_i|u_1, \dots, u_{i-1})$, $\forall i = 2, \dots, d$.

Under this definition, the null hypothesis $H_0 : C \in \mathcal{C}_0$ can be rewritten as H_{0R} : $(e_1, \dots, e_d)^\top \sim \Pi$. The first test based on the Rosenblatt transform exploits information, that under H_0 transformed observations should be exactly uniform distributed and independent, which is not the case, as those variables as not mutually independent and only approximately uniform. Nevertheless, two tests use Anderson–Darling test statistics, see Breymann et al. (2003), and are constructed as

$$T_n = -n - \sum_{i=1}^{n} \frac{2i-1}{n} [\log G_{(i)} + \log\{1 - G_{(n+1-i)}\}]$$

where G_i might be constructed in two ways. In the first possibility

$$G_{i,Gamma} = \Gamma_d \left\{ \sum_{j=1}^{d} (-\log e_{ij}) \right\},$$

where $\Gamma_d(\cdot)$ is the Gamma distribution with shape d and scale 1. The second way takes

$$G_{i,\chi^2} = \chi^2_d \left[\sum_{j=1}^{d} \{\Phi^{-1}(e_{ij})\}^2 \right],$$

where χ^2_d refers to the Chi-squared distribution with d degrees of freedom and Φ is standard normal distribution. Another possibility compares the variables not via the Anderson–Darling test statistics, but by purely deviations between estimated density functions, as in Patton et al. (2004), where the test statistics is constructed by

$$C_n^{Ch} = \frac{n\sqrt{h}\hat{J}_n - c_n}{\sigma}$$

with c_n and σ are normalization factors and $\hat{J}_n = \int_0^1 \{\frac{1}{n} \sum_{i=1}^{n} K_h(w, G_{i,\chi^2}) - 1\}^2 dw$.

As discussed by Dobrić and Schmid (2007), the problem with those tests is that they have almost no power and even do not capture the type 1 error. Much better power have tests, that work directly on the copulas of the Rosenblatt transformed data, see Genest et al. (2009). The idea is to compute Cramer–von Mises statistics of the following form

$$S_n = n \int_{[0,1]^d} \{D_n(u) - \Pi(u)\}^2 du$$

$$S_n^{(C)} = n \int_{[0,1]^d} \{D_n(u) - \Pi(u)\}^2 dD_n(u)$$

where the empirical distribution function

$$D_n(u) = D_n(u_1, \dots, u_d) = \frac{1}{n} \sum_{i=1}^{n} \prod_{j=1}^{d} I(e_{ij} \le u_j)$$

should be "close" to product copula Π under H_0.

Different from previous test are those based on the kernel density estimators, and just to mention one, let us consider test developed by Scaillet (2007), where the test statistics is given through

$$J_n = \int_{[0,1]^d} \{\hat{c}(u) - K_H * c(u; \hat{\theta})\} w(u) du,$$

with "$*$" being a convolution operator and $w(u)$ a weight function. The kernel function $K_H(y) = K(H^{-1}y)/\det(H)$ where K is the bivariate quadratic kernel with the bandwidth $H = 2.6073n^{-1/6}\widehat{\Sigma}^{1/2}$ and $\widehat{\Sigma}$ being a sample covariance matrix. The copula density is estimated non-parametrically as

$$\hat{c}(u) = n^{-1} \sum_{i=1}^{n} K_H[u - \{\widehat{F}_1(X_{i1}), \dots, \widehat{F}_d(X_{id})\}^\top],$$

where $\widehat{F}_j$ refers to an estimated marginal distribution, $j = 1, \dots, d$. The most recent goodness of fit test for copulas have been proposed recently by Zhang et al. (2016), where one compares the two-step pseudo maximum likelihood:

$$\hat{\theta} = \operatorname*{argmax}_{\theta \in \Theta} \sum_{i=1}^{n} \mathcal{L}\{\widehat{F}_1(X_{i1}), \dots, \widehat{F}_d(X_{id}); \theta\}.$$

with the delete-one-block pseudo maximum likelihood $\hat{\theta}_{-b}$, $1 \leq b \leq B$:

$$\hat{\theta}_{-b} = \operatorname*{argmax}_{\theta \in \Theta} \sum_{b' \neq b}^{B} \sum_{i=1}^{m} \mathcal{L}\{\widehat{F}_1(X_{i1}), \dots, \widehat{F}_d(X_{id}); \theta\}, \quad b = 1, \dots, B.$$

Further, "in-sample" and "out-of-sample" pseudo-likelihoods are compared with the following test statistic:

$$T_n(m) = \sum_{b=1}^{B} \sum_{i=1}^{m} \left[\mathcal{L}\{\widehat{F}_1(X_{i1}), \dots, \widehat{F}_d(X_{id}); \hat{\theta}\} - \mathcal{L}\{\widehat{F}_1(X_{i1}), \dots, \widehat{F}_d(X_{id}); \hat{\theta}_{-b}\} \right].$$

This leads to some challenges, like computation of $[\frac{n}{m}]$ dependence parameters, but Zhang et al. (2016) proposed an asymptotically equivalent test statistics based on variability and sensitivity matrices. As most of the above mentioned tests, have complicated asymptotic distributions, p-values of the tests can be performed via the parametric bootstrap sketched in the subsequent procedure:

Step 1 Generate bootstrap sample $\left\{\epsilon_i^{(k)}, i = 1, \ldots, n\right\}$ from copula $C(u; \hat{\theta})$ under H_0 with $\hat{\theta}$ and estimated marginal distribution $\widehat{F}$ obtained from original data;

Step 2 Based on $\left\{\epsilon_i^{(k)}, i = 1, \ldots, n\right\}$ from Step 1, estimate θ of the copula under H_0, and compute test statistics under consideration, say R_n^k;

Step 3 Repeat Steps (1–2) N-times and obtain N statistics $R_n^k, k = 1, \ldots, N$;

Step 4 Compute an empirical p-value as $p_e = N^{-1} \sum_{k=1}^{N} I\left(|R_n^k| \geq |R_n|\right)$ with R_n being the test statistic estimated from original data.

13.6 Empirical Study

Value-at-Risk (VaR) is an important measure in risk management. The traditional models for VaR estimation assume that the assets returns in a portfolio are jointly normally distributed. However, numerous empirical studies show that Gaussian based models are not sufficient to describe data characteristics, especially when extreme events happen such as financial crisis. The weak points of the Gaussian based models include the lack of asymmetry and tail dependence. Therefore copula methods come into the focus.

Twelve different copulas are used in this study to construct dependence structures. The employed families include the Gaussian copula, t-copula, Archimedean copulas (Clayton, Gumbel, Joe), HAC (Gumbel, Clayton, Frank), C- and D-vine structures and two factor copulas linked individually by a bivariate Gumbel and Clayton copula.

The data set utilized in this study includes five time series of stock close prices containing ADI (Analog Devices, Inc.), AVB (Avalonbay Communities Inc.), EQR (Equity Residential), LLY (Eli Lilly and Company) and TXN (Texas Instruments Inc.), from Yahoo finance. Here, ADI and TXN belong to high-tech industry, AVB and EQR to real estate industry and LLY to pharmacy industry. The time window spans from 20070113 to 20160116.

Let $w = (w_1, \ldots, w_d)^\top \in \mathbb{R}^d$ denote the long position vector of a d-dimensional portfolio, $S_t = (S_{1,t}, \ldots, S_{d,t})^\top$ stand for the vector of asset prices at time $t \in \{1, \ldots, T\}$ and $X_{i,t} = \log(S_{i,t}/S_{i,t-1})$ for the one period log-return of the i-th asset at time t. Then, $L_t = \sum_{i=1}^{d} w_i X_{i,t}$ denotes the portfolio return. The distribution function of the univariate random variable L_t is denoted by $F_{L_t}(x) = P(L_t \leq x)$ and the Value-at-Risk at level α for the portfolio is defined as the inverse of $F_{L_t}(x)$, namely $\text{VaR}_t(\alpha) = F_{L_t}^{-1}(\alpha)$.

Copula Performance in Risk Management

From the above formulations can be concluded that the idiosyncratic dependence of the log-return process $\{X_t\}_{t=1}^{T}$ is crucial for the appropriate estimation of the VaR. To remove temporal dependence from X_t, the single log-return processes are filtered through GARCH(1, 1) processes,

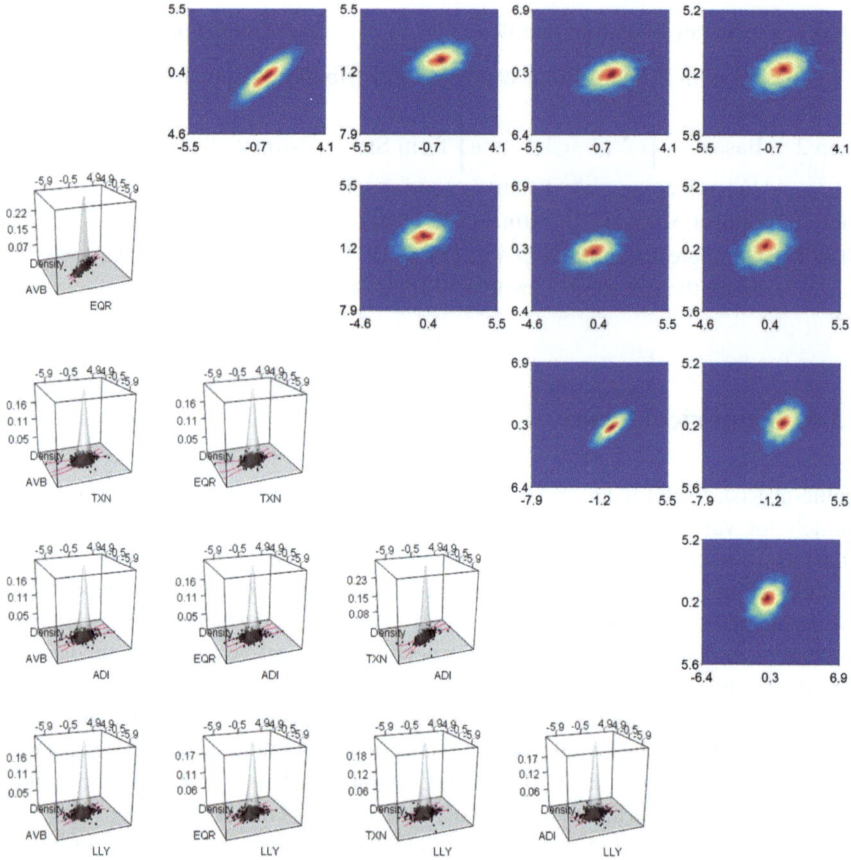

Fig. 13.3 The *lower* triangular plots give 2-dimensional kernel density estimations containing scatter plots of pairwise GARCH(1, 1)-filtered log-returns with quantile regressions under 0.05, 0.5, 0.95 quantiles. The *upper* triangular plots give pairwise contours of five variables

Table 13.1 Pairwise dependence measures including Pearson's correlation (*left*), Kendall's correlation (*center*) and Spearman's correlation (*right*)

	AVB	EQR	TXN	ADI	AVB	EQR	TXN	ADI	AVB	EQR	TXN	ADI
EQR	0.867				0.686				0.866			
TXN	0.359	0.375			0.260	0.264			0.376	0.381		
ADI	0.384	0.399	0.752		0.277	0.285	0.583		0.398	0.410	0.770	
LLY	0.358	0.370	0.358	0.362	0.268	0.260	0.272	0.270	0.390	0.376	0.393	0.391

Table 13.2 Exceeding ratios based on $\alpha \in \{0.05,\ 0.01,\ 0.005,\ 0.001\}$

Copula	$\alpha = 0.05$	$\alpha = 0.01$	$\alpha = 0.005$	$\alpha = 0.001$
Gaussian	**0.050**	0.018	0.009	0.004
t	0.048	0.014	0.011	0.005
Clayton	0.047	0.017	0.011	**0.002**
Gumbel	0.048	0.025	0.013	0.005
Joe	0.065	0.032	0.030	0.023
C-Vine	0.045	0.019	0.015	0.008
D-Vine	0.044	0.018	0.012	0.007
HAC-Clayton	0.044	**0.013**	**0.008**	0.003
HAC-Frank	0.055	0.033	0.026	0.016
HAC-Gumbel	0.070	0.036	0.028	0.017
Factor-Frank	0.046	0.026	0.017	0.015
Factor-Gumbel	0.086	0.042	0.032	0.024

$$X_{i,t} = \mu_{i,t} + \sigma_{i,t}\epsilon_{i,t}, \tag{13.21}$$

$$\sigma_{i,t}^2 = a_i + \alpha_i(X_{i,t-1} - \mu_{i,t-1})^2 + \beta_i\sigma_{i,t-1}^2. \tag{13.22}$$

The GARCH(1, 1)-filtered log-returns are illustrated in Fig. 13.3. Obviously, assets coming from the same sector have high correlation according to the GARCH residuals. For example, the AVB-EQR and TXN-ADI pairs have strong correlation coming from real estate industry and high technology industry respectively. The strong correlation is also observed in Table 13.1 presenting three dependence measures for pairs of AVB-EQR and TXN-ADI. LLY is from pharmacy industry and shows weak correlation with the other four companies according to the scatter-plots and the contours.

The performance of different copulas utilized for VaR estimation is evaluated via backtesting based on the exceeding ratio

$$\mathrm{ER}^{\alpha} = (T - w)^{-1} \sum_{t=w}^{T} \mathbf{1}\{l_t < \widehat{\mathrm{VaR}}_t(\alpha)\}, \tag{13.23}$$

where w is the sliding window size and l_t is the realization of L_t. For the twelve copulas, Table 13.2 presents the ERs which is optimal if it equals α. The Gaussian copula performs best for $\alpha = 0.05$, the HAC-Clayton copula has reached the most appropriate ER for $\alpha \in \{0.01, 0.005\}$ and the Clayton copula for $\alpha = 0.001$. The Factor-Gumbel copula provides the worst ER values for all values of α. Vines perform neither outstanding good nor bad. It deserves to be mentioned that copulas exhibiting upper-tail dependence show higher ER values, for instance, Joe copula, HAC-Gumbel copula and Factor-Gumbel copula. Even though some copulas are based on more parameters and thus, offer more flexibility, the increase of parameters does not essentially improve the ER (see Fig. 13.4).

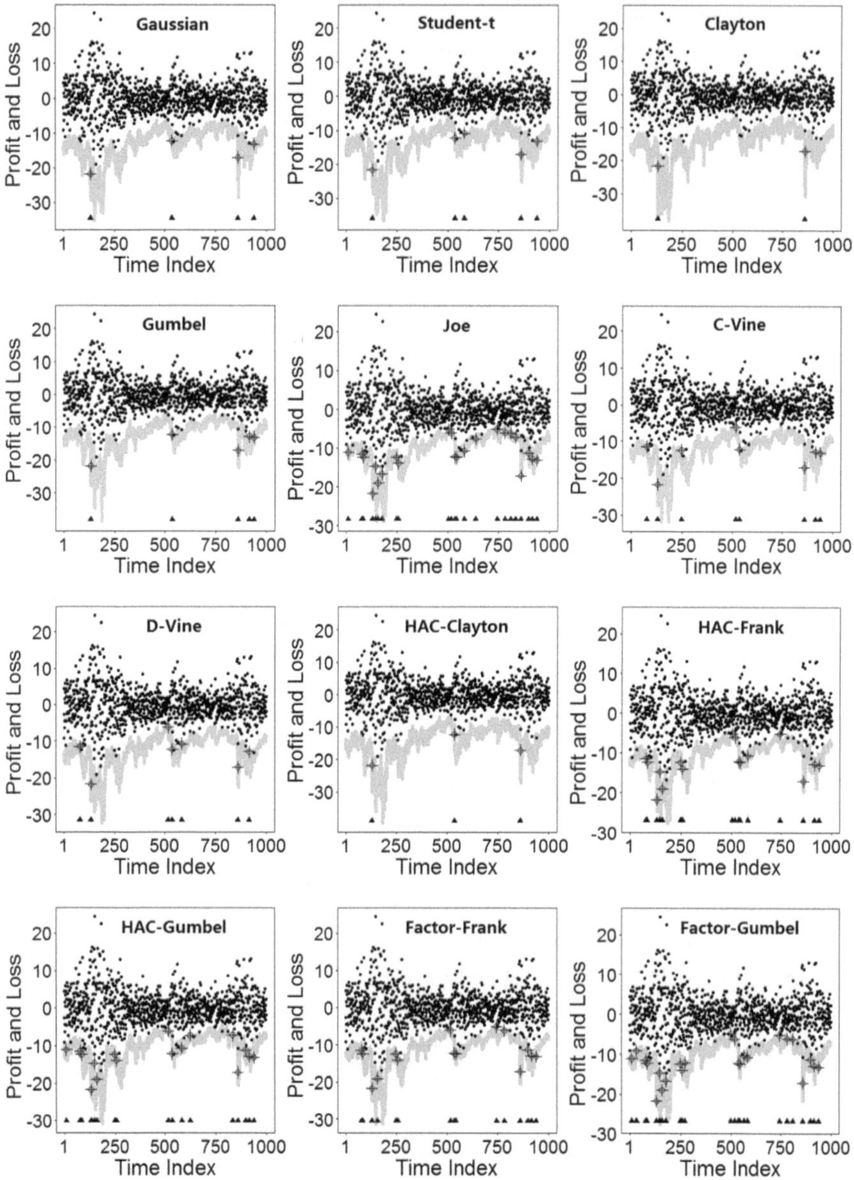

Fig. 13.4 VaRs for $\alpha = 0.001$ are constructed based on 1000 back-testing points with copulas of Gaussian, *t*, Clayton, Gumbel, Joe, C-Vine, D-Vine, HAC-Clayton, HAC-Frank, HAC-Gumbel, Factor-Frank, Factor-Gumbel, illustrated by row. XFGCHD_VaR_CVine, https://github.com/QuantLet/XFG3/tree/master/XFGCHD_VaR_CVine, https://github.com/QuantLet/XFG3/tree/master/XFGCHD_VaR_Clayton, https://github.com/QuantLet/XFG3/tree/master/XFGCHD_VaR_DVine, https://github.com/QuantLet/XFG3/tree/master/XFGCHD_VaR_Gaussian, https://github.com/QuantLet/XFG3/tree/master/XFGCHD_VaR_Gumbel, https://github.com/QuantLet/XFG3/tree/master/XFGCHD_VaR_Joe, https://github.com/QuantLet/XFG3/tree/master/XFGCHD_VaR_StuT, https://github.com/QuantLet/XFG3/tree/master/XFGCHD_VaR_hacClayton, https://github.com/QuantLet/XFG3/tree/master/XFGCHD_VaR_hacFrank, https://github.com/QuantLet/XFG3/tree/master/XFGCHD_VaR_hacGumbel

13.7 Conclusion

This work discusses bivariate copula and focuses on three high dimensional copula models including the hierarchical Archimedean copula, the factor copula and the vine copula. The three models are developed in-depth with their advantages in modeling high dimensional data for diverse research fields. For the sake of comparison, an empirical study from risk management is presented. In this study, the estimation of Value-at-Risk is performed under 12 different copula models including the discussed state-of-art copulas as well as some classical benchmarks such as some of the elliptical and Archimedean family. Considered in toto, the hierarchical Archimedean copula with Clayton generator performs better than the alternatives in terms of the exceeding ratios measure.

References

Aas, K., Czado, C., Frigessi, A., & Bakken, H. (2009). Pair-copula constructions of multiple dependence. *Insurance: Mathematics and Economics*, *44*(2), 182–198.

Andersen, L., Sidenius, J., & Basu, S. (2003). Credit derivatives: All your hedges in one basket. *Risk*, *16*, 67–72.

Barbe, P., Genest, C., Ghoudi, K., & Rémillard, B. (1996). On Kendalls's process. *Journal of Multivariate Analysis*, *58*, 197–229.

Bedford, T., & Cooke, R. M. (2001). Probability density decomposition for conditionally dependent random variables modeled by vines. *Annals of Mathematical and Artificial Intelligence*, *32*, 245–268.

Bedford, T., & Cooke, R. M. (2002). Vines - a new graphical model for dependent random variables. *Annals of Statistics*, *30*(4), 1031–1068.

Brechmann, E. C. (2014). Hierarchical kendall copulas: Properties and inference. *Canadian Journal of Statistics*, *42*(1), 78–108.

Breymann, W., Dias, A., & Embrechts, P. (2003). Dependence structures for multivariate high-frequency data in finance. *Quantitative Finance*, *1*, 1–14.

Chen, S. X., & Huang, T. (2007). Nonparametric estimation of copula functions for dependence modeling. *The Canadian Journal of Statistics*, *35*(2), 265–282.

Chen, X., & Fan, Y. (2006). Estimation and model selection of semiparametric copula-based multivariate dynamic models under copula misspesification. *Journal of Econometrics*, *135*(1–2), 125–154.

Chen, X., Fan, Y., & Tsyrennikov, V. (2006). Efficient estimation of semiparametric multivariate copula models. *Journal of the American Statistical Association*, *101*(475), 1228–1240.

Cherubini, U., Luciano, E., & Vecchiato, W. (2004). *Copula methods in finance*. New York: Wiley.

Choroś-Tomczyk, B., Härdle, W. K., & Okhrin, O. (2013). Valuation of collateralized debt obligations with hierarchical Archimedean copulae. *Journal of Empirical Finance*, *24*(C), 42–62.

Dobrić, J., & Schmid, F. (2007). A goodness of fit test for copulas based on Rosenblatt's transformation. *Computational Statistics & Data Analysis*, *51*(9), 4633–4642.

Durante, F., Fernández-Sánchez, J., & Sempi, C. (2012). A topological proof of Sklar's theorem. *Applied Mathematical Letters*, *26*, 945–948.

Durante, F., Fernández-Sánchez, J., & Sempi, C. (2013). Sklar's theorem obtained via regularization techniques. *Nonlinear Analysis: Theory, Methods & Applications*, *75*(2), 769–774.

Durante, F., & Sempi, C. (2005). *Principles of copula theory*. Boca Raton: Chapman and Hall/CRC.

Fermanian, J.-D. (2005). Goodness-of-fit tests for copulas. *Journal of Multivariate Analysis*, *95*(1), 119–152.
Fermanian, J.-D., & Scaillet, O. (2003). Nonparametric estimation of copulas for time series. *Journal of Risk*, *5*, 25–54.
Gaensler, P., & Stute, W. (1987). *Seminar on empirical processes*. Boca Raton: Springer Basel AG.
Genest, C., Ghoudi, K., & Rivest, L.-P. (1995). A semi-parametric estimation procedure of dependence parameters in multivariate families of distributions. *Biometrika*, *82*(3), 543–552.
Genest, C., Quessy, J.-F., & Rémillard, B. (2006). Goodness-of-fit procedures for copula models based on the probability integral transformation. *Scandinavian Journal of Statistics*, *33*, 337–366.
Genest, C., & Rèmillard, B. (2008). Validity of the parametric bootstrap for goodness-of-fit testing in semiparametric models. *Annales de l'Institut Henri Poincaré, Probabilités et Statistiques*, *6*(44), 1096–1127.
Genest, C., Rémillard, B., & Beaudoin, D. (2009). Goodness-of-fit tests for copulas: A review and a power study. *Insurance: Mathematics and Economics*, *44*, 199–213.
Genest, C., & Rivest, L.-P. (1989). A characterization of Gumbel family of extreme value distributions. *Statistics & Probability Letters*, *8*(3), 207–211.
Genest, C., & Rivest, L.-P. (1993). Statistical inference procedures for bivariate Archimedean copulas. *Journal of the American Statistical Association*, *88*(3), 1034–1043.
Górecki, J., Hofert, M., & Holeňa, M. (2016). An approach to structure determination and estimation of hierarchical Archimedean copulas and its application to bayesian classification. *Journal of Intelligent Information Systems*, *46*(1), 21–59.
Haff, I. H. (2013). Parameter estimation for pair-copula constructions. *Bernoulli*, *19*(2), 462–491.
Härdle, W. K., & Simar, L. (2015). *Applied multivariate statistical analysis*. Berlin: Springer.
Hering, C., Hofert, M., Mai, J.-F., & Scherer, M. (2010). Constructing hierarchical Archimedean copulas with Lévy subordinators. *Journal of Multivariate Analysis*, *101*(6), 1428–1433.
Hofert, M. (2011). Efficiently sampling nested Archimedean copulas. *Computational Statistics & Data Analysis*, *55*(1), 57–70.
Hofert, M., & Scherer, M. (2011). CDO pricing with nested Archimedean copulas. *Quantitative Finance*, *11*(5), 775–87.
Hull, J., & White, A. (2004). Valuation of a CDO and an n-th to default CDS without Monte Carlo simulation. *Journal of Derivatives*, *12*(2), 8–23.
Joe, H. (1996). Families of m-variate distributions with given margins and $m(m-1)/2$ bivariate dependence parameters. In L. Rüschendorf, B. Schweizer, & M. Taylor (Eds.), *Distribution with fixed marginals and related topics*., IMS Lecture Notes - Monograph Series Institute of Mathematical Statistics.
Joe, H. (1997). *Multivariate models and dependence concepts*. London: Chapman & Hall.
Joe, H. (2005). Asymptotic efficiency of the two-stage estimation method for copula-based models. *Journal of Multivariate Analysis*, *94*(2), 401–419.
Joe, H. (2014). *Dependence modeling with copulas*. Boca Raton: Chapman and Hall/CRC.
Johnson, R. A., & Wichern, D. W. (2013). *Applied multivariate statistical analysis* (6th ed.). Harlow: Pearson.
Jouini, M., & Clemen, R. (1996). Copula models for aggregating expert opinions. *Operation Research*, *3*(44), 444–457.
Kendall, M. (1970). *Rank correlation methods*. London: Griffin.
Krupskii, P., & Joe, H. (2013). Factor copula models for multivariate data. *Journal of Multivariate Analysis*, *120*, 85–101.
Kurowicka, M., & Cooke, R. M. (2006). *Uncertainty analysis with high dimensional dependence modelling*. New York: Wiley.
Laurent, J.-P., & Gregory, J. (2005). Basket default swaps, CDO's and factor copulas. *Journal of Risk*, *7*(4), 103–122.
Li, D. X. (2000). On default correlation: A copula function approach. *Journal of Fixed Income*, *9*, 43–54.

Luo, X., & Shevchenko, P. V. (2010). The t-copula with multiple parameters of degrees of freedom: Bivariate characteristics and application to risk management. *Quantitative Finance*, *10*, 1039–1054.

Mai, J.-F., & Scherer, M. (2013). What makes dependence modeling challenging? Pitfalls and ways to circumvent them. *Statistics & Risk Modeling*, *30*(4), 287–306.

Nelsen, R. B. (2006). *An introduction to copulas*. New York: Springer.

Oakes, D. (1994). Multivariate survival distributions. *Journal of Nonparametric Statistics*, *3*(3–4), 343–354.

Oh, D. H., & Patton, A. (2015). *Modelling dependence in high dimensions with factor copulas, Finance and Economics Discussion Series 2015–2051*. Washington: Board of Governors of the Federal Reserve System.

Oh, D. H., & Patton, A. J. (2013). Simulated method of moments estimation for copula-based multivariate models. *Journal of the American Statistical Association*, *108*(502), 689–700.

Okhrin, O., Okhrin, Y., & Schmid, W. (2013a). On the structure and estimation of hierarchical Archimedean copulas. *Journal of Econometrics*, *173*(2), 189–204.

Okhrin, O., Okhrin, Y., & Schmid, W. (2013b). Properties of hierarchical Archimedean copulas. *Statistics & Risk Modeling*, *30*(1), 21–54.

Okhrin, O., & Ristig, A. (2014). Hierarchical Archimedean copulae: The HAC package. *Journal of Statistical Software*, *58*(4), 1–20.

Patton, A., Fan, Y. & Chen, X. (2004). Simple tests for models of dependence between multiple financial time series, with applications to u.s. equity returns and exchange rates, Working paper.

Patton, A. J. (2012). A review of copula models for economic time series. *Journal of Multivariate Analysis*, *110*, 4–18.

Rachev, S., Stoyanov, S., & Fabozzi, F. (2008). *Advanced stochastic models, risk assessment, and portfolio optimization: The ideal risk, uncertainty, and performance measures*. New York: Wiley.

Radulovic, J.-D. F. D., & Wegkamp, M. (2004). Weak convergence of empirical copula processes. *Bernoulli*, *10*(5), 847–860.

Rezapour, M. (2015). On the construction of nested Archimedean copulas for d-monotone generators. *Statistics & Probability Letters*, *101*, 21–32.

Rosenblatt, M. (1952). Remarks on a multivariate transformation. *Annals of Mathematical Statistics*, *23*, 470–472.

Savu, C., & Trede, M. (2010). Hierarchies of Archimedean copulas. *Quantitative Finance*, *10*(3), 295–304.

Scaillet, O. (2007). Kernel-based goodness-of-fit tests for copulas with fixed smoothing parameters. *Journal of Multivariate Analysis*, *98*(3), 533–543.

Schreyer, M., Paulin, R., & Trutschnig, W. (2017). On the exact region determined by Kendall's τ and Spearman's ρ, to appear in: Journal of the Royal Statistical Society: Series B (Statistical Methodology).

Segers, J., & Uyttendaele, N. (2014). Nonparametric estimation of the tree structure of a nested Archimedean copula. *Computational Statistics and & Analysis*, *72*, 190–204.

Shih, J. H., & Louis, T. A. (1995). Inferences on the association parameter in copula models for bivariate survival data. *Biometrics*, *51*(4), 1384–1399.

Sklar, A. (1959). Fonctions de répartition à n dimension et leurs marges. *Publications de l'Institut de Statistique de l'Université de Paris*, *8*, 299–231.

Wang, W., & Wells, M. (2000). Model selection and semiparametric inference for bivariate failure-time data. *Journal of the American Statistical Association*, *95*(449), 62–76.

Whelan, N. (2004). Sampling from Archimedean copulas. *Quantitative Finance*, *4*(3), 339–352.

Zhang, S., Okhrin, O., Zhou, Q. M., & Song, P. X.-K. (2016). Goodness-of-fit test for specification of semiparametric copula dependence models. *Journal of Econometrics*, *193*(1), 215–233.

Zhu, W., Wang, C. -W. & Tan, K. S. (2016). Structure and estimation of Lévy subordinated hierarchical Archimedean copulas (LSHAC): Theory and empirical tests, Journal of Banking & Finance.

Chapter 14
Measuring and Modeling Risk Using High-Frequency Data

Wolfgang Karl Härdle, N. Hautsch and U. Pigorsch

Abstract Measuring and modelling financial volatility is the key to derivative pricing, asset allocation and risk management. The recent availability of high-frequency data allows for refined methods in this field. In particular, more precise measures for the daily or lower frequency volatility can be obtained by summing over squared high-frequency re- turns. In turn, this so called realized volatility can be used for more accurate model evaluation and description of the dynamic and distributional structure of volatility. Moreover, non-parametric measures of systematic risk are attainable, that can straightforwardly be used to model the commonly observed time-variation in the betas. The discussion of these new measures and methods is accompanied by an empirical illustration using high-frequency data of the IBM incorporation and of the DJIA index.

14.1 Introduction

Volatility modelling is the key to the theory and practice of pricing financial products. Asset allocation and portfolio as well as risk management depend heavily on a correct modelling of the underlying(s). This insight has spurred extensive research in

W.K. Härdle (✉)
C.A.S.E.-Center of Applied Statistics and Economics, Humboldt-Universität zu Berlin,
Unter den Linden 6, 10099 Berlin, Germany
e-mail: haerdle@hu-berlin.de

W.K. Härdle
Ladislaus von Bortkiewicz Chair of Statistics, School of Business and Economics,
Humboldt-Universität zu Berlin,
Unter den Linden 6, 10099 Berlin, Germany

N. Hautsch
Department of Statistics and Operations Research, University of Vienna as well as Center for Financial Studies(CFS), Frankfurt, Germany
e-mail: nikolaus.hautsch@univie.ac.at

U. Pigorsch
Schumpeter School of Business and Economics, University of Wuppertal, Wuppertal, Germany
e-mail: pigorsch@uni-wuppertal.de

W.K. Härdle et al. (eds.), *Applied Quantitative Finance*, Statistics and Computing,
DOI 10.1007/978-3-662-54486-0_14

financial econometrics and mathematical finance. Stochastic volatility models with separate dynamic structure for the volatility process have been in the focus of the mathematical finance literature, see Heston (1993) and Bates (2000), while parametric GARCH-type models for the returns of the underlying(s) have been intensively analyzed in financial econometrics.

The validity of these models in practice though depends upon specific distributional properties or the knowledge of the exact (parametric) form of the volatility dynamics. Moreover, the evaluation of the predictive ability of volatility models is quite important in empirical applications. However, the latent character of the volatility poses a problem. To what measure should the volatility forecasts be compared to? Conventionally, the forecasts of daily volatility models, such as GARCH-type or stochastic volatility models, have been evaluated with respect to absolute or squared daily returns. In view of the excellent in-sample performance of these models, the forecasting performance, however, seems to be disappointing.

The availability of ultra-high-frequency data opens the door for a refined measurement of volatility and model evaluation. An often used and very flexible model for logarithmic prices of speculative assets is the (continuous time) stochastic volatility model:

$$dY_t = (\mu + \beta\sigma_t)dt + \sigma_t dW_t, \tag{14.1}$$

where σ_t^2 is the instantaneous (spot) variance, μ denotes the drift, β is the risk premium, and W_t defines the standard Wiener process. The object of interest is the amount of variation accumulated in a time interval Δ (e.g., a day, week, month etc.). If $n = 1, 2, \ldots$ denotes a counter for the time intervals of interest, then the term

$$\sigma_n^2 = \int_{(n-1)\Delta}^{n\Delta} \sigma_t^2 dt \tag{14.2}$$

is called the actual volatility, see Barndorff-Nielsen and Shephard (2002b). The actual volatility is the quantity that reflects the market risk structure (scaled in Δ) and is the key element in pricing and portfolio allocation. Actual volatility (measured in scale Δ) is of course related to the integrated volatility:

$$V(t) = \int_0^t \sigma_s^2 ds \tag{14.3}$$

It is worth noting that there is a small notational confusion here: the mathematical finance literature would denote σ_t as "volatility" and σ_t^2 as "variance", see Nelson and Foster (1994). For example, an important result is that $V(t)$ can be estimated from Y_t via the quadratic variation:

$$[Y_t]_M = \sum (Y_{t_j} - Y_{t_{j-1}})^2, \tag{14.4}$$

where $t_0 = 0 < t_1 < \ldots < t_M = t$ is a sequence of partition points and $\sup_j |t_{j+1} - t_j| \to 0$. Andersen and Bollerslev (1998) have shown that

$$[Y_t]_M \xrightarrow{p} V(t), \quad M \to \infty. \tag{14.5}$$

This observation leads us to consider in an interval Δ with M observations

$$RV_n = \sum_{j=1}^{M} (Y_{t_j} - Y_{t_{j-1}})^2 \tag{14.6}$$

with $t_j = \Delta\{(n-1) + j/M\}$. Note that RV_n is a consistent estimator of σ_n^2 and is called *realized volatility*. Barndorff-Nielsen and Shephard (2002b) point out that $RV_n - \sigma_n^2$ is approximately mixed Gaussian and provide the asymptotic law of

$$\sqrt{M}(RV_n - \sigma_n^2). \tag{14.7}$$

The realized volatility turns out to be very useful in the assessment of the validity of volatility models. For instance, reconciling evidence in favor of the forecast accuracy of GARCH-type models is observed when using realized volatility as a benchmark rather than daily squared returns. Moreover, the availability of the realized volatility measure initiated the development of a new and quite accurate class of volatility models. In particular, based on the ex-post observability of the realized volatility measure, volatility is now treated as an observed rather than a latent variable to which standard time series procedures can be applied.

The remainder of this chapter is structured as follows. We first discuss the practical problems encountered in the empirical construction of realized volatility which are due to the existence of market microstructure noise. Section 14.3 presents the stylized facts of realized volatility, while Sect. 14.4 reviews the most popular realized volatility models. Section 14.5 illustrates the usefulness of the realized volatility concept for measuring time-varying systematic risk within a conditional asset pricing model (CAPM).

14.2 Market Microstructure Effects

The consistency of the realized volatility estimator builds on the notion that prices are observed in continuous time and without measurement error. In practice, however, the sampling frequency is inevitably limited by the actual quotation or transaction frequency. Since high-frequency prices are subject to market microstructure noise, such as price-discreteness, bid-and-ask bounce effects, transaction costs etc., the true price is unobservable. Market microstructure effects induce a bias in the realized volatility measure, which can straightforwardly be illustrated in the following simple discrete-time setup. Assume that the logarithmic high-frequency prices are observed with noise, i.e.,

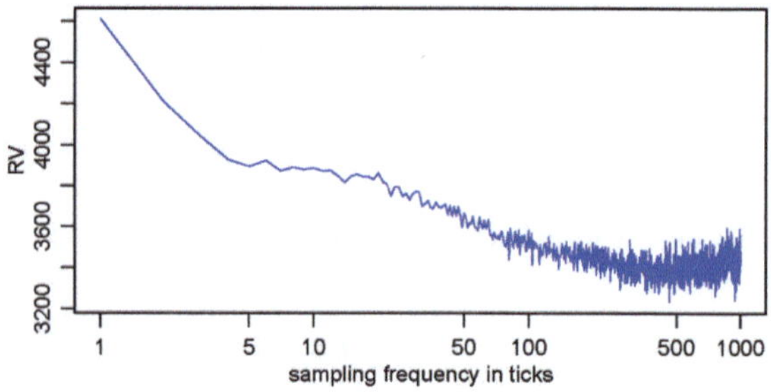

Fig. 14.1 Volatility signature plot for IBM, 2001–2006. Average time between trades: 6.78 s. XFGsignature

$$Y_{t_j} = Y^*_{t_j} + \varepsilon_{t_j}, \tag{14.8}$$

where $Y^*_{t_j}$ denotes the latent true price. Moreover, the microstructure noise ε_{t_j} is assumed to be iid distributed with mean zero and variance η^2, and is independent of the true return. Let $r^*_{t_j}$ denote the efficient return, then the high-frequency continuously compounded returns

$$r_{t_j} = r^*_{t_j} + \varepsilon_{t_j} - \varepsilon_{t_{j-1}} \tag{14.9}$$

follow an MA(1) process. Such a return specification is well established in the market microstructure literature and is usually justified by the existence of the bid-ask bounce effect, see, e.g., Roll (1984). In this model, the realized volatility is given by

$$RV_n = \sum_{i=1}^{M} (r^*_{t_j})^2 + 2\sum_{j=1}^{M} r^*_{t_j}(\varepsilon_{t_j} - \varepsilon_{t_{j-1}}) + \sum_{j=1}^{M} (\varepsilon_{t_j} - \varepsilon_{t_{j-1}})^2. \tag{14.10}$$

with

$$\mathsf{E}[RV_n] = \mathsf{E}[RV^*_n] + 2M\eta^2. \tag{14.11}$$

If the sampling frequency goes to infinity, we know from the previous section that RV^*_n consistently estimates σ^2_n and, thus, the realized volatility based on the observed price process is a biased estimator of the actual volatility with bias term $2M\eta^2$. Obviously, for $M \to \infty$, RV_n diverges.

This diverging behavior can also be observed empirically in so called volatility signature plots. Figure 14.1 shows the volatility signature for one stock of the IBM incorporation over the period ranging from January 2, 2001 to December 29, 2006. The plot depicts the average annualized realized volatility over the full sample period constructed at different frequencies measured in number of ticks (depicted in log scale). Obviously, the realized volatility is large at the very high frequency, but decays for lower frequencies and stabilizes around a sampling frequency of 300 ticks, which corresponds approximately to a 30 min sampling frequency, given that the average duration between two consecutive trades is around 6.78 s.

Thus, sampling at a lower frequency, such as every 10, 15 or 30 min, seems to alleviate the problem of market microstructure noise and has thus frequently been applied in the literature. This so-called *sparse sampling*, however, comes at the cost of a less precise estimate of the actual volatility. Alternative methods have been proposed to solve this bias-variance trade-off for the above simple noise assumption as well as for more general noise processes, allowing also for serial dependence in the noise and/or for dependence between the noise and the true price process, which is sometimes referred to as endogenous noise. A natural approach to reduce the market microstructure noise effect is to construct the realized volatility measure based on prefiltered high-frequency returns, using, e.g., an MA(1) model.

In the following we briefly present two more elaborate and under specific noise assumptions consistent procedures for estimating actual volatility. Both have been theoretically considered in several papers. The subsampling approach originally suggested by Zhang et al. (2005) builds on the idea of averaging over various realized volatilities constructed from different high-frequency subsamples. For the ease of exposition we focus again on one time period, e.g., one day, and denote the full grid of time points at which the M intradaily prices are observed by $\mathcal{G}_t = \{t_0, \ldots, t_M\}$. The realized volatility that makes use of all observations in the full grid is denoted by $RV_n^{(all)}$. Moreover, the grid is partitioned into L nonoverlapping subgrids $\mathcal{G}^{(l)}$, $l = 1, \ldots, L$. A simple way for selecting such a subgrid may be the socalled regular allocation, in which the l-th subgrid is given by $\mathcal{G}^{(l)} = \{t_{l-1}, t_{l-1+L}, \ldots, t_{l-1+M_l L}\}$ for $l = 1, \ldots, L$, and M_l denoting the number of observations in each subgrid. E.g., consider 5-min returns that can be measured at the time points 9:30, 9:35, 9:40, ..., and at the time points 9:31, 9:36, 9:41, ...and so forth. In analogy to the full grid, the realized volatility for subgrid l, denoted by $RV_n^{(l)}$, is constructed from all data points in subgrid l. Thus, $RV_n^{(l)}$ is based on sparsely sampled data.

The actual volatility is then estimated by:

$$RV_n^{(ZMA)} = \frac{1}{L}\sum_{l=1}^{L} RV_n^{(l)} - \frac{\overline{M}}{M} RV_n^{(all)}, \tag{14.12}$$

where $\overline{M} = \frac{1}{L}\sum_{l=1}^{L} M_l$. The latter term on the right-hand side is included to bias-correct the averaging estimator $\frac{1}{L}\sum_{l=1}^{L} RV_n^{(l)}$. As the estimator (14.12) consists of a component based on sparsely sampled data and one based on the full grid of price observations, the estimator is also called the *two-timescales estimator*.

Given the similarity to the problem of estimating the long-run variance of a stationary time series in the presence of autocorrelation, it is not surprising that kernel-based methods have been developed for estimating the realized volatility. Most recently, Barndorff-Nielsen et al. (2008) proposed the flat-top realized kernel estimator

$$RV_n^{(BHLS)} = RV_n + \sum_{h=1}^{H^*} K\left(\frac{h-1}{H^*}\right)(\widehat{\gamma}_h + \widehat{\gamma}_{-h}) \tag{14.13}$$

with

$$\widehat{\gamma}_h = \frac{M}{M-h}\sum_{j=1}^{M} r_{t_j} r_{t_{j-h}}, \tag{14.14}$$

and $K(0) = 1$, $K(1) = 0$. Obviously, the summation term on the righthand side is the realized kernel correction of the market microstructure noise. Zhou (1996), who was the first to consider realized kernels, proposed (14.13) with $H = 1$, while Hansen and Lunde (2006) allowed for general H but restricted $K(x) = 1$. Both of these estimators, however, have been shown to be inconsistent. Barndorff-Nielsen et al. (2008) instead propose several consistent realized kernel estimators with an optimally chosen H^*, such as the Tukey-Hanning kernel, i.e. $K(x) = \{1 - \cos \pi(1-x)^2\}/2$, which performs also very well in terms of efficiency as illustrated in a Monte Carlo analysis. They further show, that these realized kernel estimators are robust to market microstructure frictions that may induce endogenous and dependent noise terms.

14.3 Stylized Facts of Realized Volatility

Figure 14.2 shows kernel density estimates of the plain and logarithmic daily realized volatility in comparison to plots of a correspondingly fitted (log) normal distribution based on the IBM data, 2001–2006. The pictures in the top of Fig. 14.2 show the unconditional distribution of the (plain) realized volatility in contrast to a fitted normal distribution. As also confirmed by the corresponding descriptive statistics displayed by Table 14.1, we observe that realized volatility reveals severe right-skewness and excess kurtosis. This result might be surprising given that the realized volatility consists of the sum of squared intra-day returns and thus central limit theorems should apply. However, it is a common finding that intra-day returns are strongly serially dependent requiring significantly higher intra-day sampling frequencies to observe convergence to normality. In contrast, the unconditional distribution of the logarithmic realized volatility is well approximated by a normal distribution. The sample kurtosis is strongly reduced and is close to 3. Though slight right-skewness and deviations from normality in the tails of the distribution are still observed, the underlying distribution is remarkably close to that of a Gaussian distribution.

A common finding is that financial returns have fatter tails than the normal distribution and reveal significant excess kurtosis. Though GARCH models can explain excess kurtosis, they cannot completely capture these properties in real data. Consequently, (daily) returns standardized by GARCH-induced volatility, typically still show clear deviations from normality. However, a striking result in recent literature is that return series standardized by the square root of realized volatility, $r_n/\sqrt{RV_n}$, are quite close to normality. This result is illustrated by the plots in the bottom of

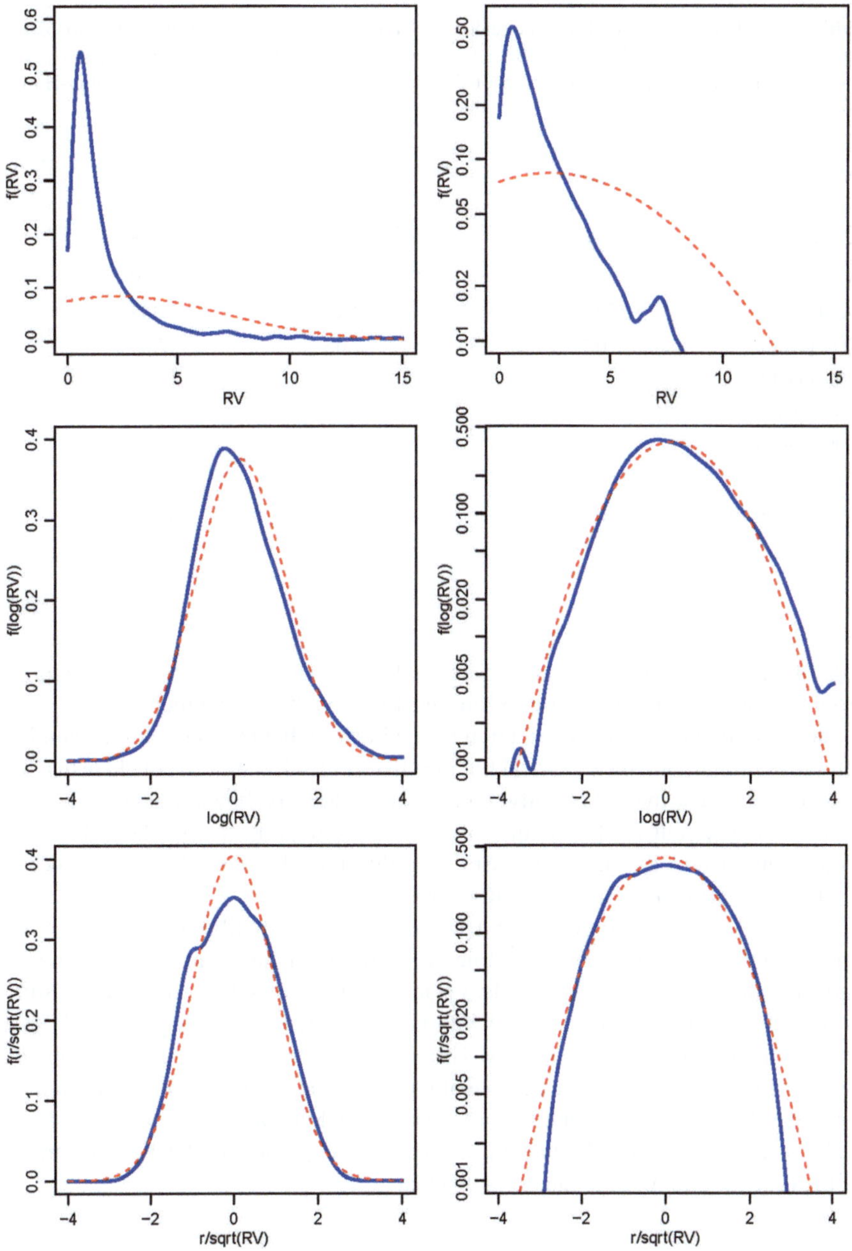

Fig. 14.2 Kernel density estimates of the (logarithmic) realized volatility and of correspondingly standardized returns for IBM, 2001–2006. The *dotted line* depicts the density of the correspondingly fitted normal distribution. The *left* column depicts the kernel density estimates based on a log scale. XFGkernelcom

Table 14.1 Descriptive statistics of the realized volatility, log realized volatility and standardized returns, IBM stock, 2001–2006. LB (40) denotes the Ljung-Box statistic based on 40 lags. The last row gives an estimate of the order of fractional integration based on the Geweke and Porter-Hudak estimator

	RV_n	$\ln RV_n$	$r_n/\sqrt{RV_n}$
Mean	2.26	0.14	−0.000
Median	1.05	0.05	−0.013
Skewness	9.93	0.42	0.035
Variance	22.57	1.13	0.979
Kurtosis	150.47	3.43	2.349
1%-quantile	0.13	−2.03	−1.980
5%-quantile	0.24	−1.41	−1.558
95%-quantile	7.58	2.00	1.628
99%-quantile	17.66	2.87	2.141
LB(40)	2140.48	14213.07	39.780
p-value LB(40)	0.00	0.00	0.480
$\widehat{d}$	0.38	0.62	–

Fig. 14.2 and the descriptive statistics in Table 14.1. Though we observe deviations from normality for returns close to zero resulting in a kurtosis which is even below 3, the fit in the tails of the distribution is significantly better than that for plain log returns. Summarizing the empirical findings from Fig. 14.2, we can conclude that the unconditional distribution of daily returns is well described by a lognormal-normal mixture. This confirms the mixture-of-distribution hypothesis by Clark (1973) as well as the idea of the basic stochastic volatility model, where the log variance is modelled in terms of a Gaussian AR(1) process.

Figure 14.3 shows the evolvement of daily realized volatility over the analyzed sample period and the implied sample autocorrelation functions (ACFs). As also shown by the corresponding Ljung-Box statistics in Table 14.1, the realized volatility is strongly positively autocorrelated with high persistence. This is particularly true for the logarithmic realized volatility. The plot shows that the ACF decays relatively slowly providing hints on the existence of long range dependence. Indeed, a common finding is that the realized volatility processes reveal long range dependence which is well captured by fractionally integrated processes. In particular, if RV_n is integrated of the order $d \in (0, 0.5)$, it can be shown that

$$\mathsf{Var}\left[\sum_{j=1}^{h} RV_{n+j}\right] \approx ch^{2d+1}, \tag{14.15}$$

with c denoting a constant. Then, plotting $\ln \mathsf{Var}\left[\sum_{j=1}^{h} RV_{n+j}\right]$ against ln h should result in a straight line with slope $2d + 1$. Most empirical studies strongly confirm

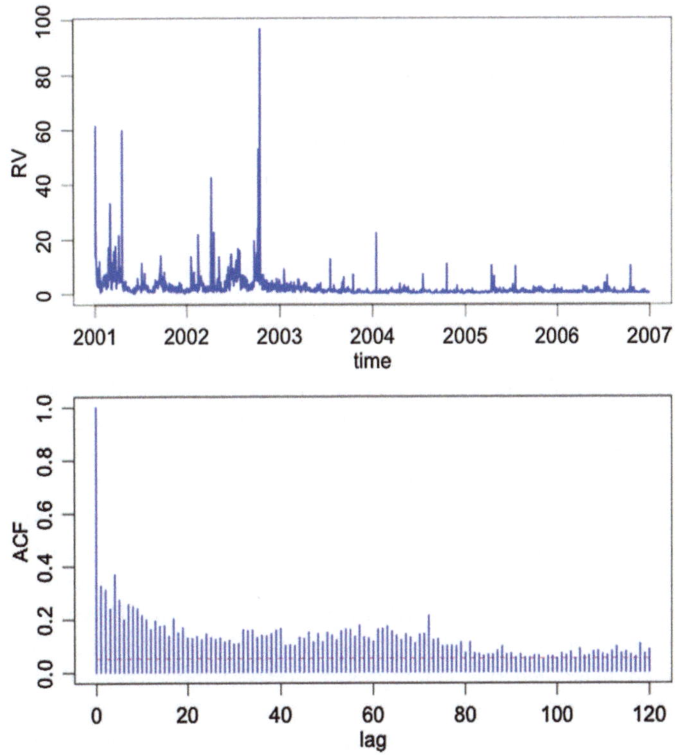

Fig. 14.3 Time evolvement and sample autocorrelation function of the realized volatility for IBM, 2001–2006. XFGrvtsacf

this relationship and find values for d between 0.35 and 0.4 providing clear evidence for long range dependence. Estimating d using the Geweke and Porter-Hudak estimator, we obtain $\widehat{d} = 0.38$ for the series of realized volatilities and $\widehat{d} = 0.62$ for its logarithmic counterpart. Hence, for both series we find clear evidence for long range dependence. However, the persistence in logarithmic realized volatilities is remarkably high providing even hints on non-stationarity of the process.

Summarizing the most important empirical findings, we can conclude that the unconditional distributions of logarithmic realized volatility and of correspondingly standardized log returns are well approximated by normal distributions and that realized volatility itself follows a long memory process. These results suggest (Gaussian) ARFIMA models as valuable tools to model and to predict (log) realized volatility.

14.4 Realized Volatility Models

As illustrated above, realized volatility models should be able to capture the strong persistence in the sample autocorrelation function. While this seemingly long-memory pattern is widely acknowledged, there is still no consensus on the mechanism generating it. One approach is to assume that the long memory is generated by a fractionally integrated process as originally introduced by Granger and Joyeux (1980) and Hosking (1981). In the GARCH literature this has lead to the development of the fractionally integrated GARCH model as, e.g., proposed by Baillie et al. (1996). For realized volatility the use of a fractionally integrated autoregressive moving average (ARFIMA) process was advocated, for example, by Andersen et al. (2003). The ARFIMA (p, q) model is given by

$$\phi(L)(1-L)^d(y_n-\mu)=\psi(L)u_n, \tag{14.16}$$

with $\phi(L)=1-\phi_1 L-\ldots-\phi_p L^p,\ \psi(L)=1+\psi_1 L+\ldots\psi_q L^q$, and d denoting the fractional difference parameter. Moreover, u_n is usually assumed to be a Gaussian white noise process, and yn denotes either the realized volatility (see Koopman et al. 2005) or its logarithmic transformation. Several extensions of the realized volatility ARFIMA model have been proposed, accounting, for example, for leverage effects (see Martens et al. 2004), for non-Gaussianity of (log) realized volatility or for time-variation in the volatility of realized volatility (see Corsi et al. 2008). Generally the empirical results show significant improvements in the point forecasts of volatility when using ARFIMA rather than GARCH-type models.

An alternative model for realized volatility has been suggested by Corsi (2009). The so-called heterogeneous autoregressive (HAR) model of realized volatility approximates the long-memory pattern by a sum of multi-period volatility components. The simulation results in Corsi (2009) show, that the HAR model can quite adequately reproduce the hyperbolic decay in the sample autocorrelation function of realized volatility even if the number of volatility components is small. For the HAR model, let the $k$period realized volatility component be defined by the average of the single-period realized volatilities, i.e.,

$$RV_{n+1-k:n}=\frac{1}{k}\sum_{j=1}^{k}RV_{n-j}. \tag{14.17}$$

The HAR model with the so-defined daily, weekly and monthly realizedvolatility components, is given by

$$\begin{aligned}\log RV_n &= \alpha_0+\alpha_d \log RV_{n-1}+\alpha_w \log RV_{n-5:n-1}\\ &+\alpha_m \log RV_{n-21:n-1}+u_n,\end{aligned} \tag{14.18}$$

with u_n typically being a Gaussian white noise. The HAR model has become very popular due to its simplicity in estimation and its excellent in-sample fit and predictive ability (see e.g. Andersen et al. 2003; Corsi et al. 2008). Several extensions exist and deal, for example, with the inclusion of jump measures (see Andersen et al. 2003) or non-linear specifications based on neural networks (see Hillebrand and Medeiros 2007).

Alternative realized volatility models have been proposed in, e.g., Barndorff-Nielsen and Shephard (2002a), who consider a superposition of Ornstein Uhlenbeck processes, and in Deo et al. (2006), who specify a long-memory stochastic volatility model. A recent and comprehensive review on realized volatility models can also be found in McAleer and Medeiros (2008b).

14.5 Time-Varying Betas

So far, our discussion focused on the measurement and modeling of the volatility of a financial asset using high-frequency transaction data. From a pricing perspective, however, systematic risk is most important. In this section, we therefore discuss, how high-frequency information can be used for the evaluation and modeling of systematic risk. A common measure for the systematic risk is given by the so-called (market) beta, which represents the sensitivity of a financial asset to movements of the overall market. As the beta plays a crucial role in asset pricing, investment decisions, and the evaluation of the performance of asset managers, a precise estimate and forecast of betas is indispensable. While the unconditional capital asset pricing model implies a linear and stable relationship between the asset's return and the systematic risk factor, i.e., the return of the market, empirical results suggest that the beta is time-varying, see, for example, Bos and Newbold (1984), and Fabozzi and Francis (1978). Similar evidence has been found for multi-factor asset pricing models, where the factor loadings seem to be time-varying rather than constant. A large amount of research has therefore been devoted to conditional CAPM and APT models, which allow for time-varying factor loadings, see, for example, Dumas and Solnik (1995), Ferson and Harvey (1991), Ferson and Harvey (1993), and Ferson and Korajczyk (1995).

14.5.1 The Conditional CAPM

Below we consider the general form of the conditional CAPM. A similar discussion for multi-factor models can be found in Bollerslev and Zhang (2003). Assume that the continuously compounded return of a financial asset i from period n to $n+1$ is generated by the following process

$$r_{i;n+1} = \alpha_{i;n+1|n} + \beta_{i;n+1|n} r_{m;n+1} + u_{i;n+1}, \tag{14.19}$$

with $r_{m;n+1}$ denoting the excess market return and $\alpha_{n+1|n}$ denoting the intercept that may be time-varying conditional on the information set available at time n, as indicated by the subscript. The idiosyncratic risk u_{n+1} is serially uncorrelated, $\mathsf{E}_n(u_{n+1}) = 0$, but may exhibit conditionally time-varying variance. Note that $\mathsf{E}_n(\cdot)$ denotes the expectation conditional on the information set available at time n. Moreover, we assume that $\mathsf{E}(r_{m;n+1}u_{n+1}) = 0$ for all n. The conditional beta coefficient of the CAPM regression (14.19) is defined as

$$\beta_{i;n+1|n} = \frac{\mathsf{Cov}(r_{i;n+1}, r_{m;n+1})}{\mathsf{Var}(r_{i;n+1})}. \tag{14.20}$$

Now, assume that lending and borrowing at a one-period risk-free rate $r_{f;n}$ is possible. Then, the arbitrage-pricing theory implies that the conditional expectation of the next period's return at time n is given by

$$\mathsf{E}_\mathsf{n}(r_{i;n+1}) = r_{f;n} + \beta_{i;n+1|n}\mathsf{E}_\mathsf{n}(r_{m;n+1}). \tag{14.21}$$

Thus, the computation of the future return of asset i requires to specify how the beta coefficient evolves over time.

The most common approach to allow for time-varying betas is to re-run the CAPM regression in each period based on a sample of 3 or 5 years. We refer to this as the rolling regression (RR) method. More elaborate estimates of the beta can be obtained using the Kalman-filter, which builds on a statespace representation of the conditional CAPM or by specifying a dynamic model for the covariance matrix between the return of asset i and the market return.

14.5.2 Realized Betas

The evaluation of the in-sample fit and predictive ability of various beta models is also complicated by the unobservability of the true beta. Consequently, model comparisons are usually conducted in terms of implied pricing errors, i.e., $e_{i,n+1} = \widehat{r}_{i,n+1} - r_{i,n+1}$, with $\widehat{r}_{i,n+1} = r_{f;n} + \widehat{\beta}_{i;n+1|n}\mathsf{E}_n(r_{m;n+1})$. Owing to the discussion on the evaluation of volatility models, the question arises, whether high-frequency data may also be useful for the evaluation of competing beta estimates. The answer is a clear "yes". In fact, high-frequency based estimates of betas are quite informative for the dynamic behavior of systematic risk. The construction of so-called *realized betas* is straightforward and builds on realized covariance and realized volatility measures. In particular, denote the realized volatility of the market by $RV_{m;n}$ and the realized covariance between the market and asset i by $RCov_{m,i;n} = \sum_{j=1}^{M} r_{i,t_j} r_{m,t_j}$, where r_{i,t_j} and r_{m,t_j} denote the j-th high-frequency return of the asset and the market, respectively, during day n. The realized beta is then defined as

$$\widehat{\beta}_{HF;i;n} = \frac{RCov_{m,i;n}}{RV_{m;n}}. \tag{14.22}$$

Barndorff-Nielsen and Shephard (2004) show that the realized beta converges almost surely for all n to the integrated beta over the time period from $n-1$ to n, i.e., the daily systematic risk associated with the market index. Note that the realized beta can also be obtained from a simple regression of the highfrequency returns of asset i on the high-frequency returns of the market, see, e.g., Andersen et al. (2006). The preciseness of the realized beta estimator can easily be assessed by constructing the $(1-\alpha)$-percent confidence intervals, which have been derived in Barndorff-Nielsen and Shephard (2004) and are given by

$$\widehat{\beta}_{HF;i;n} \pm z_{\alpha/2} \sqrt{\left(\sum_{j=1}^{M} r_{m,t_j}^2\right)^{-2} \widehat{g}_{i;n}}, \tag{14.23}$$

where $z_{\alpha/2}$ denotes the $(\alpha/2)$-quantile of the standard normal distribution,

$$\widehat{g}_{i;n} = \sum_{j=1}^{M} x_{i;j}^2 - \sum_{j=1}^{M-1} x_{i;j} x_{i;j+1}, \tag{14.24}$$

and

$$x_{i;j} = r_{i,t_j} r_{m,t_j} - \widehat{\beta}_{HF;i;n} r_{m,t_j}^2. \tag{14.25}$$

The upper panel in Fig. 14.4 presents the time-evolvement of the monthly realized beta for IBM incorporation over the period ranging from 2001 to 2006. We use the Dow Jones Industrial Average Index as the market index and construct the realized betas using 30 min returns. The graph also shows the 95%-confidence intervals of the realized beta estimator. The time-varying nature of systematic risk emerges strikingly from the figure and provides once more evidence for the relevance of its inclusion in asset pricing models.

Interestingly, the sample autocorrelation function of the realized betas depicted in the lower panel of Fig. 14.4 indicates significant serial correlation over the short horizon. This dependency can be explored for the prediction of systematic risk. Bollerslev and Zhang (2003), for example, find that an autoregressive model for the realized betas outperforms the RR approach both in terms of forecast accuracy as well as in terms of pricing errors.

14.6 Summary

We review the usefulness of high-frequency data for measuring and modeling actual volatility at a lower frequency, such as a day. We present the realized volatility as an estimator of the actual volatility along with the practical problems arising in the implementation of this estimator. We show that market microstructure effects induce a bias to the realized volatility and we discuss several approaches for the alleviation of this problem. The realized volatility is a more precise estimator of the actual volatility than the conventionally used daily squared returns, and thus provides more

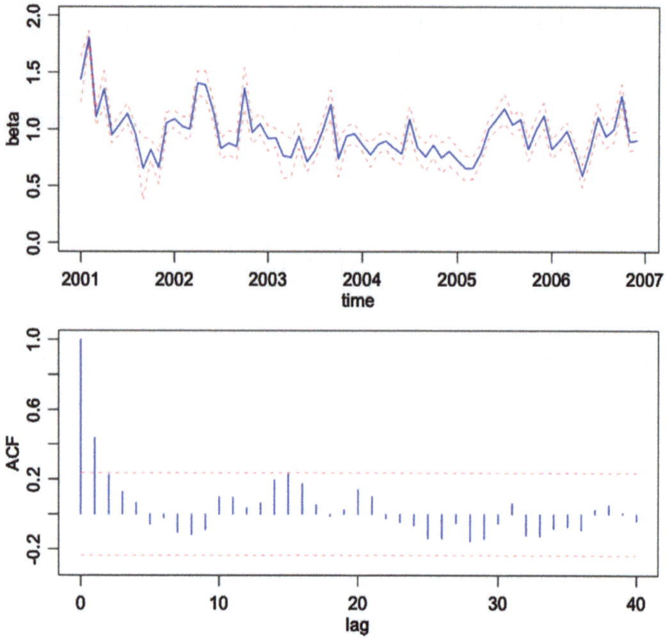

Fig. 14.4 Time evolvement and sample autocorrelation function of the realized volatility for IBM, 2001–2006. XFGbetatsacf

accurate information on the distributional and dynamic properties of volatility. This is important for many financial applications, such as asset pricing, portfolio allocation or risk management. As a consequence, several modeling approaches for realized volatility exist and have been shown to usually outperform traditional GARCH or stochastic volatility models, both in terms of in-sample as well as out-of-sample performance. We further demonstrate the usefulness of the realized variance and covariance estimator for measuring and modeling systematic risk. For the empirical examples provided in this chapter we use tick-by-tick transaction data of one stock of the IBM incorporation and of the DJIA index.

References

Andersen, T. G., & Bollerslev, T. (1998). Answering the skeptics: Yes standard volatility models do provide accurate forecasts. *International Economic Review*, *39*, 885–905.

Andersen, T. G., Bollerslev, T., Diebold, F. X., & Labys, P. (2003). Modeling and forecasting realized volatility. *Econometrica*, *71*, 579–625.

Andersen, T. G., Bollerslev, T., Diebold, F. X., & Wu, J. (2006). Realized beta: Persistence and predictability. In T. Fomby (Ed.), *Advances in econometrics: econometric analysis of economic and financial time series, volume B* (pp. 1–40).

Baillie, R. T., Bollerslev, T., & Mikkelsen, H. O. (1996). Fractionally integrated generalized autoregressive conditional heteroskedasticity. *Journal of Econometrics, 74*, 3–30.
Barndorff-Nielsen, O. E., & Shephard, N. (2002a). Econometric analysis of realised volatility and its use in estimating stochastic volatility models. *Journal of the Royal Statistical Society, Series B, 64*, 253–280.
Barndorff-Nielsen, O. E., & Shephard, N. (2002b). Estimating quadratic variation using realized variance. *Journal of Applied Econometrics, 4*(5), 457–477.
Barndorff-Nielsen, O. E., & Shephard, N. (2004). Econometric analysis of realized covariation: high frequency based covariance. *Regression, and Correlation in Financial Economics, Econometrica, 72*, 885–925.
Barndorff-Nielsen, O. E., Hansen, P. R., Lunde, A., & Shephard, N. (2008). Designing realised kernels to measure the ex-post variation of equity prices in the presence of noise. *Econometrica*, forthcoming.
Bates, D. S. (2000). Post-'87 crash fears in the S&P 500 futures option. *Journal of Econometrics, 94*(1–2), 181–238.
Bollerslev, T., & Zhang, B. Y. B. (2003). Measuring and modeling systematic risk in factor pricing models using high-frequency data. *Journal of Empirical Finance, 10*, 533–558.
Bos, T., & Newbold, P. (1984). An empirical investigation of the possibility of stochastic systematic risk in the market model. *Journal of Business, 57*, 35–41.
Clark, P. K. (1973). A subordinated stochastic process model with finite variance for speculative prices. *Econometrica, 41*, 135–156.
Corsi, F. (2009). *A simple long memory model of realized volatility. Journal of Financial Econometrics,7*, 174–196.
Corsi, F., Mittnik, S., Pigorsch, C., & Pigrosch, U. (2008). The volatility of realized volatility. *Econometric Reviews, 27*, 46–78.
Deo, R., Hurvich, C., & Lu, Y. (2006). Forecasting realized volatility using a long-memory stochastic volatility model: Estimation, prediction and seasonal adjustment. *Journal of Econometrics, 131*(1–2), 29–58.
Dumas, B., & Solnik, B. (1995). The world price of exchange rate risk. *Journal of Finance, 50*, 445–480.
Fabozzi, F. J., & Francis, J. C. (1978). Beta as a random coefficient. *Journal of Financial and Quantitative Analysis, 13*, 101–116.
Ferson, W. E., & Harvey, C. R. (1993). The risk and predictability of international equity returns. *Review of Financial Studies, 6*, 527–566.
Ferson, W. E., & Harvey, C. R. (1991). The variation of economic risk premiums. *Journal of Poltical Economy, 99*, 385–415.
Ferson, W. E., & Korajczyk, R. A. (1995). Do arbitrage pricing models explain the predictability of stock returns? *Journal of Business, 68*, 309–349.
Granger, C. W. J. (1980). Long memory relationships and the aggregation of dynamic models. *Journal of Econometrics, 14*, 227–238.
Granger, C. W. J., & Joyeux, R. (1980). An introduction to long-range time series models and fractional differencing. *Journal of Time Series Analysis, 1*, 15–30.
Granger, C. W. J., & Teräsvirta, T. (1999). A simple nonlinear time series model with misleading linear properties. *Economic Letters, 62*, 161–165.
Hansen, P. R., & Lunde, A. (2006). Realized variance and market microstructure noise. *Journal of Business & Economic Statistics, 24*, 127–161.
Heston, S. L. (1993). A closed-form solution for options with stochastic volatility with applications to bond and currency options. *The Review of Financial Studies, 6*(2), 327–343.
Hillebrand, E., & Medeiros, M. (2007). *Forecasting realized volatility models: The benefits of bagging and non-linear specifications*. Louisana State University, Working Paper.
Hosking, J. R. M. (1981). Fractional differencing. *Biometrika, 68*, 165–176.

Koopman, S. J., Jungbacker, B., & Hall, E. (2005). Forecasting daily variability of the S&P 100 stock index using historical, realised and implied volatility measurements. *Journal of Empirical Finance*, *12*(3), 445–475.

Martens, M., & Zein, J. (2004). Predicting financial volatility: High-frequency time-series forecasts vis-à-vis implied volatility. *Journal of Futures Markets*, *11*, 1005–1028.

Martens, M., van Dijk, D., & dePooter, M. (2004). *Modeling and forecasting S&P 500 volatility: Long memory, structural breaks and nonlinearity*. Erasmus University Rotterdam, Working Paper.

McAleer, M., & Medeiros, M. (2008a). A multiple regime smooth transition heterogenous autoregressive model for long memory and asymmetries. *Journal of Econometrics, 147*, 104–119.

McAleer, M., & Medeiros, M. (2008b). Realized volatility: A review. *Econometric Reviews*, *26*, 10–45.

Müller, U. A., Dacorogna, M. M., Dav, R. D., Olsen, R. B., Pictet, O. V., & von Weizsäcker, J. E. (1997). Volatilities of different time resolutions-analyzing the dynamics of market components. *Journal of Empirical Finance*, *4*(2–3), 213–239.

Nelson, D. B., & Foster, D. P. (1994). Asymptotic filtering theory for univariate ARCH models. *Econometrica*, *62*, 1–41.

Roll, R. (1984). A simple implicit measure of the effective bid-ask spread in an efficient market. *Journal of Finance*, *39*, 1127–1139.

Zhang, L., Mykland, P. A., & Ait-Sahalia, Y. (2005). A tale of two time scales: Determining integrated volatility with noisy high-frequency data. *Journal of the American Statistical Association*, *100*(472), 1394–1411.

Zhou, B. (1996). High-frequency data and volatility in foreign-exchange rates. *Journal of Business & Economic Statistics*, *14*, 45–52.

Chapter 15
Measuring Financial Risk in Energy Markets

S. Žiković

Abstract We investigate the relative performance of a wide array of Value at risk (VaR) and Expected Tail Loss (ETL) risk models in the energy commodities markets. The risk models are tested on a sample of daily spot prices of WTI oil, Brent oil, natural gas, heating oil, coal and uranium yellow cake during the recent global financial crisis. The analysed sample includes periods of backwardation and contango. After obtaining the VaR and ETL estimates we proceed to evaluate the statistical significance of the differences in performance of the analysed risk models. We employ a novel methodology for comparing VaR performance allowing us to rank competing models. Our simulation results show that for a significant number of different VaR models there is no statistical difference in the performance.

15.1 Introduction

Energy commodities are constantly at a centre stage of the global financial and geopolitical interest. The multiplicative effect of energy commodities on electricity, agricultural and industrial production makes protecting against commodity risk associated energy prices a necessity. This applies not only to energy producers and users but also to financial institutions and a wide spectrum of players from different industries. Looking from the financial modelling perspective it is hard to treat energy commodities as a single asset class since their specificities and respective markets differ significantly. The main differences refer to the influence of geopolitics, ecological issues, storage costs, safety issues, spatial distance between production and consumption sites and geographical dispersion. For these reasons energy commodities usually display higher volatility, fatter tails and skewness compared to classical financial assets. Hedging against energy price changes is equally important to buyers and sellers/producers of energy, which protect their businesses from rising and/or falling energy prices as it is to the financial sector, where commodities serve as an alternative investment vehicle. In order to protect the company's business against

S. Žiković (✉)
Faculty of Economics, University of Rijeka, Rijeka, Croatia
e-mail: szikovic@efri.hr

W.K. Härdle et al. (eds.), *Applied Quantitative Finance*, Statistics and Computing,
DOI 10.1007/978-3-662-54486-0_15

commodity risks the first step would be to correctly evaluate the market risk of energy commodities. A reliable energy risk forecasting model is essential for this task.

Value at Risk (VaR) and Expected Tail Loss (ETL) have established themselves as an essential risk management tool in the financial industry. Same as with other asset classes, VaR/ETL can be used to quantify the market risk of energy commodities associated with the specific probability level. Mining and energy companies undertake natural hedges but it usually not sufficient and a proactive approach to hedging and risk management is required. With the use of VaR and/or ETL it is possible to differentiate between risks which are negligible and those that require hedging. In light of the dramatic and protracted fall in the prices of fossil fuels we will focus on the risks facing energy producers i.e. risks from holding a long position in energy.

The issue of energy hedging has been well studied in the energy economics literature. Among others, Agnolucci (2009) studied the market volatility of WTI and found that asymmetric GARCH models outperform implied volatility models in terms of predictive accuracy. Cheong (2009) investigated the out-of-sample performance of four GARCH models under three loss functions, finding that the simplest and most parsimonious GARCH model provides a superior fit to Brent oil data. On the other hand, a complex FIAPARCH out-of-sample WTI oil forecasts provided superior performance. Wei et al. (2010) claim that no model can outperform all of the other models for Brent and WTI markets across different loss functions. They find the nonlinear GARCH models, which are capable of capturing long-memory and asymmetric volatility, exhibit solid forecasting accuracy, especially in over longer time periods. Mohammadi and Su (2010) considered oil spot prices in eleven markets and compared the forecasting accuracy of four GARCH-class models under two loss functions.

As opposed to the energy commodity volatility that has been widely studied, there is only a limited number of papers dealing with energy price risk management. Hung et al. (2008) highlight the importance of selecting the appropriate distribution in a GARCH volatility context. They found that the VaR of crude oil and oil products is adequately captured by fat-tailed distributions. Marimoutou et al. (2009) found that extreme value based models perform well in the oil markets and that they offer a major improvement over the traditional (non-parametric and parametric) methods. Bunn et al. (2013) showed that a structural linear quantile regression model outperforms skewed t GARCH and CAViaR models regarding the accuracy of out-of-sample forecasts. A number of authors found long range memory in energy returns and report as their top VaR performers models based on this characteristic (Aloui 2008; Mabrouk 2011). In recent studies Žiković et al. (2015) and Žiković and Tomas Žiković (2016) analysed the statistical significance of the differences in performance of a wide range of VaR models by employing a simulation-based methodology. VaR/ETL model performance was tested on Natural gas, Brent, WTI, coal, uranium yellow cake and heating oil contracts. They found that for a large number of different VaR models there is no statistical difference in performance. Overall the findings reported in the VaR and ETL literature on the topic of energy commodities are not conclusive. A similar situation and findings can also be found in the electricity price forecasting literature.

We add to previous research on energy risk measurement by investigating whether there are some identifiable common model features that yield consistently superior results under both risk metrics and at the same time investigate whether there is any significant statistical difference in performance of analysed VaR/ETL models.

The rest of the paper is organized in the following manner: Sect. 15.2 presents the data and the methodology, with emphasis on risk ranking procedure that is used in our analysis. Section 15.3 presents and discusses the VaR and ETL backtesting results. Section 15.4 concludes.

15.2 Methodology and Data

We analyze the performance of 10 VaR and 7 ETL models with their definitions summarized in Table 15.1. Tested VaR models are: simple moving average (VCV), the RiskMetrics approach, historical simulation (HS 100, 250 and HS 500; the number indicates the window length used to compute VaR), mirrored historical simulation (MHS 100, 250 and MHS 500), BRW (Boudoukh, Richardson, Whitelaw) simulation with the usually used decay factors of 0.97 and 0.99 and the approach proposed by Žiković and Aktan (2011) with individually optimized decay factors, GARCH model, filtered historical simulation (FHS), Hull and White (1998) approach and the conditional EVT approach (EVT GARCH) using the generalized Pareto distribution (GPD). Tested ETL models are: VCV (Gumbel and Frchet distribution), RiskMetrics (Gumbel distribution), bootstrapped historical simulation, bootstrapped mirrored historical simulation, bootstrapped BRW and FHS approach and conditional extreme value (EVT-GARCH) approach. For validation purposes we employ the log-daily spot returns on natural gas (NG1 Henry Hub), Brent, WTI, uranium 5% yellow cake (UXA1), heating oil (HO1 NYMEX) and US low sulphur coal - Big Sandy Barge Fob (COALBGSD).

In the risk management arena there are several approaches to testing whether a risk model is superior to others. Some of them are: Diebold and Mariano (1995) Equal Predictive Ability (EPA), White (2000) Reality Check test (RC) and Hansen (2005) Superior Predictive Ability (SPA). All of them investigate whether any alternative forecast is better than the benchmark, or in another way, whether the best alternative forecasting model is better than the benchmark. This question can be addressed by testing the hypothesis that the benchmark is not inferior to any alternative forecast. Using such tests is useful for exploring if there is a better forecasting model than the one currently used. We employ the methodology for comparing VaR model performance developed by Žiković and Filer (2013) allowing for consistent ranking of competing VaR models based on several general assumptions.

To implement the forecast evaluation proposed by Žiković and Filer (2013) it is necessary to specify the loss function. A number of loss functions have been proposed in the risk management literature. A very intuitive, simple and symmetric loss function was proposed by Lopez (1999). It allows for the sizes of tail losses to influence the models final rating. Risk model that generates the same number of

Table 15.1 Definitions of analysed VaR and ETL models

Model	VaR	ETL	Description
Historical simulation	$VaR^{cl} = F^{-1}(cl) = X_{(i)}$	$ETL^{cl} = \left(\sum_{i=[ncl]}^{n} X_{n(i)}\right)/(n-[ncl])$	$F_n(t) = \frac{1}{n}\sum_{i=1}^{n} I(X_i \le t)$
Mirrored HD	$Y_i = \lvert X_i \rvert$ $VaR_t^{cl} = F^{-1}(cl) = Y_{(i)}$	$ETL^{cl} = \left(\sum_{i=[ncl]}^{n} Y_{n(i)}\right)/(n-[ncl])$	
BRW simulation	$G(x;t,N) = \sum_{i=1}^{N} 1_{\{r_{t-i} \le x\}} w_{t-i}$ $VaR_t^{cl} = (r \in \{r_{t-1}, ..., r_{t-1-N}\})\lvert$ $G(r;t,N) \ge cl$	$ETL_t^{cl} = \left(\sum_{i=[ncl]}^{n} X_{n(i)}\right)/(n-[ncl])$	$\{w\} = \frac{1-\lambda}{1-\lambda^N}, ..., \left(\frac{1-\lambda}{1-\lambda^N}\right)\lambda^{N-1}$
VCV	$VaR_t^{cl} = \mu_t + \sigma_t \alpha_{cl}$	$ETL^{cl} = \mu_t + \sigma_t E[Z \lvert Z < z_{cl}]$	$\sigma_t = \sqrt{\frac{1}{T}\sum_{t=1}^{T}(r_t - \overline{r})}$
RiskMetrics	$VaR_t^{cl} = \mu_t + \sigma_t \alpha_{cl}$	$ETL^{cl} = \mu_t + \sigma_t E[Z \lvert Z < z_{cl}]$	$\sigma_t = \sqrt{0.94\sigma_{t-1}^2 + 0.06\varepsilon_t^2}$
GARCH	$VaR_t^{cl} = \mu_t + \sigma_t \alpha_{cl}$	$ETL^{cl} = \mu_t + \sigma_t E[Z \lvert Z < z_{cl}]$	$\sigma_t^2 = \alpha_0 + \sum_{i=1}^{q} \alpha_i \varepsilon_{t-1}^2 + \sum_{i=1}^{p} \beta_i \sigma_{t-1}^2$
FHS	$VaR^{cl} = (r \in \{r_{t-1}, ..., r_{t-1-N}\})\lvert$ $G(r;t,N) \ge cl$	$ETL_t^{cl} = \left(\sum_{i=[ncl]}^{n} \hat{Z}_{n(i)}\right)/(n-[ncl])$	$z_t = \varepsilon_t/\sigma_t \quad \hat{z}_{t+1} = z_t \times \hat{\sigma}_{t+1}$ $\hat{r}_{t+1} = \alpha_0 + \sum_{i=1}^{p} \alpha_i r_{t-i+1} +$ $\sum_{i=1}^{q} \theta_i \hat{z}_{t-i+1} + \hat{z}_{t+1}$
Unconditional GPD	$VaR^{cl} = q^{cl}(F) =$ $u + \frac{\sigma}{\xi}\left(\left(\frac{1-cl}{\overline{F}(u)}\right)^{-\xi} - 1\right)$	$ETL_t^{cl} = \frac{1}{1-cl}\int_{cl}^{1} q_x(F)dx =$ $\frac{VaR_{cl}}{1-\xi} + \frac{\sigma-\xi}{1-\xi}$	
Conditional GPD	$VaR_t^{cl} = \mu_t + \sigma_t VaR(Z)_{cl}$	$ETL_t^{cl} = \mu_t + \sigma_t ES(Z)^{cl}$ $ETL(Z)^{cl} = \frac{VaR_{cl}}{1-\xi} + \frac{\sigma-\xi u}{1-\xi}$	$\sigma_t^2 = \alpha_0 + \sum_{i=1}^{q} \alpha_i \varepsilon_{t-1}^2 + \sum_{i=1}^{p} \beta_i \sigma_{t-1}^2$ $Z = \left(\frac{x_{t-n+1} - \mu_{t-n+1}}{\sigma_{t-n+1}}, ..., \frac{x_t - \mu_t}{\sigma_t}\right)$ $VaR(Z)^{cl} = u_Z + \frac{\sigma_Z}{\xi_Z}\left(\left(\frac{1-cl}{\overline{F}(u_Z)}\right)^{-\xi_Z} - 1\right)$

errors but higher tail losses than an alternative one would generate higher values under this size adjusted loss function. The ranking procedure consists of five steps:

- Fitting an ARMA-GARCH model to the analysed time series in order to obtain IID observations. Estimating the empirical CDF of the non-tail distributional regions using a suitable kernel (e.g. Epanechnikov kernel). Kernel smooths the CDF estimates, eliminating the staircase pattern of unsmoothed sample CDFs.
- Finding the upper and lower thresholds such that someF percentage of the residuals is reserved for each tail. Fitting the generalized Pareto distribution (GPD) to the extreme residuals in each tail.
- Generating N simulated paths for the residuals from the obtained semi-parametric distribution (each path is T observations long) and adding the ARMA-GARCH model to the residuals to obtain N x T simulated returns.
- Calculating VaR for each of the N x T simulated returns for each VaR model and N Lopez scores for each of the N simulated return pairs, for every tested VaR model.
- Comparing if the mean values of the Lopez scores for different VaR models differ significantly. For this purpose a non-parametric Kruskal-Wallis test is employed.

Kruskal-Wallis test makes only mild assumptions about the data and is appropriate when the distribution of the data is non-normal. The assumption behind this test is that the measurements come from a continuous distribution. The test is based on an analysis of the variance using the ranks, not the individual observations themselves. The limitation of this approach is the assumption that the description of the central mass and the tails of the process distribution are adequate i.e. that the underlying process is well described by the recorded realization. This is not an unusual assumption and is made in all the models that are used in practice. By simulating the data generating process in the above described way, stochastic randomness is allowed in the data set. Limitations are not stricter than the ones usually used.

Returns are collected from the Bloomberg website for the period January 1st, 2005 through January 1st 2016. The analysed period is divided into two parts: the period from January 2005 to January 2012 was used to calculate distributional/volatility parameters and VaR/ETL starting values. The second period, consisting of 1.000 trading days, from January 2012 to January 1st 2016, was used to perform out-of-the-sample backtesting. The only exception to this rule was applied to uranium UXA1 series since it starts on May 7th 2007. VaR and ETL figures are calculated for a one-day ahead long position and 99% confidence level. ETL model performance is evaluated according to the root mean squared error (RMSE) and Blanco and Ihle (1998) loss function. The analysed VaR models are tested by using: Kupiec (1995) test, Christoffersen (1998) independence (IND) test and Lopez size adjusted tests. In the applied two-stage backtesting procedure, the best performing VaR model must first satisfy both the Kupiec (1995) and Christoffersen (1998) IND tests and then provide the minimal deviation from the expected value of losses by minimizing the Lopez (1999) error statistics.

15.3 Backtesting Results

All of the analysed energy commodities' time series show asymmetry, leptokurtosis and heteroskedasticity, with pronounced autoregression indicating periodicity in the daily returns. Based on the AIC and BIC results the best GARCH representation of volatility (predominantly GED and Students t distribution) was used to capture the dynamics of data-generating processes of each energy commodity. The asymmetry parameter in GARCH models was found to be significant only for WTI oil. In the case of WTI oil spot prices the asymmetry parameter, controlling the asymmetric impact of positive and negative shocks on the conditional variance, indicates significantly higher conditional volatility after positive shocks. For the correct application of extreme value theory based models (EVT) the crucial point is the estimation of the tail index. Estimation of the tail index is tightly linked to the threshold value u that the modeller defines as the level above/bellow which returns are considered extreme. The threshold value for each index was determined by comparing the Hill estimator with the mean excess plot and the quantile-quantile (QQ) plot. The same procedure is applied to IID innovations required for the implementation of the EVT-GARCH model (McNeil and Frey 2000). The Hill estimator, QQ and mean excess plots, as well as the maximum likelihood estimates indicate that the tail indexes for energy commodities are equal to or greater than zero. This means that energy commodities are characterized by significant leptokurtosis and that the GPD fitted to the tails belong to Gumbel domain of attraction.

Out of sample VaR model performance according to Kupiec (1995) test, Christoffersen (1998) IND tests, Lopez size adjusted score and average VaR values for natural gas (NG), Brent, WTI, heating oil (HO), uranium (UR) and US coal at 99% confidence level is presented in Tables 15.2, 15.3 and 15.4.

In Table 15.5, grey cells represent VaR models with lowest average VaR values irrespective of their backtesting performance (more than one figure is reported when the statistics do not differ significantly). Yellow cells represent the VaR model with lowest average VaR values which satisfy the Kupiec (1995) and the Christoffersen (1998) independence test.

The VaR backtesting results from Tables 15.2 to 15.5 are to a large degree consistent. A wide range of models satisfied the Kupiec (1995) coverage criteria (MHS, BRW and EVT-GARCH models). Christoffersen (1998) IND test i.e. errors do not bunch together, making their exceptions IID, was problematic in cases of uranium and coal, where almost all of the models failed, with the exception of MHS and EVT GARCH models. Overall only the MHS250 and EVT-GARCH satisfied both the Kupiec and Christoffersen test across all the commodities. The worst performers were the VCV and RiskMetrics. Coal and uranium presented the biggest challenge in forecasting VaR, a fact that can be attributed to their low liquidity and stale prices. Although independence of VaR errors is not required under any regulatory rules, in practice this characteristic is important. Dependence of errors is crucial for the financial stability since bunched errors can deplete capital reserves much faster than the simple underestimations of risk. From the security standpoint EVT based models

Table 15.2 Kupiec backtesting results at 99% confidence level, period: 1.000 days up to January 1st 2016

	NG	BRENT	WTI	HO	UR	COAL
HS 100	0.01	0.00	0.00	0.00	0.00	0.00
HS 250	0.02	0.25	0.11	0.14	0.00	0.05
HS 500	0.17	0.85	0.73	0.66	0.66	0.08
MHS 100	0.22	0.27	0.34	0.28	0.03	0.45
MHS 250	0.60	0.91	0.68	0.89	0.33	0.67
MHS 500	0.94	0.96	0.83	0.91	0.82	0.82
BRW $\lambda = 0.97$	0.00	0.05	0.05	0.03	0.05	0.03
BRW $\lambda = 0.99$	0.12	0.54	0.54	0.40	0.51	0.36
BRW $\lambda =$opt	1.00	1.00	1.00	1.00	0.78	0.62
VCV	0.04	0.03	0.08	0.03	0.00	0.00
Risk Metrics	0.21	0.01	0.00	0.03	0.00	0.01
GARCH	0.41	0.05	0.05	0.03	0.08	0.08
FHS	0.97	0.54	0.51	0.81	1.00	0.07
HW EWMA	0.00	0.00	0.00	0.00	0.00	0.13
Gumbel GARCH	1.00	0.81	1.00	1.00	1.00	0.39
Frechet GARCH	1.00	1.00	1.00	1.00	1.00	1.00

Table 15.3 Christoffersen independence (IND) backtesting results at 99% confidence level, period: 1.000 days up to January 1st 2016

	NG	BRENT	WTI	HO	UR	COAL
HS 100	0.33	0.34	0.31	0.30	0.00	0.32
HS 250	0.24	0.58	0.52	0.47	0.02	0.41
HS 500	0.45	0.72	0.82	0.70	0.04	0.08
MHS 100	0.59	0.55	0.55	0.51	0.04	0.37
MHS 250	0.70	0.81	0.67	0.72	0.14	0.60
MHS 500	0.86	0.83	0.80	0.89	0.14	0.62
BRW $\lambda = 0.97$	0.24	0.32	0.17	0.25	0.02	0.41
BRW $\lambda = 0.99$	0.51	0.72	0.55	0.55	0.04	0.54
BRW $\lambda =$opt	0.90	0.88	0.90	0.89	0.04	0.73
VCV	0.14	0.32	0.42	0.53	0.00	0.00
Risk Metrics	0.51	0.50	0.26	0.30	0.02	0.00
GARCH	0.69	0.31	0.32	0.35	0.19	0.08
FHS	0.87	0.79	0.60	0.70	0.88	0.02
HW EWMA	0.16	0.22	0.26	0.73	0.00	0.30
Gumbel GARCH	NaN	0.68	NaN	0.82	NaN	1.00
Frechet GARCH	NaN	0.91	NaN	0.90	NaN	0.00

Table 15.4 Lopez test scores at 99% confidence level, period: 1.000 days up to January 1st 2016

	NG	BRENT	WTI	HO	UR	COAL
HS 100	9.74	10.67	9.33	9.21	19.33	8.31
HS 250	6.73	1.02	3.65	2.16	9.29	6.30
HS 500	2.13	-4.42	-1.22	-2.87	-2.87	4.27
MHS 100	2.34	2.52	0.06	0.15	6.24	-0.81
MHS 250	-1.76	-5.45	-1.67	-3.87	0.19	-1.86
MHS 500	-4.55	-7.04	-3.55	-4.89	-4.94	-2.90
BRW $\lambda = 0.97$	9.65	6.85	7.93	10.18	11.22	4.21
BRW $\lambda = 0.99$	3.43	-1.86	-0.92	0.13	0.20	1.20
BRW λ =opt	-7.00	-9.03	-8.99	-7.95	-2.85	-0.83
VCV	5.77	5.96	4.17	1.17	16.36	11.40
Risk Metrics	2.33	12.11	10.22	11.18	23.39	15.47
GARCH	0.09	9.44	7.17	6.17	14.31	11.33
FHS	-7.04	-0.95	-0.92	-2.88	-7.99	7.30
HW EWMA	10.04	10.67	8.15	16.20	10.09	3.23
Gumbel GARCH	-10.00	-3.05	-10.00	-8.14	-10.00	0.07
Frechet GARCH	-10.00	-8.12	-10.00	-9.00	-10.00	-10.00

Table 15.5 Average VaR values (%) at 99% confidence level, period: 1.000 days up to January 1st 2016

	NG	BRENT	WTI	HO	UR	COAL
HS 100	4.92	3.19	3.69	3.44	2.38	2.24
HS 250	5.44	3.77	4.37	3.98	2.71	2.78
HS 500	6.91	4.97	5.36	4.57	3.88	5.98
MHS 100	6.05	3.75	4.18	3.77	3.42	3.94
MHS 250	6.97	4.08	4.70	3.98	3.85	4.63
MHS 500	8.99	5.02	6.22	5.21	5.53	7.84
BRW $\lambda = 0,97$	5.23	3.46	3.78	3.98	2.72	2.86
BRW $\lambda = 0,99$	6.65	3.79	4.21	4.05	3.18	3.75
BRW $\lambda =$ opt	8.48	4.53	5.65	4.93	3.51	3.02
VCV	6.37	3.41	3.90	3.52	1.97	2.06
Risk Metrics	5.89	3.12	3.14	3.34	1.87	1.64
GARCH	6.00	3.10	3.16	3.32	1.84	2.18
FHS	6.85	3.61	4.34	4.01	4.87	2.44
HW EWMA	7.54	3.87	4.72	5.02	5.26	6.53
Gumbel GARCH	8.02	4.25	5.23	4.95	2.52	2.86
Frechet GARCH	10.27	6.55	7.96	7.02	3.77	4.54

Table 15.6 RMSE ETL backtesting results at 99% confidence level. period: 1.000 days up to January 1st 2016

	WTI	BRENT	NG	HO	COAL	UR
Frechet GARCH RM	0.17	0.09	0.20	0.07	0.05	0.06
Gumbel GARCH RM	0.07	0.04	0.11	0.03	0.04	0.03
Bootstrap FHS	0.02	0.03	0.09	0.03	0.03	0.04
Frechet VCV	0.21	0.08	0.15	0.08	0.03	0.04
Gumbel VCV	0.06	0.05	0.09	0.04	0.03	0.05
Bootstrap HS250	0.04	0.04	0.06	0.03	0.03	0.05
Bootstrap HS500	0.04	0.04	0.05	0.03	0.05	0.06
Bootstrap MHS250	0.03	0.06	0.04	0.05	0.04	0.05
Bootstrap MHS500	0.05	0.05	0.05	0.04	0.05	0.07
Bootstrap BRW	0.04	0.05	0.08	0.05	0.03	0.06
RM Gumbel	0.05	0.04	0.09	0.05	0.04	0.04

Table 15.7 Modified Blanco-Ihle ETL backtesting results at 99% confidence level. period: 1.000 days up to January 1st 2016

	WTI	BRENT	NG	HO	COAL	UR
Frechet GARCH RM	2.78	1.68	1.57	1.28	0.78	1.02
Gumbel GARCH RM	0.43	0.42	0.57	0.41	0.31	0.31
Bootstrap FHS	0.14	0.19	0.48	0.20	0.35	0.72
Frechet VCV	2.96	1.72	1.43	1.05	0.63	0.96
Gumbel VCV	0.49	0.53	0.52	0.38	0.31	0.31
Bootstrap HS250	0.25	0.17	0.50	0.30	0.70	0.37
Bootstrap HS500	0.22	0.17	0.38	0.22	0.70	0.29
Bootstrap MHS250	0.19	0.31	0.16	0.38	0.38	0.37
Bootstrap MHS500	0.19	0.28	0.19	0.30	0.25	0.21
Bootstrap BRW	0.25	0.29	0.38	0.31	0.40	0.45
RM Gumbel	0.40	0.43	0.50	0.41	0.32	0.41

are the only acceptable ones, sonly due to uranium IID errors issue. Results of the Lopez test favour the MHS (100 and 250), BRW (0.99 and optimized) and FHS models. Similar results are also found for the average VaR values, with BRW (optimized), MHS100 and FHS being the highest ranked models. EVT-GARCH model performed very well with regards to security, providing very safe and independent VaR forecasts, but it is overestimating the risk and thus tends to hurt the lowest cost/profitability criteria.

The results of the overall ETL model performance, at 99% confidence level, for the selected energy commodities are presented in Tables 15.6 and 15.7.

The results of the ETL model comparison at 99% level are somewhat more conclusive than the VaR figures. According to the RMSE statistic, bootstrapped FHS ETL model provided a superior fit to the extreme losses. The worst performers are the bootstrapped VCV, EVT-GARCH model with Frchet distribution (except for uranium). According to the Blanco-Ihle statistic the best performers are the bootstrapped FHS and the bootstrapped MHS500 model. The worst performers are the bootstrapped VCV and EVT-GARCH models.

From the obtained results we can conclude that the more advanced semiparametric VaR/ETL models, using conditional volatility and extreme tails, are required to capture the true level of risk in the energy markets. These results are, to an extent, in line with the results of Hung et al. (2008). They also emphasise the need of integrating fat tails into VaR forecasting. With regards to the performance of EVT based models we find similar results to Marimoutou et al. (2009) that reports that EVT performs superiorly over the traditional VaR models.

To further analyse the true performance of VaR in the energy markets we apply the methodology by Žiković and Filer (2013) to test whether there is any significant difference in the performance of the tested VaR models. The data is simulated based on the distribution of returns in the period 2012–2016. For each commodity we perform 3.000 simulations with the window length of 1.000 days. Lopez size adjusted score used in the evaluation gives the best performance to the model whose score is closest to zero. After obtaining 3.000 Lopez size adjusted scores for each VaR model and for each of six energy commodities we test for the existence of differences among the tested VaR models by using a Kruskal-Wallis test. Simulation results are reported in Table 15.8.

In case when the simulated mean value of the VaR model score lies outside of the 95% confidence bands of all the other tested VaR models, it is ranked according to its relative performance. If a model is not significantly different from the other models it shares the same ranking. From Table 15.8 we see that for a large number of VaR models there is no statistically significant difference when measured by the selected loss function. When looking at the overall performance for the tested commodities the best performing VaR models that are statistically significant are the filtered historical simulation (FHS) and semiparametric BRW model with optimized λ. These models are followed by the conditional EVT GARCH model and simple nonparametric MHS model and BRW simulation with fixed λ. The worst performance, measured by the distance from expected losses is recorded for Hull-White, RiskMetrics, VCV and HS model with shortest window length. Although we tested ten VaR models (with 16 combinations) on a sample of six energy commodities the number of statistically different VaR models never surpasses five.

A second finding that can be pointed out is that there is consistency in the performance of risk models under VaR and ETL metrics. According to VaR performance we can point out four superior models: FHS, BRW, EVT-GARCH and MHS. ETL estimation yielded only two superior models: MHS and FHS. It is obvious that best performing VaR and ETL models overlap. Both MHS and FHS models can be viewed as semiparametric model i.e. models not making a priori parametrical assumptions about the distribution of returns, but using the empirical returns. These findings are,

Table 15.8 Lopez size adjusted score ranking of simulated VaR model performance (3.000 simulations, 1.000 days forecasting horizon)

	NG	BRENT	WTI	HO	UR	COAL	Total
HS 100	5	5	5	5	5	5	**7**
HS 250	3	2	3	2	4	3	**4**
HS 500	2	2	2	2	4	3	**4**
MHS 100	4	5	5	3	4	2	**6**
MHS 250	2	3	3	3	3	2	**4**
MHS 500	1	2	2	3	3	3	**3**
BRW $\lambda = 0.97$	4	3	4	4	3	4	**6**
BRW $\lambda = 0.99$	2	1	2	1	1	2	**2**
BRW $\lambda = \text{opt}$	1	1	1	1	1	2	**1**
VCV	5	5	5	4	5	4	**7**
Risk Metrics	5	5	5	4	4	4	**7**
GARCH	1	4	5	3	3	3	**5**
FHS	1	1	1	1	1	3	**1**
HW EWMA	5	5	5	5	4	4	**7**
Gumbel GARCH	1	2	2	1	1	1	**1**
Frechet GARCH	3	3	3	2	1	2	**3**

to a significant extent, in line with the findings of Žiković et al. (2015) and Žiković and Tomas Žiković (2016).

Similar results regarding the performance of FHS model can be found in Costello et al. (2008) which report that FHS performs superiorly to wide spread models, mainly due to relaxed distribution assumptions and treatment of volatility clustering. In the case of MHS the key to its success is in creating a bootstrapped empirical series and using order statistics, with the only assumption being that the past will be similar to the future. In the case of FHS it is a more elaborate scheme of using GARCH volatility and EVT innovations to rescale the empirical returns. In regards to GARCH volatility modelling and using GARCH volatility as part of a parametric risk model Fan et al. (2008) claim superior performance, a result which we cannot confirm. During the dynamic volatility modelling phase we also tested the performance of fractionally integrated parametric GARCH (FIGARCH) models but although providing excellent fit to the in-sample volatility, out-of-the-sample performance was unremarkable, a finding which is contrary to the results obtained by Aloui and Mabrouk (2010) and Mabrouk (2011) about the superiority of FIGARCH models.

15.4 Conclusion

We investigate common risk model features that result in superior forecasting results under both VaR and ETL metrics in the energy commodities markets. Our goal was not only to find the models that accurately forecast VaR figures but also give the best approximation to the tail losses i.e. minimize the deviation between ETL forecasts and extreme losses. VaR backtesting results showed that only MHS250 and EVT-GARCH satisfied both the Kupiec and Christoffersen test across all the tested commodities. These findings confirm the findings of Marimoutou et al. (2009) regarding the EVT performance. The worst performers were the VCV and RiskMetrics models failing both tests. Coal and uranium presented the biggest challenge in forecasting VaR at higher confidence level, a fact that can be attributed to the low liquidity and stale prices of these commodities. Results of the Lopez test clearly favour the MHS, BRW and FHS models. Similar results are also found for the average VaR values, a finding that confirms the Costello et al. (2008) results. EVT-GARCH performed very well with regards to security, providing very safe and independent VaR forecasts, but sometimes overestimates the risk and thus tends to hurt the lowest cost/profitability criteria. This finding is in line with the conclusions by Žiković et al. (2015) which find that advanced models based on conditional EVT, as well as the nonparametric models, yield very robust and consistent results.

The simulation study shows that for a large number of different VaR models there is no statistical difference measured by Lopez size adjusted loss function. Overall, statistically significant top performers are FHS model, semiparametric BRW simulation, conditional EVT GARCH and nonparametric MHS model. Simpler parametric models, e.g. VCV, Hull-White and RiskMetrics were the worst performers in our comparison. It is also interesting to note that although historical simulation based models are clearly theoretically inferior to EVT in risk estimation, their empirical track record is impressive. This finding may suggest that during the analysed period there were a large number of extreme events that allowed the simpler nonparametric models to correctly asses the true level of risk. After performing our simulation based test we conclude that there is sufficient statistically significant evidence that models like FHS, BRW, EVT-GARCH and MHS perform superiorly to other models. Advanced VaR models based on conditional EVT, FHS and BRW simulation as well as the very simple, nonparametric models such as MHS yield very robust and consistent results. Models that do not fall into these groups have shown poor performance in the energy markets.

The results of the ETL model comparison are similar to VaR backtesting results. According to RMSE and Blanco-Ihle statistics bootstrapped FHS and MHS models provided the closes fit to the expected value of extreme losses. The worst performers under both statistics were the VCV and Frchet EVT-GARCH model. Both VaR and ETL results show that advanced semiparametric models, with conditional volatility and extreme tails, are required to capture the true level of risk for energy commodities. There is significant overlapping in the performance of tested models under both risk measures. The most consistent top performers under both risk measure, FHS and

MHS models, do not a priori assume the parametrical distribution of returns but use instead the empirical returns. In both cases the common factor is the use of empirical distribution without a priori parameterization of return distribution.

References

Agnolucci, P. (2009). Volatility in crude oil futures: A comparison of the predictive ability of GARCH and implied volatility models. *Energy Economics*, *31*(2), 316–321.

Aloui, C. (2008). Value-at-risk analysis for energy commodities: Long-range dependencies and fat tails in return innovations. *Journal of Energy Markets*, *1*(1), 31–63.

Aloui, C., & Mabrouk, S. (2010). Value-at-risk estimations of energy commodities via long-memory, asymmetry and fat-tailed GARCH models. *Energy Policy*, *38*(5), 2326–2339.

Blanco, C. & Ihle, G. (1998, August). How Good is Your VaR Using Backtesting to Assess System Performance. Financial Engineering News, pp. 1-2.

Bunn, D., et al. (2013). Analysis and forecasting of electricity price risks with quantile factor models. London Business School Working Paper. http://www.ceem-dauphine.org/assets/dropbox/Derek_BUNN.pdf.

Cheong, C. W. (2009). Modeling and forecasting crude oil markets using ARCH-type models. *Energy Policy*, *37*(6), 2346–2355.

Christoffersen, P. F. (1998). Evaluating interval forecasts. *International Economic Review*, *39*(4), 841–862.

Costello, A., Asem, E., & Gardner, E. (2008). Comparison of historically simulated VaR: Evidence from oil prices. *Energy Economics*, *30*(5), 2154–2166.

Diebold, F. X., & Mariano, R. (1995). Comparing predictive accuracy. *Journal of Business and Economic Statistics*, *13*, 253–263.

Fan, Y., et al. (2008). Estimating Value at Risk of crude oil price and its spillover effect using the GED-GARCH approach. *Energy Economics*, *30*(6), 3156–3171.

Hansen, P. R. (2005). A test for superior predictive ability. *Journal of Business and Economic Statistics*, *23*(4), 365–380.

Hull, J., & White, A. (1998). Incorporating volatility updating into the historical simulation method for value at risk. *Journal of Risk*, *1*(Fall), 1–19.

Hung, J. C., Lee, M. C., & Liu, H. C. (2008). Estimation of value-at-risk for energy commodities via fat-tailed GARCH models. *Energy Economics*, *30*(3), 1173–1191.

Kupiec, P. (1995). Techniques for verifying the accuracy of risk management models. *Journal of Derivatives*, *3*(2), 73–84.

Lopez, A. J. (1999). Methods for evaluating value-at-risk estimates - federal reserve bank of New York. *Economic Policy Review*, *2*, 3–17.

Mabrouk, S. (2011). Value-at-risk and expected shortfall estimations based on GARCH-type models: Evidence from energy commodities. *Journal of Energy and Development*, *35*(1), 279–314.

Marimoutou, V., Raggad, B., & Trabelsi, A. (2009). Extreme value theory and value at risk: Application to oil market. *Energy Economics*, *31*(4), 519–530.

McNeil, A. J., & Frey, R. (2000). Estimation of tail-related risk measures for heteroscedastic financial time series: An extreme value approach. *Journal of Empirical Finance*, *7*, 271–300.

Mohammadi, H., & Su, L. (2010). International evidence on crude oil price dynamics: Applications of ARIMA-GARCH models. *Energy Economics*, *32*(5), 1001–1008.

Wei, Y., Wang, Y., & Huang, D. (2010). Forecasting crude oil market volatility: Further evidence using GARCH-class models. *Energy Economics*, *32*(6), 1477–1484.

White, H. (2000). A reality check for data snooping. *Econometrica*, *68*, 1097–1126.

Žiković, S., & Aktan, B. (2011). Decay factor optimisation in time weighted simulation - evaluating VaR performance. *International Journal of Forecasting*, *27*(4), 1147–1159.

Žiković, S., & Filer, R. K. (2013). Ranking of VaR and ES models: Performance in developed and emerging markets. *Czech Journal of Economics and Finance*, *63*(4), 327–359.
Žiković, S., & Tomas Žiković, I. (2016). Two sides of the same coin, risk measures in the energy markets. *Journal of Energy Markets*, *9*(2), 51–68.
Žiković, S., Weron, R., & Tomas Žiković, I. (2015). Evaluating the performance of VaR models in energy markets. *Stochastic models, statistics and their applications*, Springer proceedings in mathematics and statistics (Vol. 19(122), pp. 479–487). Berlin: Springer.

Chapter 16
Risk Analysis of Cryptocurrency as an Alternative Asset Class

L. Guo and X.J. Li

Abstract The purpose of this study is to analyze the risk of cryptocurrencies, as an alternative investment. In particular, we find the wealth distribution of the cryptocurrency, evaluate its corresponding effects on the market and analyze other risk factors resulting in the death of altcoins. The paper concludes that the closer the right tail of wealth distribution approaching the Power-Law model, the more stable the market will be. This result is quite useful for investors to make decisions when investing in cryptocurrencies.

16.1 Introduction

As a representative of cryptocurrency, Bitcoin was developed by an anonymous hacker in 2009. Within 4 years' development, the price of Bitcoin had reached higher than $1,000 by the end of 2013. What's more, the total number of Bitcoins that can be mined has been limited within 21 million while it appears to be a more complicated question to calculate the amount of gold that can be mined. As a result, during the period when the gold price has collapsed, Bitcoin appears to be a better store of value than gold for investors.

Besides, Bitcoin can be used to make online purchases via mobile phones or other devices. Popular with the techno tribe, the currency is regarded as being beyond the reach of government regulation – the anonymous founder of Bitcoin introduced the idea of a distributed block chain to prevent the counterfeiting of Bitcoin (Lee et al. 2014). The block chain, also known as the public ledger, is a technical innovation that solves a 20-year-old problem called the General Byzantine problem (Lam et al. 2014), which is a problem all distributed systems face. For instance, how to reach

L. Guo (✉)
Lee Kong Chian School of Business, Singapore Management University,
50 Stamford Road, Singapore 178899, Singapore
e-mail: liguo.2014@pbs.smu.edu.sg

X.J. Li
College of Letters and Science, University of California, Berkeley, CA, USA
e-mail: lxjpub@gmail.com

W.K. Härdle et al. (eds.), *Applied Quantitative Finance*, Statistics and Computing,
DOI 10.1007/978-3-662-54486-0_16

consensus in a system without any central authorities instructions or how to prevent the double spending of digital currency.

In 2014, the Bitcoin Central partnered with a French bank becoming a registered Payment Services Provider (PSP) under the European Union Law. It means that Bitcoins now can offer debit cards, account insurance and other banking facilities to the Bitcoin owners. This phenomenon became a breaking news because the amount of Bitcoin value is becoming infinite due to the excess demand of market which changed drastically from its original value. Nowadays Bitcoin has already gained worldwide attention, as people can sell products or services overseas by using Bitcoins and make profits immediately. There are more than twelve million Bitcoin users including digital miners, traders and small business owners.

Meanwhile, similar cryptocurrencies or alternative cryptocurrencies (aka. altcoins) are proliferating, and there are now over 400 active altcoins in the market (Lee et al. 2014). Examples of popular cryptocurrencies include Bitcoin, Ripple, Litecoin and Dogecoin (Coinmarketcap.com, 2014). However, many of the coins are ephemeral and become inactive shortly after they are launched. Such coins are known as dead coins e.g. Auroracoin (AUR), Alcohoin (ALC), 2chcoin (2ch), 66coin (66). Digital currencies can potentially play a major role in lowering the cost of financial services and enable financial institutions to reach out to the unbanked banking and the under-banked (Ignacio et al. 2014). As a payment system, digital currency can contribute to the banking and achieve the goal of financial inclusion for being advocated by 90 countries in the Maya Declaration as well as the Bill and Melinda Gates Foundation (gatesfoundation.org, 2014). Therefore, it is important to investigate the factors that determine the success of a coin as we can then avoid similar pitfalls in the future when constructing a new coin, which can benefit the less privileged and those at the bottom of the wealth pyramid (XiangJun et al. 2014). To this end, we have decided to compare the different characteristics of Auroracoin and Bitcoin to figure out those risk factors leading to the death of Auroracoin but the success of Bitcoin. The present paper adopts a complete empirical methodology for detecting Power-Laws introduced by (Clauset et al. 2009). To verify whether the whole range of the upper tails of wealth distributions obeys the Power-Law model. We estimate both the Power-Law exponent and the lower bound on the Power-Law behavior.

The paper is organized as follows: Sect. 16.2 shortly describes our data sets drawn from the original blockchain and other sources. Section 16.3 presents the statistical framework introduced by Clauset et al. which is used for measuring and analyzing Power-Law behavior in empirical data. Section 16.4 is the empirical analysis while Sect. 16.5 serves as the conclusion.

16.2 Data Collection

Data are collected mainly through the following four methods:

16.2.1 *Parse the Balance Information of Each Address from the Downloaded Block Chain Using C++*

A source code written in C++ by John W. Ratcliff was used and modified. Basically, this program provides us the balance information of each address which can be used to find the wealth distribution of both Bitcoin and Auroracoin.

16.2.2 *Parse Other Fundamental Variables of Bitcoin*

Blockchain.info contains all the fundamental variables of Bitcoin market except for the balance information. The data include market price, transaction volume, developer's revenue, etc. All the data were downloaded in CSV format and R 3.1.2 was used to group the data together and calculate the aggregate where appropriate.

16.2.3 *Historical Price Data for Auroracoin Are from a Data Provider Named Myip*

Myip is a data provider that stores the historical price and transaction volumes of Auroracoin. Additionally, different from Bitcoin, the block chain explorer of Auroracoin doesn't have the historical price, so we have to use this data provider to collect the historical price of Auroracoin.

16.2.4 *Parse Other Fundamental Variables of Auroracoin from Online Block Chain Explorer Using Python*

We obtained the data of other fundamental variables of Auroracoin from the Block Chain Explorer. Figure 16.1 shows how the webpage looks like.

16.3 Methodology

In the beginning, we believe it is necessary for a coin's wealth distribution to follow a pareto optimal distribution. The reason is that, in the initial stage, we expect to see "Top few" users to develop the market and their wealth of the coin takes large position of the overall market. Indeed, the existence of "Top few" users are necessary for a coin to survive and gain popularity, hence shaping the overall wealth distribution to follow a power law. So in the paper, the first hypothesis we want to test is that

AuroraCoin 123101

Short Link: http://explorer.auroracoin.eu/b/12dF1oi8os

Hash: bd4055064262d2f7742a42472e5c350aa0064416935d5aa57ad857fc0b1c6165
Previous Block: 83f8a8b3dff6dee0642aebcc47858d40f6167d95ea9b48c46c808594ad77f548
Next Block: d093fff0f67f0366a8a84031ccf903a3e265604cade9cdaf47a693fdcb96b002
Height: 123101
Version: 1
Transaction Merkle Root: 612e55180bc4507e087252045411c751be13a78d2d89de379e6c1fc5cce9e7ba
Time: 1431477552 (2015-05-13 00:39:12)
Difficulty: 440.196 (Bits: 1c0094e1)
Cumulative Difficulty: 19 550 578.013
Nonce: 23539883
Transactions: 2
Value out: 50
Transaction Fees: 0
Average Coin Age: 128.82 days
Coin-days Destroyed: 12.58998842
Cumulative Coin-days Destroyed: 70.7386%

Fig. 16.1 Original Data from Blockchain

whether the wealth distribution is one of the key factors that determine the success of a crypto currency. In fact, by plotting the wealth distribution of both Bitcoin and Auroracoin, we find the right tail of both wealth distribution seem to follow the Power-Law model. Hence, in the paper, we fit the wealth distribution using the Power-Law model. There's no denying that there are some other candidates to fit the wealth distribution, such as Log-normal distribution, which has a similar pattern as Power-Law. However, as a preliminary study, we do not focus on the comparison of density functions.

In order to find the Power-Law behavior in wealth distributions we use a toolbox proposed by Clauset et al. (2009). A density of continuous Power-Law model is given by

$$p(x) = \frac{\alpha - 1}{x_{min}}(\frac{x}{x_{min}})^{-\alpha} \tag{16.1}$$

The maximum likelihood estimator (MLE) of the Power-Law exponent, $\hat{\alpha}$, is

$$\hat{\alpha} = 1 + n\{\sum_{i=1}^{\infty} \log \frac{x_i}{x_{min}}\} \tag{16.2}$$

where $x_i; i = 1, 2 \ldots n$ are independent observations such that $x_i > x_{min}$. In the meantime, x_{min} can be found by minimizing the well-known Kolmogorov–Smirnov (KS) statistic, which can be defined as follow:

$$KS = \max_{x \geq x_{min}} |S(x) - P(x)| \tag{16.3}$$

In the equation above, $S(x)$ stands for the CDF of the data for the observations with value at least x_{min} while $P(x)$ represents the CDF for the Power-Law model that best fits the data in the region $x \geq x_{min}$. Hence, the lower bound on the Power-Law, x_{min}:

$$x_{min} = \underset{x_{min}}{\operatorname{argmin}}\, KS \tag{16.4}$$

The next step in measuring Power-Law involves testing goodness of fit. A positive result of such a test allows us to conclude that a Power-Law model is consistent with a given data set. Following Clauset et al. again, we start with fitting a Power-Law model to data using the MLE for α and the KS-based estimator for x_{min}. Meanwhile, we have the KS statistic for this MLE fitting. Next, we generate the synthetic data sets with scaling parameter $\hat{\alpha}$ and lower bound x_{min} from previous step. To be more specific, the synthetic data sets have the Power-Law model above the estimated x_{min} and have the same non-Power-Law distribution as the original data set below $\hat{x}_{min}$. Then, Power-Law models are fitted to each of the generated data sets with the KS statistics calculated. Finally, we define the p-value of the test as the fraction of data sets for which their own KS statistics are larger than the KS found in the empirical data set. Hence, the Power-Law hypothesis is rejected if this p-value is smaller than the chosen threshold. In the reference (Clauset et al. 2009), Clauset et al. rules out the Power-Law model if the estimated p-value for the test is smaller than 0.1.

16.4 Empirical Results

In this section, we compared the wealth distribution of two different altcoins, Bitcoin and Auroracoin to illustrate the importance of achieving a Pareto optimal distribution. After that, we further test the predicting power of wealth distribution, defined as the frequency distribution of public addresses of the digital currency under study. In particular, we examine the following hypothesis that the wealth distribution within the system has predictive power over its lifespan and price. On top of that, we also study the different characteristics of both coins and document the important features that lead to the survivorship of the cryptocurrency.

16.4.1 Data Visualization

We first plot a histogram of frequency of public addresses of Bitcoin and Auroracoin, as we have shown in Figs. 16.2 and 16.3.

It seems that Auroracoin does not appear to follow a distinct Power-Law distribution while the distribution of Bitcoin does. In the meantime, although the wealth

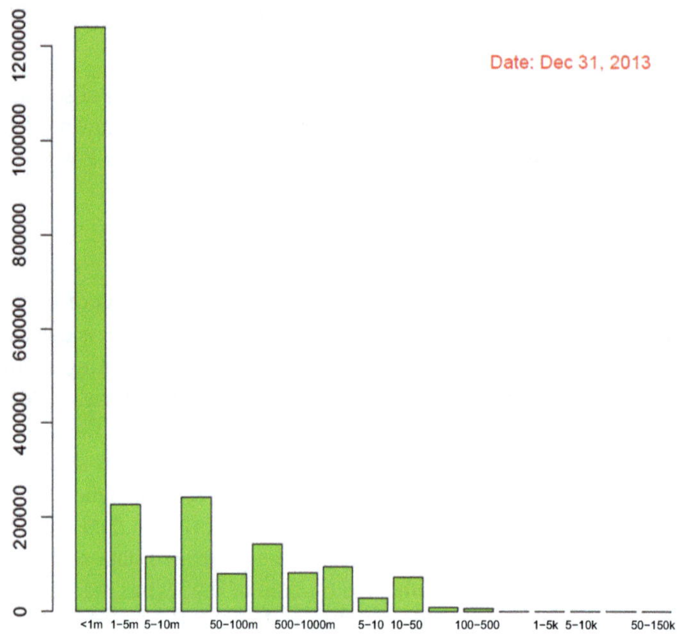

Fig. 16.2 Bitcoin Histogram. XFGHistWealthD

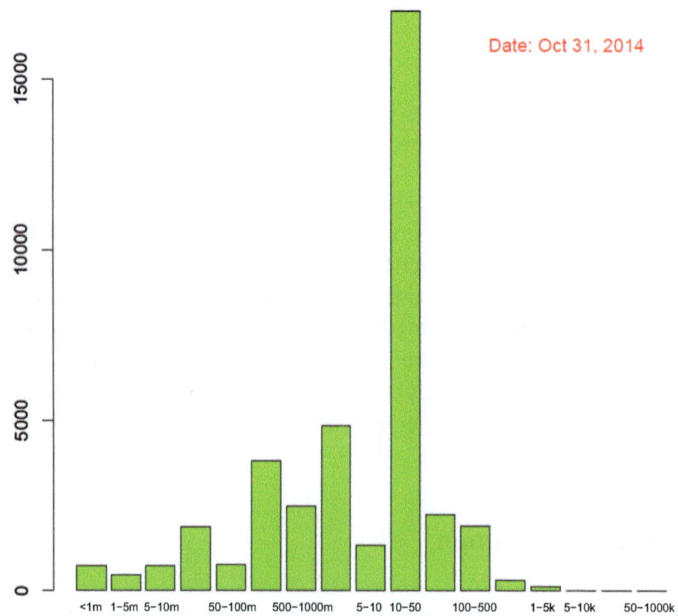

Fig. 16.3 Auroracoin Histogram. XFGHistWealthD

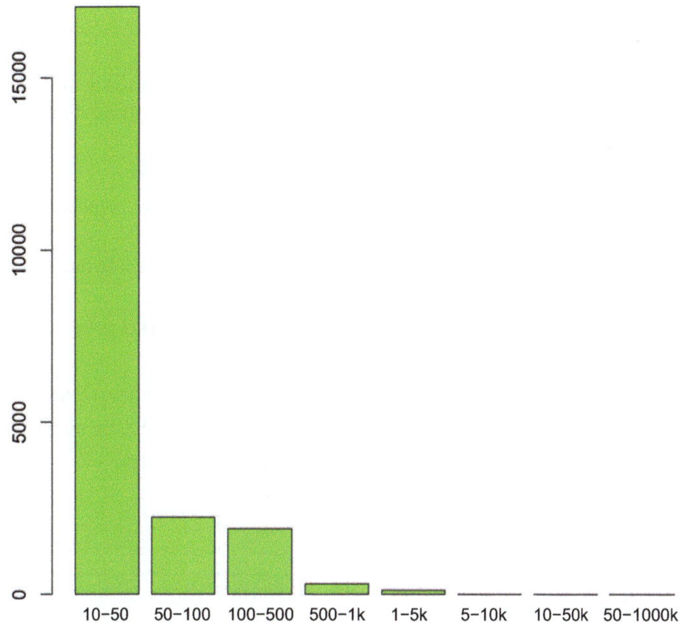

Fig. 16.4 Right Tail Auroracoin Histogram. XFGHistWealthD

distribution in Auroracoin does not exhibit Power-Law distribution on the whole, the tail part of the distribution does seem to follow a Power-Law distribution (shown in Fig. 16.4). Therefore, when calculating the α, x_{min} is set free and is automatically determined by the programme to minimize the Kolmogorov–Smirnov statistics. The motivation is to investigate if the Power-Law parameters for the right side of the distribution have any explanatory power for those fundamental variables of cryptocurrencies. Table 16.1 lists the fundamental variables treated as dependent variables in the following regression.

16.4.2 Power-Law Estimation and Empirical Analysis

In this section, we fit the wealth distribution of Bitcoin and Auroracoin using the Power-Law model. For the Auroracoin, only the right tail seems to follow the Power-Law pattern so the x_{min} is optimally selected by minimizing the KS statistic in the Auroracoin case.

Shown in Figs. 16.5 and 16.6, α of Bitcoin increases smoothly while α of Auroracoin goes up and down. What's more, the α shows no significant predicting power on those fundamental variables in terms of Bitcoin while for the Auroracoin, its significant predicting power is not only limited to the price movements but also applicable to other fundamental variables with average R-square = 0.65, much larger than that

Table 16.1 Variable list

Variable list	Definition
Days destroyed	A measure of the transaction volume of Cryptocurrency. If someone has 100 BTC that they received a week ago and they spend it then 700 bitcoin days have been destroyed
MB.1	The total size of all block headers and transactions
Difficulty	A measure of how difficult it is to find a new block compared to the easiest it can ever be
Hashrate	The estimated number of billions of hashes per second the bitcoin network is performing
Market cap	Total number of bitcoins in circulation * the market price in USD
Market price	Price of Cryptocurrency
Miners revenue	(Number of bitcoins mined per day + transaction fees) * market price
Network deficit	Difference between transaction fees and cost of bitcoin mining
No. of deals	Total number of unique bitcoin transactions per day
Ratio	Transaction volume/USD exchange volume

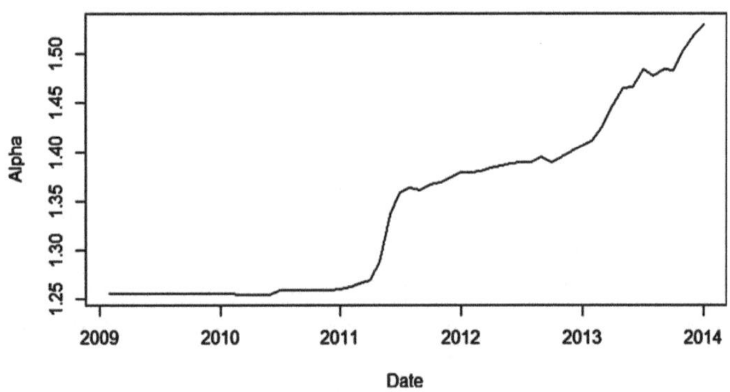

Fig. 16.5 Bitcoin Power-Law Estimation using whole sample. XFGPowerLawAlpha

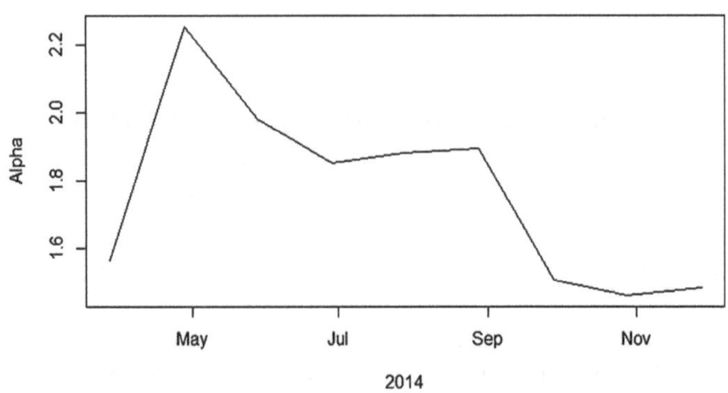

Fig. 16.6 Auroracoin Power-Law Estimation (Right Tail). XFGPowerLawAlpha

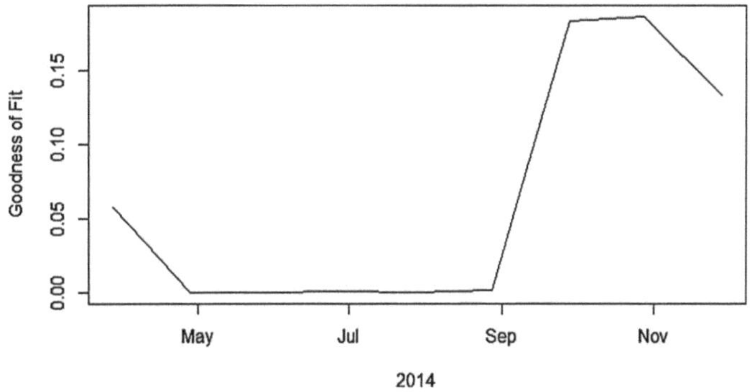

Fig. 16.7 Goodness of Fit of Auroracoin (Right Tail). XFGPowerLawP

of Bitcoin. In addition, all the fundamental variables have been taken first-order difference so that the variables put into regressions are stationary. We are surprised about these findings as it contradicts to our expectations. Auroracoin is short lived after it is launched while Bitcoin is one of successful crytos in the market. Why the distribution of a dead coin play an important role in determining the market of a deadcoin but has no effects on Bitcoin? To answer this question, we further test the goodness of fit on wealth distributions of Bitcoin and Auroracoin respectively. Results suggest that none of the models survives – for Bitcoin, all p-values across the whole sample period are 0, indicating the whole sample doesn't follow Power-Law. Similarly, with respect to the Auroracoin, p-values above 10% level only occur in 3 months, suggesting that only 3 of them can be fitted using the Power-Law distribution (see Fig. 16.7).

Given the goodness of fit, we know that the $\hat{\alpha}$ in both cases cannot reflect the wealth distribution very well. In Bitcoin case, $\hat{\alpha}$ has no prediction power over the market which is not due to the wealth distributions lacking impact on the market but because the $\hat{\alpha}$ cannot stand for the wealth distribution of Bitcoin market. As for the Auroracoin case, it becomes an another story – although the fitted α cannot reflect the wealth distribution of Auroracoin, it shows significant prediction power on those fundamental variables. Looking into the regression, we note that only 8 observations are included in the regression, so the estimation results may not be so convincing. Later, daily data instead of monthly data should be tested in order to expand the samples. Running the Power-Law model for a long time made us neglect the checking of regression results using daily data.

Keeping in mind that we have already selected the optimal x_{min} for Auroracoin to fit the Power-Law distribution but not for Bitcoin, it is safe to conclude that wealth distribution of Auroracoin market doesn't follow the Power-Law distribution and for the Bitcoin market, using the whole sample to fit the wealth distribution is inappropriate. Therefore, we try to improve the model by analyzing the right tail wealth distribution of Bitcoin with x_{min} optimally selected in order to improve the goodness of fit of the Power-Law model.

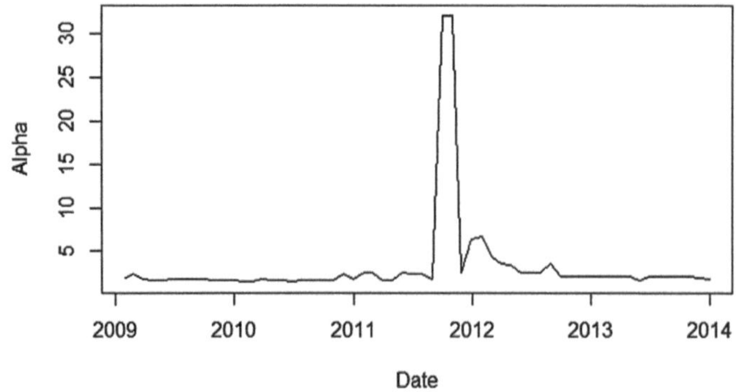

Fig. 16.8 Bitcoin Power-Law Estimation (Right Tail). XFGPowerLawAlpha

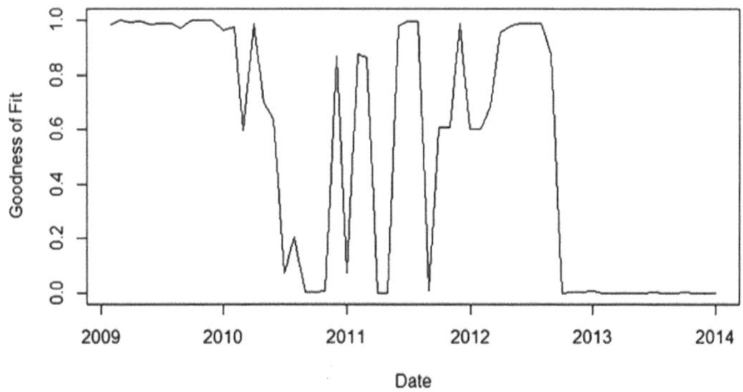

Fig. 16.9 Goodness of Fit of Bitcoin (Right Tail). XFGPowerLawP

Figures 16.8 and 16.9 suggest that although the overall wealth distribution of Bitcoin doesn't follow Power-Law, its right tail perfectly fits the Power-Law distribution. Besides, the p-value varies dramatically before the end of 2012, while the majority of p-values are above 10% level, indicating that the right tail wealth distribution of Bitcoin market follows the Power-Law distribution well. However, the p-value drops below 5% level after September 2012, implying that the wealth distribution deviates a lot from the Power-Law distribution. This is quite consistent with the sharp increase of price from the end of 2012. It is believed that when the price explodes, the Bitcoin market will begin to deviate from its previous state due to the extraordinary amount of investors in the market. As we know, there are multiple big events happening during that time. On 15th Nov, 2012, Wordpress as one of the 25 most popular domains on the web, its move paved the way for later retail ventures of Bitcoin. On 25th March, 2013, the Eurogroup, the European Commission, the European Central Bank and the International Monetary Fund orchestrated the 10 billion bailout to fortify the flagging Cypriot economy. As a result, the increasing trading volume broke Mt. Gox in April. Then on 18 Nov, 2013, US Senate held a hearing of Bitcoin. Afterwards and most importantly, Bitcoin was accepted in China. Chinese people were free to participate in the Bitcoin market finally. BTC China achieved a trading volume more than twice of the second place in Mt. Gox. Within one year, the Bitcoin price jumped from \$11.04 to \$1075.16. Of course, all these events exerted profound effects on the Bitcoin market and thus causing the wealth distribution deviating from its previous status.

In terms of α value, we note that it jumps to 32.06 in Sep 2011. That is because earlier that month, Mt. Gox was hacked. A copy of the users' database was leaked and was used to launch attacks against accounts held by users of the MyBitcoin online wallet service, because they shared the same password on both sites. The attack resulted in thefts of over 4,019 BTC from about 600 wallets. Consequently, the Bitcoin market experienced a downward trend in the following months. Even large Bitcoin holders began to sell the coins, increasing the diversification of the Bitcoins. As the α parameter stands for the diversification of the wealth distribution – a higher α means that wealth is more diversified. As a matter of fact, Alpha increases a lot in the following months. However, the wealth distribution of that time still follows the Power-Law. We believe that this event exerted a great influence on the market, but it is not strong enough to disrupt the whole market, which is different from the case when the market price exploded to \$1000 with big events (Tables 16.2 and 16.3).

However, the Auroracoin doesn't follow the Power-Law distribution even considering the right tail of wealth distribution. In the wealth distribution plots in Fig. 16.3, the air-dropped amount is seldom spent by the recipients for Auroracoin. The value of Auroracoin is thus severely undermined. What's worse, since the coins are acquired for free as opposed to arduous processes such as mining or trading, people do not show appreciation for the coin, which leads to its death. Nonetheless, for Bitcoin, as mining is required, people do think the coin worth a certain amount of value. Over time, the wealth distribution of Bitcoin edges towards Pareto distribution (in previous analysis, we have already concluded that wealth distribution of Bitcoin followed Power-Law and so in this part we mainly refer to the optimal Power-Law distribution indicated by the α). Pareto is a mathematical model that the wealth distribution in the real world follows, and more and more people start to use it for various reasons, such as analyzing international fund transfer. It can be easily verified – excluding in high volatile periods when big events happen and the price explodes. The Power-Law parameter, α, has been increasing evidenced by Fig. 16.8. On the contrary, the wealth distribution of Auroracoin has not shown many changes during 2014 and certainly does not follow the Power-Law distribution (Fig. 16.7). Therefore for a coin to survive and gain popularity, attaining Pareto distribution is absolutely helpful.

However, using the truncated sample, the $\hat{\alpha}$ still does not show any significant predicting power over those fundamental variables (Table 16.4). The reason is that, for most months, the x_{min} is too large to be considered in the majority of observations. For example, in some months, less than 100 observations are counted when fitting the Power-Law distribution while the whole sample size for that month amounts to 20 million. In other words, α has lost some generality if the truncated sample size is used. That is why the diversification of wealth, treating α as a proxy, has no predicting power over fundamental variables, while the right tail of wealth distribution follows the Power-Law model.

Table 16.2 Predicting power of Bitcoin wealth distribution

	Days destroyed $(*10^6)$	MB.1	Difficulty $(*10^6)$	Hashrate $(*10^6)$	Market cap $(*10^6)$	Market price	Miners revenue $(*10^6)$	Network deficit $(*10^6)$	No. of transactions $(*10^6)$	Ratio
(Intercept)	2, 934.07***	6, 594.04***	506.68*	4.52*	5, 146.12*	424.53*	1.93*	−1.92*	15, 63***	152.65***
	(534.17)	(1, 333.22)	(275.29)	(2.47)	(2, 946.43)	(242.76)	(1.07)	(1.06)	(3.04)	(19.95)
D_Alpha	5, 786.83	11, 592.00	1, 134.81	9.79	13, 487.32	1, 129.65	5.08	−5.04	26.88	246.75
	(5, 483.28)	(13, 685.48)	(2, 825.86)	(25.33)	(30, 245.12)	(2, 491.93)	(10.95)	(10.88)	(31.23)	(204.80)
R^2	0.02	0.01	0.00	0.00	0.00	0.00	0.00	0.00	0.01	0.02
Adj. R^2	0.00	−0.00	−0.01	−0.01	−0.01	−0.01	−0.01	−0.01	−0.00	0.01
Num. obs.	59	59	59	59	59	59	59	59	59	59
RMSE	4,088.67	10,204.73	2,107.13	18.89	22,552.61	1858.13	8.16	8.11	23.29	152.71

$^{***}p < 0.01$, $^{**}p < 0.05$, $^{*}p < 0.1$

XFGPowerLawEst1

Table 16.3 Predicting power of Auroracoin wealth distribution

	Days destroyed ($*10^6$)	Cost. transaction	Difficulty	Transaction volume ($*10^6$)	Market price	No. of transactions	Transaction.fees
(Intercept)	0.01	1.26	28.01	50.17	−2.37*	1, 376.34	0.00
	(0.06)	(5.89)	(194.34)	(292.66)	(1.12)	(5, 544.57)	(0.00)
D_Alpha	0.06**	67.63**	2, 083.06**	3, 198.36**	−13.83***	61, 054.08**	0.03**
	(0.02)	(19.53)	(644.66)	(970.80)	(3.70)	(18, 392.50)	(0.01)
R^2	0.60	0.67	0.64	0.64	0.70	0.65	0.62
Adj. R^2	0.53	0.61	0.57	0.58	0.65	0.59	0.56
Num. obs.	8	8	8	8	8	8	8
RMSE	17104.18	16.63	548.99	826727991.81	3.15	15662.94	0.01

$^{***}p < 0.01$, $^{**}p < 0.05$, $^{*}p < 0.1$

XFGPowerLawP

Table 16.4 Predicting power of Bitcoin right tail wealth distribution

	Days destroyed (∗10⁶)	MB.1	Difficulty (∗10⁶)	Hashrate	Market cap (∗10⁶)	Market price	Miners revenue (∗10⁶)	Network deficit (∗10⁶)	No. of transactions	Ratio
(Intercept)	1, 915.90***	1, 896.68***	1.93*	15, 159.10*	99.66	10.32	0.08	−81, 039.97	5.35***	92.70***
	(410.62)	(576.45)	(0.99)	(7, 985.15)	(105.59)	(15.36)	(0.16)	(0.16)	(1.45)	(23.54)
D_Alpha	35.60	66.73	0.28*	477.30	−23.21	−3.67*	−34, 639.12	3, 4505.21	0.18	−5.14
	(57.63)	(80.90)	(0.14)	(1, 120.63)	(14.82)	(2.16)	(23, 107.99)	(22, 949.44)	(0.20)	(3.30)
R^2	0.01	0.02	0.11	0.01	0.07	0.08	0.06	0.06	0.02	0.07
Adj. R^2	−0.02	−0.01	0.08	−0.02	0.04	0.05	0.04	0.04	−0.01	0.04
Num. obs.	35	35	35	35	35	35	35	35	35	35
RMSE	2,429.22	3,410.29	5.88	47,239.87	624.65	90.89	0.97	967430.57	8.57	139.28

$^{***}p < 0.01$, $^{**}p < 0.05$, $^{*}p < 0.1$

XFGPowerLawP

Table 16.5 Predicting power of goodness of fit (Power-Law periods)

	Days destroyed (*10^6)	MB.1	Difficulty (*10^6)	Hashrate	Market cap (*10^6)	Market price	Miners revenue (*10^6)	Network deficit	No. of transactions (*10^6)	Ratio
(Intercept)	1, 567.09	731.37	−1.93	−32, 421.78	−505.22	−79.12	−0.67	0.66	2.54	75.43
	(1, 966.83)	(2, 767.07)	(4.96)	(37, 198.99)	(510.27)	(74.67)	(0.80)	(0.79)	(6.97)	(116.19)
P_value	402.97	1, 343.61	4.45	54, 748.14	694.53	102.68	0.86	−0.85	3.25	19.61
	(2, 211.73)	(3, 111.61)	(5.58)	(41, 830.81)	(573.81)	(83.97)	(0.90)	(0.89)	(7.83)	(130.65)
R^2	0.00	0.01	0.02	0.05	0.04	0.04	0.03	0.03	0.01	0.00
Adj. R^2	−0.03	−0.02	−0.01	0.02	0.01	0.01	−0.00	−0.00	−0.02	−0.03
Num. obs.	35	35	35	35	35	35	35	35	35	35
RMSE	2442004307.47	3435.57	6158266.21	46185.99	633553438.34	92.71	993133.43	986436.90	8649344.11	144.26

$^{***}p < 0.01$, $^{**}p < 0.05$, $^{*}p < 0.1$

XFGPowerLawP

So now we relax our hypothesis by considering another indicators, the goodness of fit and p-value. The smaller the p-value is, the more the true wealth distribution deviates from the Power-Law distribution. Then the new hypothesis becomes that whether the extent of wealth distribution approaching the optimal Power-Law distribution has its significant predicting power on the Bitcoin market.

Similar to α, we test the predicting power using sample where the goodness of fit is above 10% and the results suggest that increase in the goodness of fit has no significant impacts on the fundamental variables, evidenced by Table 16.5. Namely, whether the right tail of wealth distribution approaches the Power-Law rarely affects the Bitcoin market. The result may not be reliable since the sample we used in the regression only covers the periods that appear to follow the Power-Law. The p-values are relative stable across the time and the overall market environment. Hence, the effect of approaching Pareto optimal distribution cannot be fully reflected by the market during those periods. After that, we reestimate the predicting power of Clauset's goodness of fit by including in the non-Power-Law periods. This is another great advantage of using p-value to measure the whole market. In the case of α, we are restricted to do so since when P drops below 0.1, the wealth distribution doesn't follow Power-Law and hence α loses its ability to explain the market. Generally speaking, the goodness of fit has more general effects on the cryptocurrency market. We mainly test the following most related two hypothesis:

1. Whether a wealth distribution that follows Power-Law does improve the stability of the market.
2. Whether approaching the Power-Law distribution can significantly reduce the fluctuations of the market.

Table 16.6 shows the estimation results of the state variable (according to Clauset's criterion, we regard the periods whose goodness of fit below the 10% as non-Power-Law periods). Especially, this dummy variable exhibits significant predicting power on those fundamental variables. What's more, we note that the sign of the coefficient of this dummy variable is always opposite to the sign of the constant. Namely, when the right tail wealth distribution follows the Power-Law model, the changes of market cap, market price, transaction fees and other fundamental variables become much smaller compared to the changes during the non-Power-Law periods. This indicates that when the wealth distribution follows Power-Law, the Bitcoin market would become more stable than otherwise.

Above result is consistent with what we observed in Fig. 16.9 — evidenced by the strong price explosion starting from the 3rd quarter of 2013. We suspect that the wealth distribution doesn't follow Power-Law distribution during that period. In fact, it can be easily verified by analyzing the goodness of fit –. After the middle of 2013, p-value drops to almost 0, indicating that the previous stability has been disrupted. We may also note that the p-value drops below 10% level during the 4th quarter of 2011, which is also consistent with the Mt. Gox hacker event (Table 16.7).

From the above analysis, again, we claim that during the non-Power-Law period, the price movements or other fundamental variables can hardly be explained by the wealth distribution parameter, α, since the stability has been disrupted. However, this

Table 16.6 Predicting power of goodness of fit (Dummy)

	Days destroyed ($*10^6$)	MB.1	Difficulty ($*10^6$)	Hashrate ($*10^6$)	Market cap ($*10^6$)	Market price	Miners revenue ($*10^6$)	Network deficit ($*10^6$)	No. of transactions ($*10^6$)	Ratio
(Intercept)	4, 729.47***	13, 341.79***	1, 220.46***	10.90***	12, 211.28***	1, 001.94***	4.48***	−4.45***	30.72***	235.46***
	(780.87)	(1, 734.68)	(412.39)	(3.70)	(4, 443.42)	(366.34)	(1.61)	(1.60)	(3.99)	(27.99)
Ps	−3, 106.03***	−11, 534.05***	−1, 218.81**	−10.89**	−12, 094.98**	−988.87**	−4.37**	4.34**	−25.81***	−142.98***
	(1, 013.84)	(2, 252.22)	(535.42)	(4.80)	(5, 769.12)	(475.63)	(2.09)	(2.08)	(5.19)	(36.35)
R^2	0.14	0.32	0.08	0.08	0.07	0.07	0.07	0.07	0.30	0.21
Adj. R^2	0.13	0.30	0.07	0.07	0.06	0.05	0.05	0.05	0.29	0.20
Num. obs.	59	59	59	59	59	59	59	59	59	59
RMSE	3,825.46	8,498.14	2,020.28	18.12	21,768.22	1,794.68	7.88	7.84	19.57	137.15

$^{***}p < 0.01$, $^{**}p < 0.05$, $^{*}p < 0.1$

XFGPowerLawP

Table 16.7 Predicting power of goodness of fit

	Days destroyed (*10^6)	MB.1	Difficulty (*10^6)	Hashrate (*10^6)	Market cap (*10^6)	Market price	Miners revenue (*10^6)	Network deficit (*10^6)	No. of transactions (*10^6)	Ratio
(Intercept)	4, 561.26***	12, 718.87***	1, 164.89***	10.40***	11, 637.72***	953.34**	4.25***	−4.22***	29.31***	227.43***
	(769.83)	(1, 736.26)	(404.46)	(3.63)	(4, 356.72)	(359.25)	(1.58)	(1.57)	(3.99)	(27.78)
P_value	−3, 225.63***	−11, 981.52***	−1, 285.86**	−11.48**	−12, 717.66*	−1, 036.50*	−4.55*	4.52*	−26.79***	−147.93***
	(1, 123.74)	(2, 534.47)	(590.40)	(5.29)	(6, 359.60)	(524.40)	(2.30)	(2.29)	(5.83)	(40.55)
R^2	0.13	0.28	0.08	0.08	0.07	0.06	0.06	0.06	0.27	0.19
Adj. R^2	0.11	0.27	0.06	0.06	0.05	0.05	0.05	0.05	0.26	0.18
Num. obs.	59	59	59	59	59	59	59	59	59	59
RMSE	3,858.93	8,703.33	2,027.44	18.18	21,838.82	1,800.79	7.91	7,86	20,02	139.24

$^{***}p < 0.01$, $^{**}p < 0.05$, $^{*}p < 0.1$

XFGPowerLawP

fluctuations have been well captured by the Clauset's goodness of fit. Afterwards, we continue to test the second hypothesis that during the whole sample period, whether approaching the Power-Law distribution which is indicated by the p-value, has any predicting power over those fundamental variables. The estimation results have been shown as below:

As expected, all the fundamental variables listed above are significantly affected by the p-value and the change of directions also meet our expectation – when p-value increases or the wealth distribution approaches the Power-Law distribution, the changes of transaction fees, the price movements, market cap and other fundamental variables become much smaller than otherwise, suggesting that the Bitcoin market becomes more and more stable when the wealth distribution approaches the Power-Law distribution.

16.5 Other Risk Analysis

Apart from those shortages reflected by the Auroracoin, more reasons are required to be considered as dangers for a coin to survive. The current section provides more aspects to investigate these reasons.

To begin with, many coins died because of badly designed mechanism, especially the block reward scheme. It could be a too complicated scheme, for example, Aircoin (AIR) adjusts the block rewards in response to the exchange rate in order to target a gradually rising exchange rate. The reward halving time was supposed to be about once per five years, therefore it is hard to comprehend given the mining reward adjustments to target an exchange rate. Or like the case of EToken (ETOK) where the block reward for the latter block such as the ten thousandth block. People tended to abandon the block once it was mined, eventually leading to the death of the project.

Secondly, the developer issues. Some coins like BellaCoin (BELA) died since its developer is completely unknown. Some coins faded simply because the developer disappeared after launching the coin, for example, the Melange (SPICE). Besides, the anonymous developer of the BatCoin claimed to be attacked during the night of 3rd–4th, April in his home and hospitalized by an assailant intent on stealing the premined coins. If that's true, then clearly he was not anonymous to the assailant. If that's false, then he made a small but respectable profit on the premine. Either way, he hindered the development and support of Batcoin like a hot rock.

Thirdly, there are moral issues causing the death of coins. On one hand, pure IPO scams occurred (NeonCoin, VisaCoin, etc.) where developer just disappear with the money. On the other hand, plenty of coins are malware. For instance, Nerdcoin contained a key logger and a wallet stealer and Oreocoin contained a remote desktop exploit. Moreover, there is a keyboard recorder in the Thecoin and the developer apparently hoped to get the passwords people were using for their wallets.

To add up, the uniqueness of a coin also affects its survival. Some coins died because they are clones or forks of other coins. To name a few, FairBrix is clone of Tenebrix and FairQuark is clone of Quark. Nucoin, Nutcoin and Stop are all forks of NXT. Moreover, duplicate names can jeopardize a coins prosperity, too. Taking Aircoin (AIR) as an example, apparently there are at least two different coins both named Aircoin and both trading with the symbol AIR. One is effectively dead and the other apparently alive as of 12th, July 2014.

Last but not least, some coins are dead due to the bad listed timing, usually too early. For instance, Global Denomination (GDN) is unwisely initially listed on exchanges while its market cap was still under $5000. Muniti (MUN) went for exchanges way too quickly, even before there was any market capitalization to distinguish them from the thousands of dead coins.

16.6 Conclusion

In the paper, we are trying to figure out what characteristics are necessary for a cryptocurrency to be a good alternative investment. To be more specific, first, we believe that for a coin to survive and gain popularity, achieving a Pareto optimal distribution is absolutely helpful. Hence we start looking at the wealth distribution and characterizing it by fitting a Power-Law model. To verify the hypothesis, we consider two Cryptocurrencies in two situations – one with the whole sample size and the other with the truncated sample size by optimally selecting the right tail of the wealth distribution. We find that for Auroracoin market, although the fitted parameter α using the truncated sample size has significant predicting power on both the price movement, changes of market cap, and the other fundamental variables posted on Blockchain web. It doesn't follow Power-Law suggested by Clauset's goodness of fit. While in terms of Bitcoin market, it becomes a little tricky. We find that using the whole sample size, it doesn't follow Power-Law model at all and the fitted parameter α has no predicting power on those fundamental variables. After that, we fit the truncated sample size and find that the Power-Law fits the wealth distribution very well. Nevertheless, the parameter, α, still shows no predicting power over those fundamental variables. After further looking into the Bitcoin market, we relax the hypothesis by considering whether the wealth distribution which follows Power-Law has significant predicting power over the market and instead of using parameter α. We choose Clauset's goodness of fit which is more appropriate to measure the whole market. As expected, the predicting power is significant and the closer the wealth distribution approaches the Power-Law, the more stable of the Bitcoin market will be.

In addition, a better crypto-currency usually entails the following characteristics.

i. Known creator;
ii. Some work is required to get the coin;
iii. Coins are not distributed for free;
iv. Attains Pareto-distribution;
v. Appropriate reward scheme;
vi. Good credit of developer;
vii. Uniqueness;
viii. Good launching time.

With this knowledge in mind, countries can use this information to create a successful cryptocurrency when they ever desire to make use of digital currency in the future. Our preliminary study of two digital currencies needs to be expanded to more digital and crypto currencies in order to draw a firmer conclusion. Nevertheless, while Clauset's goodness of fit can be used and has some predictive power over the fundamental variables of a cryptocurrency, it may not be sufficient enough. We may need to combine the indicator with other explanatory variables to yield more accurate predictions. In the future, we could continue to monitor the goodness of fit in order to verify the results obtained in this research paper and to expand the study to other coins. Furthermore, weekly instead of monthly data could be used for Auroracoin in order to further validate the significant level of goodness of fit in relation with the Bitcoin market. Finally, we would compare more density functions to fit the wealth distribution instead of using only the Power-Law model.

References

Clauset, A., Shalizi, C. R., & Newman, M. E. J. (2009). Power-law distributions in empirical data. *SIAM Review*, *51*(4), 661–703.

Ignacio, M., & Lee, D. K. C. (2014). Bitcoin-like Protocols and Innovations. Sim Kee Boon Institute for Financial Economics, Singapore Management University.

Lam, P. N., & Lee, D. K. C. (2014). Introduction to Bitcoin. Sim Kee Boon Institute for Financial Economics, Singapore Management University.

Lee, D. K. C., Ong, B. C. E., Lee, T.M., & Li, G. (2014). Evaluating the Potential of Alternative Cryptocurrencies. Sim Kee Boon Institute for Financial Economics, Singapore Management University.

Li, Xiang Jun, Lee, David K. C., & Teo, Ernie. (2014). Life Cycle of Cryptocurrencies. Sim Kee Boon Institute for Financial Economics, Singapore Management University.

Chapter 17
Time Varying Quantile Lasso

Wolfgang Karl Härdle, W. Wang and L. Zboňáková

Abstract In the present chapter we study the dynamics of penalization parameter λ of the least absolute shrinkage and selection operator (Lasso) method proposed by Tibshirani (J Roy Stat Soc Series B 58:267–288, 1996) and extended into quantile regression context by Li and Zhu (J Comput Graph Stat 17:1–23, 2008). The dynamic behaviour of the parameter λ can be observed when the model is assumed to vary over time and therefore the fitting is performed with the use of moving windows. The proposal of investigating time series of λ and its dependency on model characteristics was brought into focus by Härdle et al. (J Econom 192:499–513, 2016), which was a foundation of FinancialRiskMeter. Following the ideas behind the two aforementioned projects, we use the derivation of the formula for the penalization parameter λ as a result of the optimization problem. This reveals three possible effects driving λ; variance of the error term, correlation structure of the covariates and number of nonzero coefficients of the model. Our aim is to disentangle these three effects and investigate their relationship with the tuning parameter λ, which is conducted by a simulation study. After dealing with the theoretical impact of the three model characteristics on λ, empirical application is performed and the idea of implementing the parameter λ into a systemic risk measure is presented.

W.K. Härdle · W. Wang · L. Zboňáková (✉)
C.A.S.E.-Center of Applied Statistics and Economics,
Humboldt-Universität zu Berlin, Spandauer Str. 1, 10178 Berlin, Germany
e-mail: lenka.zbonakova@hu-berlin.de

W.K. Härdle
Singapore Management University, 50 Stamford Road,
178899 Singapore, Singapore
e-mail: haerdle@hu-berlin.de

W. Wang
Department of Economics, City, University of London,
Northampton Square, London EC1V 0HB, UK
e-mail: wangwein@cms.hu-berlin.de

W.K. Härdle et al. (eds.), *Applied Quantitative Finance*, Statistics and Computing,
DOI 10.1007/978-3-662-54486-0_17

17.1 Introduction

The least absolute shrinkage and selection operator (Lasso) method as proposed by Tibshirani (1996) has been widely used and extended during recent years. The literature presents a method which simultaneously completes the task of model selection and parameter estimation, while studying its consistency. A key factor for the estimation precision is choosing a tuning parameter which controls the degree of penalization. Although there is much literature on Lasso, including a time series context, the time variation of the tuning parameter remains unexplored.

Here we explain dynamics of the penalization parameter λ and how it can be used in financial practice, particularly when dealing with systemic risk. Let us assume for the moment a linear model with a vector of responses $Y = (Y_1, Y_2, \ldots, Y_n)^\top$, a vector of parameters $\beta = (\beta_1, \ldots, \beta_p)^\top$, an $(n \times p)$ design matrix X, which might be either fixed or random, and a vector of independent identically distributed errors ε with zero mean and variance σ^2. Then the objective function of Lasso is

$$\min_{\beta} \left\{ \frac{1}{2} \sum_{i=1}^{n} \left(Y_i - X_i^\top \beta\right)^2 + \lambda \sum_{j=1}^{p} |\beta_j| \right\}, \tag{17.1}$$

with tuning parameter $\lambda \geq 0$ and X_i, $0 \leq i \leq n$, denoting row vectors of X. In (17.1) one assumes that the columns of the matrix $X = (x_{ij})_{i=1,\ldots,n, j=1,\ldots,p}$ have been standardized, i.e. $n^{-1} \sum_{i=1}^{n} x_{ij} = 0$ and $n^{-1} \sum_{i=1}^{n} x_{ij}^2 = 1$. Solving this type of penalized least squares problem with L_1-penalization allows some of the coefficients of the model to shrink to 0. This is a highly advantageous property when dealing with high-dimensional data, since variable selection and shrinkage of coefficients are performed simultaneously. Shrinking some of the coefficients to exactly 0 also improves the interpretability of the fitted model.

Modification of Lasso in quantile regression (Koenker and Basset 1978) studied by Li and Zhu (2008) and Belloni and Chernozhukov (2011) solves the optimization problem with

$$\min_{\beta} \left\{ \frac{1}{2} \sum_{i=1}^{n} \rho_\tau \left(Y_i - X_i^\top \beta\right) + \lambda \sum_{j=1}^{p} |\beta_j| \right\}, \tag{17.2}$$

where $\tau \in (0, 1)$ and $\rho_\tau(\cdot)$ is the check function

$$\rho_\tau(x) = \begin{cases} \tau \cdot x & \text{if } x > 0; \\ -(1-\tau) \cdot x & \text{otherwise.} \end{cases} \tag{17.3}$$

The Lasso models described above account for independent observations. However, there is much literature on the Lasso in time series context as well. For the univariate case we refer to Wang et al. (2007), Nardi and Rinaldo (2011) and Chen and Chan

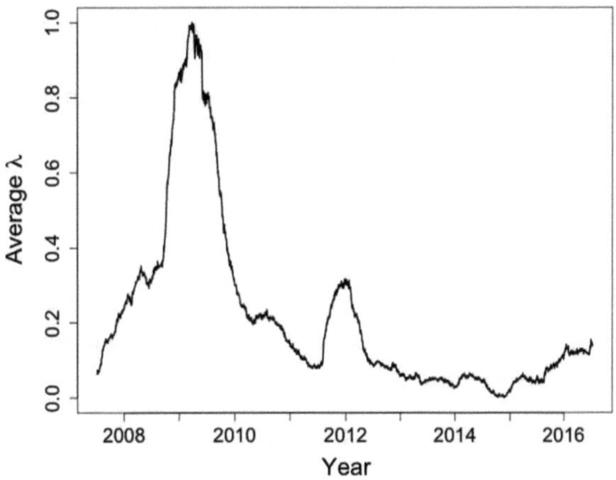

Fig. 17.1 Time series of λ taken from FinancialRiskMeter (http://frm.wiwi.hu-berlin.de), normalized to interval (0,1)

(2011). The case of multivariate time series, particularly vector autoregression, was covered by e.g. Hsu et al. (2008).

Lasso in quantile regression has been used by Härdle et al. (2016) to model tail event dependencies among U.S. financial companies. Based on the penalization parameters the FinancialRiskMeter (FRM), http://frm.wiwi.hu-berlin.de, was developed, see Fig. 17.1. The value of the averaged penalization parameter λ was elevated during the financial crises. This fact led us to the question we indicated above; what drives the penalization parameter λ and what are the dynamics of λ? We investigate this by simulation study and empirical application.

The computations included in this chapter were performed in the environment of R software developed by R Core Team (2014) and the codes are available on http://quantlet.de/d3/ia/.

17.2 Lasso Method

17.2.1 Lasso as an Optimization Problem

In this section we firstly follow Osborne et al. (2000) to derive formula for the penalization parameter λ of the Lasso method when applied in linear regression problems. Then we aim our focus on the representation of λ in penalized quantile regression.

If we treat λ as a fixed value in the objective function of the penalized regression

$$f(\beta, \lambda) = \left\{ \frac{1}{2} \sum_{i=1}^{n} \left(Y_i - X_i^\top \beta\right)^2 + \lambda \sum_{j=1}^{p} |\beta_j| \right\}, \tag{17.4}$$

then the function $f(\beta, \lambda)$ is convex in parameter β. Moreover, with diverging β we observe that $f(\beta, \lambda) \to \infty$. Hence there exists at least one minimum of the function $f(\cdot, \lambda)$. According to Osborne (1985) this minimum is attained in $\widehat{\beta}(\lambda)$ if and only if the null-vector $0 \in \mathbb{R}^p$ is an element of the subdifferential

$$\frac{\partial f(\beta, \lambda)}{\partial \beta} = -X^\top (Y - X\beta) + \lambda u(\beta), \tag{17.5}$$

where $u(\beta) = (u_1(\beta), \dots, u_p(\beta))^\top$ is defined as $u_j(\beta) = 1$ if $\beta_j > 0$, $u_j(\beta) = -1$ if $\beta_j < 0$ and $u_j(\beta) \in [-1, 1]$ if $\beta_j = 0$. Then, for $\widehat{\beta}(\lambda)$ as a minimizer of $f(\beta, \lambda)$ the following has to be satisfied

$$0 = -X^\top \{Y - X\widehat{\beta}(\lambda)\} + \lambda u(\widehat{\beta}(\lambda)), \tag{17.6}$$

Here we denote the estimator of a parameter vector β as a function of the penalization parameter λ. This dependency follows from the formulation of the penalized regression method and its objective function (17.4), where we first select λ and then search for $\widehat{\beta}(\lambda)$ which minimizes (17.4). Using the fact that $u(\beta)^\top \beta = \sum_{j=1}^p |\beta_j| = ||\beta||_1$, where $||\cdot||_1$ denotes L_1-norm of a p-dimensional vector, (17.6) can be further rewritten in the formula

$$\lambda = \frac{\{Y - X\widehat{\beta}(\lambda)\}^\top X\widehat{\beta}(\lambda)}{||\widehat{\beta}(\lambda)||_1}. \tag{17.7}$$

The identity (17.7) leads us to consider possible constituents which influence the value of parameter λ and therein its dynamics when treated in a time-dependent framework. Here we propose to study three effects which are related to the size of λ:

1. size of residuals of the model;
2. absolute size of the coefficients of the model, $||\beta||_1$;
3. singularity of a matrix $X^\top X$.

The second effect can also be translated into the effect of a number of nonzero parameters the so-called active set of the model, $q = ||\beta||_0 = \sum_{j=1}^p \mathbf{I}(\beta_j \neq 0)$, where $||\cdot||_0$ stands for L_0-norm on $\mathbb{R}^p$ and $\mathbf{I}(\cdot)$ is an indicator function. As a measure of the third structure, the condition number $\kappa(X^\top X)$ defined as the ratio $\phi_{\max}(X^\top X)/\phi_{\min}(X^\top X)$, the maximum and the minimum eigenvalue of the matrix $X^\top X$, can be used.

Similarly, one can derive formulae for the penalization parameter λ in a quantile regression problem (17.2) and (17.3). Following Li and Zhu (2008)

$$\lambda = \frac{\theta^\top X\widehat{\beta}(\lambda)}{||\widehat{\beta}(\lambda)||_1}, \tag{17.8}$$

where $\theta = (\theta_1, \dots, \theta_n)^\top$ satisfies the following

$$\theta_i = \begin{cases} \tau & \text{if } Y_i - X_i^\top \widehat{\beta}(\lambda) > 0; \\ -(1-\tau) & \text{if } Y_i - X_i^\top \widehat{\beta}(\lambda) < 0; \\ \in (-(1-\tau), \tau) & \text{if } Y_i - X_i^\top \widehat{\beta}(\lambda) = 0. \end{cases} \tag{17.9}$$

Hence, we observe that λ depends on cardinality of the active set q, which is again influenced by the correlation structure of the design matrix. Direct impact of the variance of residuals disappears and only the sign of the residuals stays in effect. However, when looking at Fig. 17.2 one can see similarities between the time series of λ and historic values of the implied volatility index (VIX) reported by the Chicago Board Options Exchange. This fact leads us to believe that the dynamics of λ is also influenced by the changes in the variance of model residuals.

17.2.2 Choosing the Penalization Parameter

In theory the equalities (17.7) and (17.8) hold for every solution of the Lasso optimization problems (17.1) and (17.2) respectively, since first λ is chosen and afterwards the model is fitted according to the given value of the penalization parameter. One of the commonly used methods of choosing estimator of λ is cross-validation in its three forms; k-fold, leave-one-out and generalized cross-validation method, see e.g. Tibshirani (1996). As pointed out in Hastie et al. (2009), cross-validation is a widely used method for estimation of prediction error. This feature is used when estimating λ in Lasso method, where, on a grid of penalization parameters λ, the one which minimizes estimated prediction error is chosen. However, as Leng et al. (2006) argued in their work, methods of choosing penalization parameter based on prediction accuracy are in general not consistent when variable selection is considered. The same argument was used by Wang et al. (2009) where they compared the asymptotic behaviour of the generalized cross-validation to the one of Akaike's

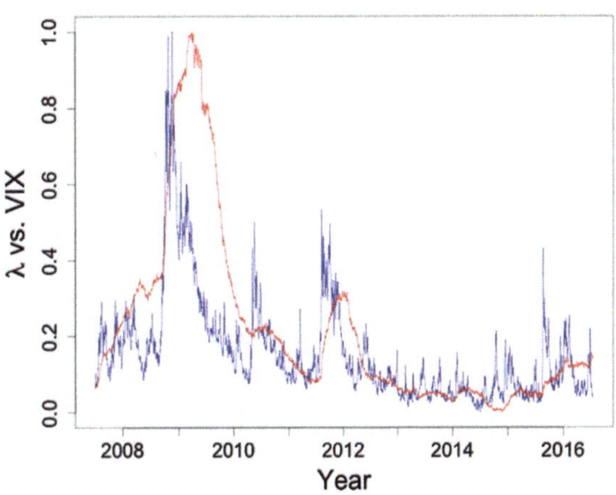

Fig. 17.2 Normalized implied volatility index (*blue*) and λ from FinancialRiskMeter (*red*). XFGTVP_LambdaVIX

information criterion (AIC); it is efficient if one is interested in the model error, but inconsistent in selecting the true model.

The second widely used method of estimating λ is the Bayesian information criterion (BIC). By $\beta_0 = (\beta_{01}, \dots, \beta_{0p})^\top$ we denote the true vector of coefficients of the regression model and q_0 defines the number of its nonzero elements, i.e. $\beta_{0j} \neq 0$ for $1 \leq j \leq q_0$ and $\beta_{0j} = 0$ for $j > q_0$. The permutation of the elements of β_0 is performed without loss of generality, so the previous notation holds. Secondly, by $\mathcal{S} = \{j_1, \dots, j_q\}$ we denote an arbitrary model with $X_\mathcal{S} = (X_{j_1}, \dots, X_{j_q}) \in \mathbb{R}^{n\times q}$ as a design matrix associated with it. Vector of coefficients of a model $\mathcal{S}$ is $\beta_\mathcal{S} = (\beta_{j_1}, \dots, \beta_{j_q})^\top$ and the model size is $|\mathcal{S}| = q$. The true model is referred to by $\mathcal{S}_0$.

Using the notation from above, the BIC is written in the following form

$$\mathrm{BIC}_\mathcal{S} = \log(\widehat{\sigma}^2_\mathcal{S}) + |\mathcal{S}|\frac{\log(n)}{n}C_n, \tag{17.10}$$

with $\widehat{\sigma}^2_\mathcal{S} = n^{-1}\mathrm{SSE}_\mathcal{S} = \inf_{\beta_\mathcal{S}}(n^{-1}||Y - X_\mathcal{S}\beta_\mathcal{S}||_2^2)$ where $||\cdot||_2$ denotes L_2-norm of a vector and C_n is some positive constant. Wang and Leng (2007) prove the consistency of (17.10) in selecting a true model also for a diverging parameter vector dimension p and a true number of nonzero coefficients q_0. This is shown in unpenalized as well as in penalized regression models.

Modification of (17.10) in terms of a tuning parameter leads to

$$\mathrm{BIC}_\lambda = \log(\widehat{\sigma}^2_\lambda) + |\mathcal{S}_\lambda|\frac{\log(n)}{n}C_n, \tag{17.11}$$

where $\widehat{\sigma}^2_\lambda = n^{-1}\mathrm{SSE}_\lambda = n^{-1}||Y - X\widehat{\beta}(\lambda)||_2^2$ and $\mathcal{S}_\lambda = \{j : \widehat{\beta}(\lambda)_j \neq 0\}$. The estimation of the tuning parameter $\widehat{\lambda}$ is then chosen by minimizing (17.11) with $C_n = \log\{\log(p)\}$ or $C_n = \sqrt{n}/p$, see Chand (2012).

Consistency of the BIC_λ selector holds for the penalized regression methods such as smoothly clipped absolute deviation (SCAD) method defined by Fan and Li (2001) and adaptive Lasso introduced by Zou (2006). For the regular Lasso method by Tibshirani (1996) the additional assumption on a design matrix X called irrepresentable condition has to be fulfilled.

The aforementioned condition was presented by Zhao and Yu (2006). Firstly they assumed that $n^{-1}X^\top X \xrightarrow{p} C$, with C a positive definite matrix

$$C = \begin{pmatrix} C_{11} & C_{12} \\ C_{21} & C_{22} \end{pmatrix}. \tag{17.12}$$

Here C_{11} is a $(q_0 \times q_0)$ matrix that corresponds to the q_0 active predictors and is assumed to be invertible. Then the formulation of the irrepresentable condition is

$$\left|\left[C_{21}C_{11}^{-1}\mathrm{sgn}(\beta_{\mathcal{S}_0})\right]_k\right| \leq 1, \quad k = 1, \dots, p - q_0. \tag{17.13}$$

Adopting the notation from above, q_0 is a number of nonzero parameters in the true model $\mathcal{S}_0$ and $\operatorname{sgn}(\beta_{\mathcal{S}_0}) = (\operatorname{sgn}(\beta_{01}), \ldots, \operatorname{sgn}(\beta_{0q_0}))^\top$ with sign function $\operatorname{sgn}(\beta_j) = 1$ if $\beta_j > 0$, $\operatorname{sgn}(\beta_j) = -1$ if $\beta_j < 0$ and $\operatorname{sgn}(\beta_j) = 0$ if $\beta_j = 0$.

Modified selection criteria for penalized quantile regression which were used by Li and Zhu (2008) are BIC for quantile regression presented by Koenker et al. (1994) and generalized approximate cross-validation criterion (GACV) introduced by Yuan (2006)

$$\text{BIC}(\lambda) = \log\left[n^{-1}\sum_{i=1}^{n}\rho_\tau\{Y_i - X_i^\top\widehat{\beta}(\lambda)\}\right] + \frac{\log(n)}{2n}\widehat{\text{df}}(\lambda), \tag{17.14}$$

$$\text{GACV}(\lambda) = \frac{\sum\limits_{i=1}^{n}\rho_\tau\{Y_i - X_i^\top\widehat{\beta}(\lambda)\}}{n - \widehat{\text{df}}(\lambda)}, \tag{17.15}$$

where $\widehat{\text{df}}(\lambda)$ stands for the estimated effective dimension of the fitted model. Li and Zhu (2008) argued that number of interpolated observations Y_i denoted by $\mathcal{E}$ is a plausible measure for this quantity, i.e. $\widehat{\text{df}}(\lambda) = |\mathcal{E}|$.

17.2.3 Algorithms to Solve Lasso

Finding a feasible solution of the optimization problems (17.1) and (17.2) can be computationally demanding, since one has to check all of the combinations of values of the tuning parameter λ and its respective model parameter estimates $\widehat{\beta}(\lambda)$. Only after all of the possible combinations are found, the particular method of choosing $\widehat{\lambda}$ can be applied.

The first algorithm for finding solution of Lasso was presented by Tibshirani (1996) in his work introducing the Lasso method itself. Then Osborne et al. (2000) developed an algorithm which works not only for the case where $p < n$ but also $n > p$. In order to make the computation more efficient, Efron et al. (2004) proposed the use of the least angle regression algorithm (LARS). The latter procedure is as efficient as a single least squares fit and can also be used in cases where number of parameters of the investigated model is much larger than the number of observations. As a selection criterion of $\widehat{\lambda}$ for LARS, Efron et al. (2004) suggested to use C_p-type selection criterion. Zou et al. (2007) then defined model selection criteria such as C_p, Akaike information criterion (AIC) and BIC suitable for the Lasso framework.

Another approaches to find a path of Lasso solutions, particularly for the quantile regression, were proposed by Belloni and Chernozhukov (2011) and Li and Zhu (2008). The second one comes into focus in this chapter, since one is interested in modeling tail event dependencies when dealing with systemic risk evaluations.

17.3 Simulation Study

As derived in the previous section, the penalization parameter λ of the Lasso regression depends on three effects. The factors driving its dynamics are variance of the error term of the model, conditionality of the matrix $X^\top X$ and absolute size of the coefficients of the model, $||\beta||_1$. In this section we conduct simulations which describe the relationships between these three effects and the parameter λ focusing mainly on a quantile regression case. Our aim is to disentangle these effects and find the way to explain behaviour of λ in dependency of the three aforementioned elements.

17.3.1 Penalty λ Dependent on Variance σ^2

Firstly we investigate the effect of the size of variance σ^2 of the error term ε on the penalty parameter λ. According to the identity (17.7) λ is supposed to rise with higher σ^2 and vice versa. This holds for the linear regression problem, and as discussed previously for the quantile regression as well. The evidence is visible from Fig. 17.2, whereas when considering the formula (17.8) this dependency is not straightforward to follow.

In our simulation study we use quantile regression model $Y = X\beta + \varepsilon$ with a vector of responses $Y = (Y_1, \ldots, Y_n)^\top$, a vector of parameters $\beta = (\beta_1, \ldots, \beta_p)^\top$, an $(n \times p)$ design matrix X and *iid* error term $\varepsilon = (\varepsilon_1, \ldots, \varepsilon_n)^\top$ such that $\mathrm{P}(\varepsilon_i \leq 0 | X_i = x) = \tau$ for almost every $x \in \mathbb{R}^p$ with $\tau \in (0, 1)$ denoting conditional quantile of Y.

The design matrix X is simulated from the p-dimensional normal distribution

$$\{X_i\}_{i=1}^n \sim \mathrm{N}_p(0, \Sigma), \tag{17.16}$$

where the elements of $(p \times p)$ covariance matrix $\Sigma = (\sigma_{ij})_{i,j=1}^p$ are defined as follows

$$\sigma_{ij} = \rho^{|i-j|} \text{ for } i, j = 1, \ldots, p, \tag{17.17}$$

with $\rho = 0.5$ as in Tibshirani (1996). Here we select $n = 600$ and $p = 100$. In order to study the effect of increased dispersion (in the error term ε) on λ, the vector of parameters is set to

$$\beta_{0(100\times 1)} = (1, 1, 1, 1, 1, 0, \ldots, 0)^\top. \tag{17.18}$$

The error term is simulated such that its variance changes after the observation $i_0 = 300$. We assume ε_i for $i = 1, \ldots, n$ to be independently distributed with asymmetric Laplace distribution

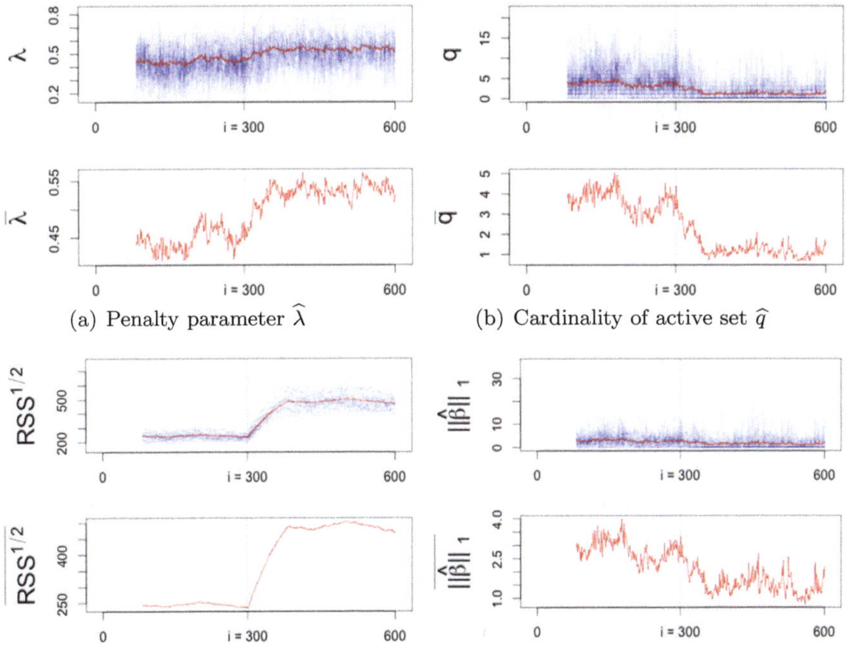

(a) Penalty parameter $\widehat{\lambda}$

(b) Cardinality of active set $\widehat{q}$

(c) L_2-norm of residuals

(d) L_1-norm of coefficients

Fig. 17.3 Time series of $\widehat{\lambda}$ (*blue*), other model characteristics and their respective averages (*red*) drawn from the 50 simulations with change of σ_i after $i_0 = 300$, moving windows of length 80. XFGTVP_LambdaSim

$$\varepsilon_i \sim \begin{cases} \text{ALD}(0, 1, 0.05), & \text{if } i \leq i_0 \\ \text{ALD}(0, 2, 0.05), & i > i_0 \end{cases}. \tag{17.19}$$

The density of asymmetric Laplace distribution is

$$f(x|\mu, \sigma, \tau) = \frac{\tau(1-\tau)}{\sigma} \exp\left\{-\frac{\rho_\tau(x-\mu)}{\sigma}\right\}, \tag{17.20}$$

with location parameter μ, scale parameter $\sigma > 0$, skewness parameter $\tau \in (0, 1)$ and the check function $\rho_\tau(\cdot)$ as defined in (17.3). The idea to use this type of distribution comes from Lee et al. (2014).

We simulate 50 scenarios using the algorithm designed by Li and Zhu (2008) and select $\widehat{\lambda}$ according to BIC (17.14). For model fitting we apply moving windows technique to capture the dynamics of the tuning parameter λ. The size of the moving window is set to be $w = 80$. Resulting values of $\widehat{\lambda}$ obtained by simulation settings above are, together with other model characteristics of interest, captured in Fig. 17.3.

As can be seen from Fig. 17.3, the values of the estimated tuning parameter $\widehat{\lambda}$ are indeed increasing with higher variation σ^2 of the error term. Number of nonzero parameters $q_0 = ||\beta_0||_0$ was set to be constant over all $n = 600$ observations and

Table 17.1 Relative and absolute change in averaged values of $\widehat{\lambda}$ before and after the change point $i_0 = 300$ with starting value of the scale parameter $\sigma_i = 1$ for $i \leq i_0$

σ_i $i > i_0$	$\frac{\bar{\hat{\lambda}}_{end}}{\bar{\hat{\lambda}}_{start}}$	$\bar{\hat{\lambda}}_{end} - \bar{\hat{\lambda}}_{start}$
1.1	1.061	0.027
1.2	1.084	0.037
1.3	1.112	0.050
1.4	1.135	0.060
1.5	1.144	0.064
1.6	1.162	0.072
1.7	1.169	0.075
1.8	1.177	0.079
1.9	1.187	0.083
2.0	1.199	0.089

also the level around which the condition number $\kappa(X^\top X)$ fluctuates stays constant. However, the L_1-norm of estimated model coefficients $||\widehat{\beta}(\widehat{\lambda})||_1$ changes with higher values of $\widehat{\lambda}$. Since that is an idea of the Lasso method itself, this can be seen as a natural effect.

In order to study the size of impact of σ^2 on λ we conducted a set of simulations, where different values of scale parameter σ were used after the change point i_0. The starting value was defined as in the previous case, $\sigma = 1$, and the relative and absolute change of average $\widehat{\lambda}$ were examined. Observed changes are noted in Table 17.1.

From Table 17.1 one can see that the penalization parameter $\widehat{\lambda}$ increases in dependency of the change in the scale parameter σ of the distribution of the error term in the assumed model. This conclusion of course corresponds to what we see from Fig. 17.2. Again we use BIC as a selection criterion. However, as discussed before, theoretically other methods yield the same dependency structure.

17.3.2 Penalty λ Dependent on Model Size q

The second effect driving the size of the penalization parameter λ is the number of nonzero parameters q. In order to study this case, the design matrix X was again set as in (17.16) and (17.17) with $\rho = 0.5$. The error term ε_i was simulated to have scale $\sigma = 1$ for all $1 \leq i \leq n$ and the change in vector of model parameters β came into focus. The number of nonzero parameters of the model was defined by setting β_0 to have the form

$$\beta_{0i} = \begin{cases} (1, 1, 1, 1, 1, 0, \ldots, 0)^\top, \; i \leq i_0 \\ (\underbrace{1, 1, \ldots, 1}_{10\times}, 0, \ldots, 0)^\top, \; i > i_0. \end{cases} \tag{17.21}$$

Thus, the first i_0 simulated observations have five active parameters and the rest has ten of them.

The paths of the values of $\widehat{\lambda}$ obtained from the aforementioned simulation settings are plotted in Fig. 17.4. Visible are also other characteristics of the model which we are interested in to examine.

As expected from (17.8) defining λ, an increasing value of $||\widehat{\beta}(\lambda)||_1$ or q results in a decreasing value of the tuning parameter λ. In this specific case $||\beta_0||_1 = q_0$. From Fig. 17.4 one can see that the value of $\widehat{\lambda}$ decreased with higher $\widehat{q}$.

To study the reaction of λ on the cardinality of the active set q, we performed simulations with different changes of q after the observation i_0, the starting value was always $q_0 = 5$. The results are summarized in Table 17.2. From Eq. (17.8) the relationship between λ and $||\widehat{\beta}(\lambda)||_0$ as well as q is inversely proportional and values in Table 17.2 correspond to this statement.

We may conclude that the cardinality of the active set q has a real impact on change in value of λ. Since in (17.8) the effect of q is captured by the effect of $||\widehat{\beta}(\lambda)||_1$, this is also of our interest. Another simulation was conducted to investigate the impact of the L_1-norm of the model coefficients. Previously the coefficients were hard thresholded, i.e. cut off abruptly and set to be zero. Now the parameters are allowed to decrease to zero more smoothly

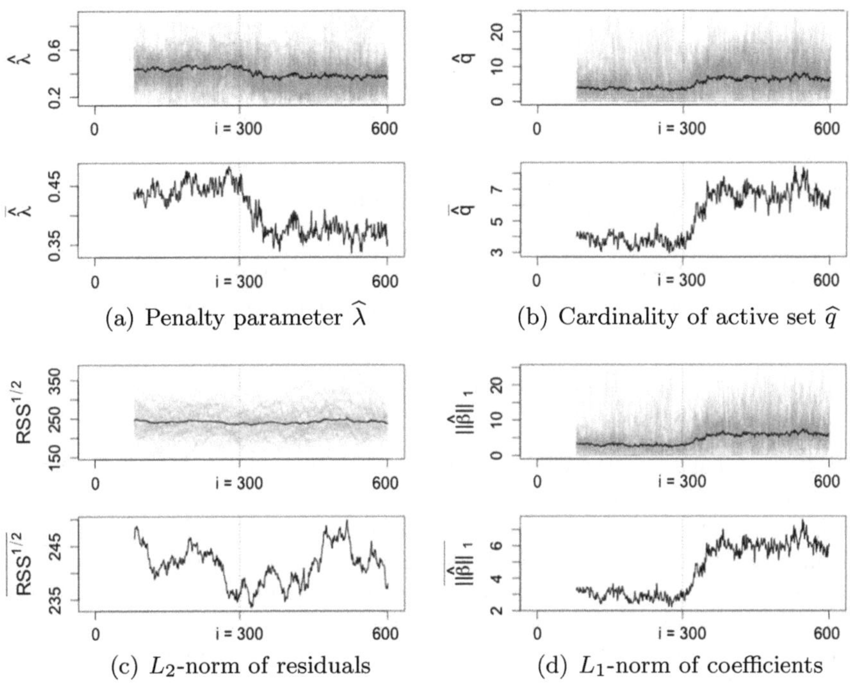

Fig. 17.4 Time series of $\widehat{\lambda}$ (*gray*), other model characteristics and their respective averages (*black*) drawn from the 50 simulations with change of q_0 after $i_0 = 300$, moving windows of length 80. XFGTVP_LambdaSim

Table 17.2 Relative and absolute change in averaged values of $\widehat{\lambda}$ before and after the change point $i_0 = 300$ with starting number of nonzero parameters $q_{0i} = 5$ for $i \le i_0$

q_{0i} $i > i_0$	$\frac{\bar{\widehat{\lambda}}_{end}}{\bar{\widehat{\lambda}}_{start}}$	$\bar{\widehat{\lambda}}_{end} - \bar{\widehat{\lambda}}_{start}$
6	0.952	−0.021
7	0.922	−0.035
8	0.905	−0.043
9	0.862	−0.062
10	0.837	−0.073
15	0.736	−0.118

$$\beta_{0i} = \begin{cases} (\underbrace{1, 1, \ldots, 1}_{10\times}, 0, \ldots, 0)^\top, & i \le i_0 \\ (1, 0.9, 0.8, \ldots, 0.2, 0.1, 0, \ldots, 0)^\top, & i > i_0, \end{cases} \tag{17.22}$$

i.e. $||\beta_{0i}||_1 = 10$ for $i \le i_0$ and $||\beta_{0i}||_1 = 5.5$ for $i > i_0$.

We put this simulation setting forward, because it seems more natural that the effect of particular covariates fades away rather than disappears. Time series of model characteristics of this case are to be found in Fig. 17.5. The relative and absolute change of average $\widehat{\lambda}$ after the point $i_0 = 300$ is 1.245 and 0.091 respectively.

17.3.3 Penalty λ Dependent on Design

We examine the dependency of the parameter λ on the design matrix X of the given model through the characteristics called condition number of a matrix:

$$\kappa(X^\top X) = \frac{\phi_{\max}(X^\top X)}{\phi_{\min}(X^\top X)},$$

where $\phi_{\max}(\cdot)$ and $\phi_{\min}(\cdot)$ are the largest and the smallest eigenvalues of a matrix. If the condition number κ is low the problem is called well-conditioned, matrices with higher κ values are referred to as ill-conditioned. The condition number can help to diagnose a multicollinearity issue. With the presence of multicollinearity, one can expect more coefficients to be incorrectly defined as significant and therefore values of q and $||\beta||_1$ to rise. This is in analogy to the situation described in the previous subsection and regarding the formula (17.8) we expect the tuning parameter λ to decrease with higher condition number of the matrix $X^\top X$.

The simulation settings are as follows; parameter β_0 as in (17.18) and the error term is *iid* with $\varepsilon_i \sim$ ALD(0, 1, 0.05) for $0 \le i \le n$. The design matrix X is simulated from (17.16) and (17.17), but here the parameter ρ is allowed to change after the

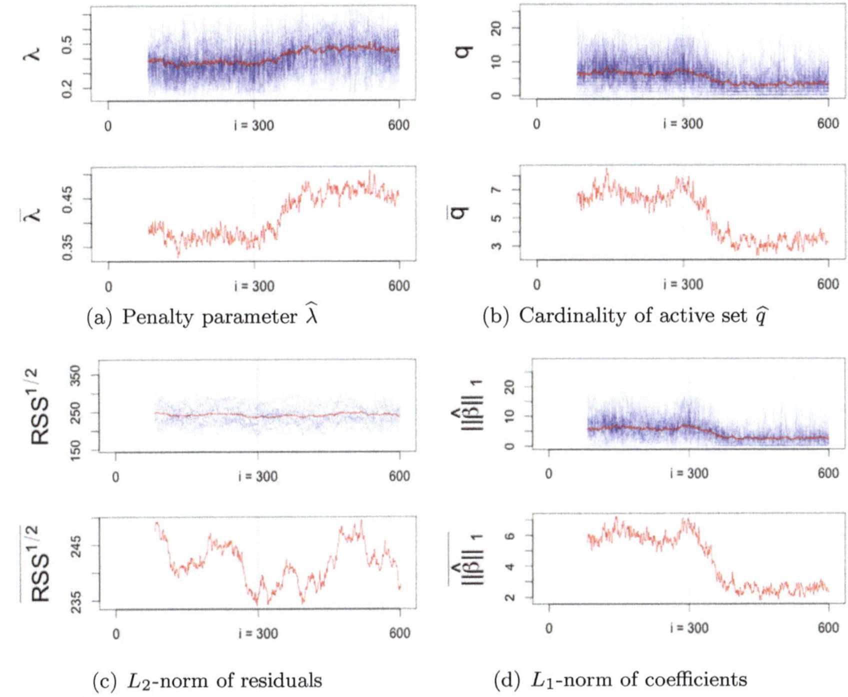

Fig. 17.5 Time series of $\widehat{\lambda}$ (*blue*), other model characteristics and their respective averages (*red*) drawn from the 50 simulations with change of $||\beta_{0i}||_1$ after $i_0 = 300$, moving windows of length 80. XFGTVP_BetaChange

Table 17.3 Relative and absolute change in averaged values of $\widehat{\lambda}$ before and after the change point $i_0 = 300$ with starting number of nonzero parameters $\rho_i = 0$ for $i \leq i_0$

$\rho_i\ i > i_0$	$\frac{\bar{\bar{\lambda}}_{end}}{\bar{\bar{\lambda}}_{start}}$	$\bar{\bar{\lambda}}_{end} - \bar{\bar{\lambda}}_{start}$
0.1	1.023	0.012
0.3	0.943	−0.028
0.5	0.890	−0.055
0.7	0.692	−0.155
0.9	0.750	−0.126

point $i_0 = 300$. The case where $\rho_i = 0$ for $i \leq i_0$ and $\rho_i = 0.5$ for $i > i_0$ is illustrated in Fig. 17.6.

Indeed, our expectations presented above hold true. Increased correlation between the covariates and with that increased condition number $\kappa(X^\top X)$ result in decreasing values of the estimated tuning parameter $\widehat{\lambda}$. This case together with other simulated changes in correlation structure between covariates are summarized in Table 17.3. Starting value of ρ from (17.17) is always 0.

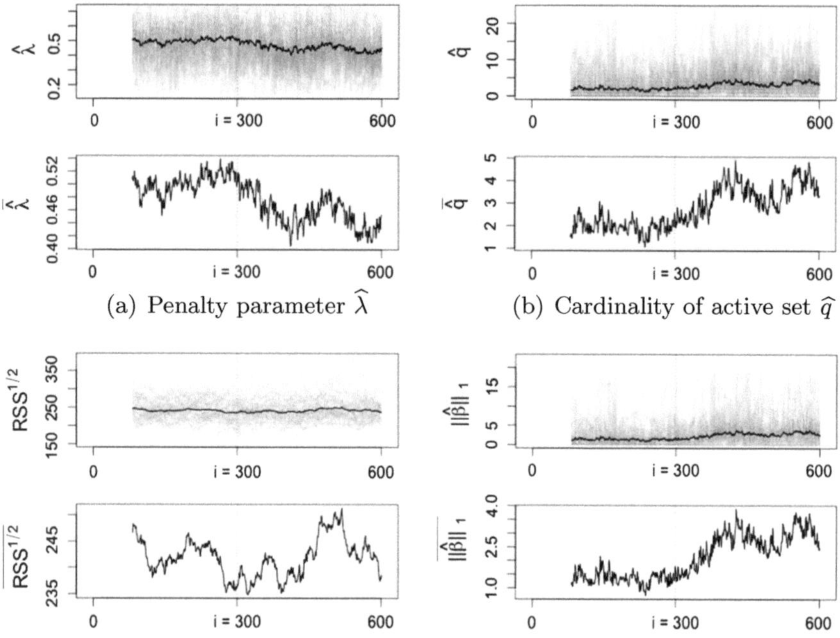

(a) Penalty parameter $\widehat{\lambda}$ (b) Cardinality of active set $\widehat{q}$

(c) L_2-norm of residuals (d) L_1-norm of coefficients

Fig. 17.6 Time series of $\widehat{\lambda}$ (*gray*), other model characteristics and their respective averages (*black*) drawn from the 50 simulations with change of ρ_i after $i_0 = 300$, moving windows of length 80. XFGTVP_LambdaSim

17.3.4 All Factors Affecting the Value of λ

So far we investigated the effect of the change in the variance of error term σ^2, in structure of the vector of parameters β and in the correlation structure of the covariates ceteris paribus. In this subsection we focus on all of the factors driving dynamics of λ at once and examine the strength of their impact when combined together.

For each of the elements driving the dynamics of the penalization parameter λ we simulated three cases. The values of interest either stayed constant, increased or

Table 17.4 Relative changes $\widehat{\lambda}_{end}/\widehat{\lambda}_{start}$ as a result of combinations of changes in a model Relative changes $\widehat{\lambda}_{end}/\widehat{\lambda}_{start}$ as a result of combinations of changes in a model

	$\sigma^2 \nearrow$			$\sigma^2 \rightarrow$			$\sigma^2 \searrow$		
	$\kappa \nearrow$	$\kappa \rightarrow$	$\kappa \searrow$	$\kappa \nearrow$	$\kappa \rightarrow$	$\kappa \searrow$	$\kappa \nearrow$	$\kappa \rightarrow$	$\kappa \searrow$
$q_0 \nearrow$	0.884	1.101	1.311	0.783	0.843	1.003	0.659	0.710	0.841
$q_0 \rightarrow$	0.992	1.198	1.425	0.854	1.001	1.191	0.719	0.843	0.998
$q_0 \searrow$	1.162	1.403	1.555	1.000	1.172	1.300	0.759	0.889	1.125

decreased after the point $i_0 = 300$. If constant, the scale parameter σ of the distribution of the error term was set to be 1. Otherwise it increased from the value of 1 to 2 or decreased from 2 to the value of 1. Number of nonzero parameters was either $q_0 = 5$ for all $n = 600$ observations or it increased to the value $q_0 = 10$ or decreased from $q_0 = 10$ to $q_0 = 5$ after the point i_0. The change of the design matrix was again defined by the change of the correlation structure between corresponding covariates, i.e. change of ρ from (17.17). For the constant case it was set to be $\rho = 0.5$, when increased it had value 0.9 after the i_0-th observation and for the decreasing case it was $\rho = 0.9$ for $i \leq i_0$ and $\rho = 0.5$ for $i > i_0$.

Results of all combinations of the changes in the factors having impact on λ are summarized in Table 17.4. There we can see that the effects can overpower each other when combined. This holds particularly for the cases, when the condition number κ is increased and number of nonzero parameters q_0 decreased and vice versa. This fact can be explained by the issue of multicollinearity as discussed before.

Empirically, when considering the situation on financial markets (particularly modeling of stock prices), increased volatility indicates elevated risk. Parameter λ is sensitive to the changes in degree of variation and therefore can be bound to the risk evaluation problem. Another aspect indicating time series of λ as a measure of systemic risk is its dependency on interconnectedness of financial institutions, which can be measured by the number of nonzero parameters in estimated model and their magnitude.

17.4 Empirical Analysis

17.4.1 Data Description

In order to be able to apply our insight to the FinancialRiskMeter (http://frm.wiwi.hu-berlin.de), we closely follow the choice of data of Härdle et al. (2016). Due to the computational efficiency, our dataset consists of daily stock returns of the first 100 largest U.S. financial companies ordered by market capitalization according to NASDAQ company list. In the FRM case it is 200. The stock returns are downloaded from Yahoo Finance and the list of the corresponding companies is to be found in Table 17.6.

As a characterization of the general state of the economy, six macroprudential variables are used as covariates in our model settings. These are implied volatility index reported by the Chicago Board Options Exchange, daily S&P500 index returns, daily Dow Jones U.S. Real Estate index returns, changes in the three-month Treasury bill rate, changes in the slope of the yield curve corresponding to the yield spread between the ten-year Treasury rate and the three-month bill rate and, finally, changes in the credit spread between BAA-rated bonds and the Treasury rate. The former three are obtained from Yahoo Finance and the latter three from the Federal Reserve Board. The macro state variables are summarized in Table 17.7. The data are downloaded with help of FRM_download_data. All of the variables are recorded in the time interval from 03 January 2007 to 17 August 2016. For the macroprudential variables we use 1 day lagged values.

17.4.2 Construction of Time Series of $\widehat{\lambda}$

In order to capture interdependencies among the companies and to reduce the dimensionality of the data set into single time series of the penalization parameter λ of the Lasso regression, we proceed as follows.

We take each of the 100 companies as a dependent variable and use the remaining 99 together with the macro variables as predictors, i.e. $p = 105$. This way we get hundred regression models, which are then fitted with help of the quantile Lasso method by Li and Zhu (2008). To record the dynamics of $\widehat{\lambda}$, we use moving windows of size 63 observations ($n = 63$) which in this case represents 3 months.

Within each window algorithm designed by Li and Zhu (2008) is used to fit the Lasso model. Then the best fit and with it also the tuning parameter $\widehat{\lambda}$ are chosen with help of the BIC criterion (17.14). We obtain time series of tuning parameters $\widehat{\lambda}_k$ for each of the hundred regressed companies. These are plotted in Fig. 17.7a together with the average over all estimated parameters $\widehat{\lambda}_k$, $k = 1, \ldots, 100$, which we are interested in.

Indeed as suggested in our previous simulation study, $\widehat{\lambda}$ is driven by characteristics of an investigated model. From Fig. 17.7 we can see that its values are higher when the residuals of the model are higher, too. There are several peaks in time series of $\widehat{\lambda}$, which correspond to time periods of financial crises. This fact drives us to the conclusion that the dynamics of $\widehat{\lambda}$ can serve as an indicator of a systemic risk.

17.4.3 $\widehat{\lambda}$ and Systemic Risk Measures

In the past decade, much attention has been paid to measuring of systemic risk, particularly after the financial crisis between 2007 and 2009. It has uncovered the cross-sectional dependencies among financial institutions to be important when determining the risk on the market. Adrian and Brunnermeier (2016), Hautsch et al. (2015) and Härdle et al. (2016), just to mention a few, dealt with evaluating systemic risk according to the relevance of each financial institution itself. This inspired us to connect the Lasso parameter λ with the systemic risk, since it depends not only on the volatility but also on the size of model parameters and the correlation structure of the design matrix. The latter two effects can be translated into the connectedness of financial institutions throughout the market.

To illustrate the connection between $\widehat{\lambda}$ computed according to the method mentioned previously and other systemic risk measures, we plotted their common time development starting from 3 April 2007 to 17 August 2016, see Fig. 17.8.

We chose VIX to show the dependency between $\widehat{\lambda}$ and volatility observed on the financial market. The Standard & Poor's 500 stock market index (S&P500) moves in opposite direction of $\widehat{\lambda}$, which can also provide some information about behaviour of $\widehat{\lambda}$ in connection to the situation on financial markets. Another systemic risk measure is CoVaR presented by Adrian and Brunnermeier (2016) and extended by Härdle et al. (2016), where a single index model for generalized quantile regression instead of linear quantile regression was employed. The data for CoVaR$_S$ were downloaded from TENET_VaR_CoVaR where only weekly data between 7 December 2007 and 4 January 2013 were available. Financial turbulence as a risk measure was proposed by Kritzman and Li (2010). Its comovement with the time series of $\widehat{\lambda}$ is visible from the Fig. 17.8d. A composite indicator of systemic risk (CISS) is an indicator of contemporaneous stress in the financial system developed by Holló et al. (2012) and computed for the area of Europe on weekly basis. Even when considering another financial market, particularly collecting data from another countries, periods where CISS was elevated correspond to the periods of higher $\widehat{\lambda}$ values. And, finally, credit spread, i.e. changes in the credit spread between BAA-rated bonds and the Treasury rate, suggested by Giglio et al. (2016), was used to relate $\widehat{\lambda}$ to systemic risk level.

From Fig. 17.8 it is visible, that $\widehat{\lambda}$ has a common trend with some of the aforementioned systemic risk measures. For CoVaR$_S$ and S&P500 index it holds, that their time development goes in opposite direction compared to $\widehat{\lambda}$.

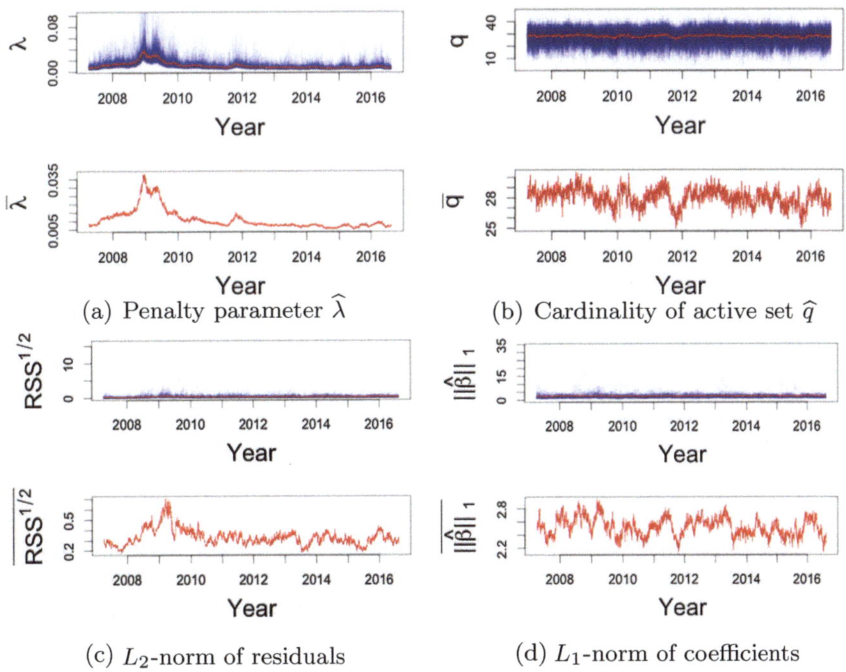

Fig. 17.7 Time series of $\widehat{\lambda}_k$ (*blue*) and other model characteristics and their respective averages (*red*) when fitted to given dataset, moving windows of length 63. XFGTVP_FRM

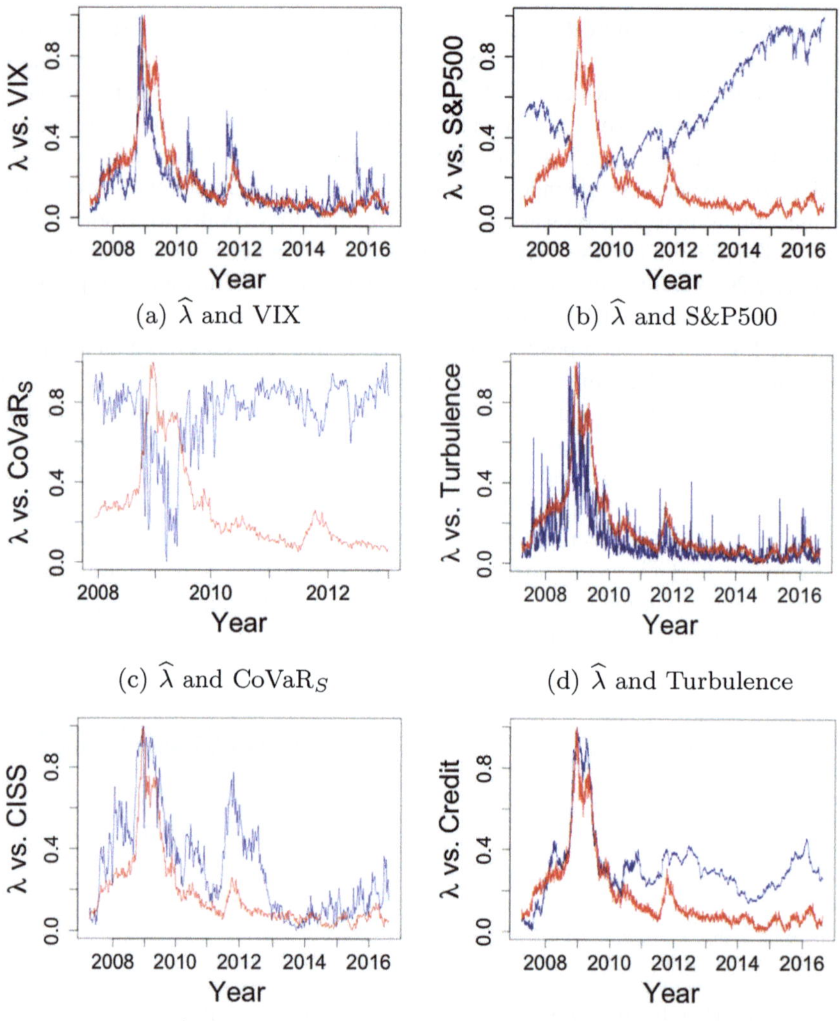

(a) $\widehat{\lambda}$ and VIX (b) $\widehat{\lambda}$ and S&P500

(c) $\widehat{\lambda}$ and CoVaR_S (d) $\widehat{\lambda}$ and Turbulence

(e) $\widehat{\lambda}$ and CISS (f) $\widehat{\lambda}$ and Credit spread

Fig. 17.8 Time series of $\widehat{\lambda}$ (*red*) and various systemic risk measures (*blue*). XFGTVP_LambdaSysRisk

In order to show there is a comovement between $\widehat{\lambda}$ and other systemic risk measures also from the statistical point of view, we conducted several cointegration tests. When looking at Fig. 17.8 one can see, that the time series of observed measures are nonstationary, however, there may exist cointegration relations between them which would make it a stationary stochastic process.

As a testing procedure we chose the Johansen (1991) test, where we used its eigenvalue type. In Table 17.5 there are stated resulting values of test statistics and their corresponding critical values on significance levels 10 and 5%. Variable r corresponds to a number of cointegration relations found between the two investigated nonstationary time series, i.e. for the valid inference we require that $r = 1$.

In Table 17.5 we included 3 more systemic risk measures. We chose also CoVaR computed with variable selection based on linear quantile regression (CoVaR_L). Another systemic risk measure is the volatility connectedness index designed by Diebold and Yilmaz (2014) and accessed from http://financialconnectedness.org. Yield slope denotes changes in the slope of the yield curve corresponding to the yield spread between the 10-year Treasury rate and the 3-month bill rate.

As we can see, many of the measures are cointegrated with the estimated Lasso parameter $\widehat{\lambda}$. The obtained results can lead to further work, such as studying and developing a theoretical model which might serve for prediction of the Lasso parameter λ. Furthermore, implementing our work into time series context might be of interest.

Table 17.5 Cointegration of $\widehat{\lambda}$ with systemic risk measures, r is number of cointegration relations in Johansen procedure, measures cointegrated with $\widehat{\lambda}$ are written in bold

	H_0	Test statistic	10 %	5 %
VIX	$r \leq 1$	4.80	7.52	9.24
	$r = 0$	87.43	13.75	15.67
S&P500	$r \leq 1$	7.59	10.49	12.25
	$r = 0$	9.20	16.85	18.96
CoVaR$_S$	$r \leq 1$	4.52	10.49	12.25
	$r = 0$	50.58	16.85	18.96
CoVaR$_L$	$r \leq 1$	4.59	10.49	12.25
	$r = 0$	57.15	16.85	18.96
Turbulence	$r \leq 1$	8.94	10.49	12.25
	$r = 0$	212.24	16.85	18.96
CISS	$r \leq 1$	6.90	10.49	12.25
	$r = 0$	31.12	16.85	18.96
Volatility connectedness	$r \leq 1$	9.48	10.49	12.25
	$r = 0$	10.51	16.85	18.96
Yield slope	$r \leq 1$	7.20	10.49	12.25
	$r = 0$	13.63	16.85	18.96
Credit spread	$r \leq 1$	5.45	10.49	12.25
	$r = 0$	42.29	16.85	18.96

Appendix

See Tables 17.6 and 17.7.

Table 17.6 List of 100 U.S. largest financial companies

WFC	Wells Fargo & Company	SEIC	SEI Investments Company
JPM	JP Morgan Chase & Co.	ETFC	E*TRADE Financial Corporation
BAC	Bank of America Corporation	AMG	Affiliated Managers Group, Inc.
C	Citigroup Inc.	RJF	Raymond James Financial, Inc.
AIG	American International Group, Inc.	UNM	Unum Group
GS	Goldman Sachs Group, Inc. (The)	NYCB	New York Community Bancorp, Inc.
USB	U.S. Bancorp	Y	Alleghany Corporation
AXP	American Express Company	SBNY	Signature Bank
MS	Morgan Stanley	CMA	Comerica Incorporated
BLK	BlackRock, Inc.	AJG	Arthur J. Gallagher & Co.
MET	MetLife, Inc.	JLL	Jones Lang LaSalle Incorporated
PNC	PNC Financial Services Group, Inc. (The)	TMK	Torchmark Corporation
BK	Bank Of New York Mellon Corporation (The)	WRB	W.R. Berkley Corporation
SCHW	The Charles Schwab Corporation	AFG	American Financial Group, Inc.
COF	Capital One Financial Corporation	SIVB	SVB Financial Group
PRU	Prudential Financial, Inc.	EWBC	East West Bancorp, Inc.
TRV	The Travelers Companies, Inc.	ROL	Rollins, Inc.
CME	CME Group Inc.	ZION	Zions Bancorporation
CB	Chubb Corporation (The)	AIZ	Assurant, Inc.
MMC	Marsh & McLennan Companies, Inc.	PACW	PacWest Bancorp
BBT	BB&T Corporation	AFSI	AmTrust Financial Services, Inc.
ICE	Intercontinental Exchange Inc.	ORI	Old Republic International Corporation
STT	State Street Corporation	PBCT	People's United Financial, Inc.
AFL	Aflac Incorporated	CACC	Credit Acceptance Corporation
AON	Aon plc	BRO	Brown & Brown, Inc.

(continued)

Table 17.6 (continued)

ALL	Allstate Corporation (The)	ERIE	Erie Indemnity Company
BEN	Franklin Resources, Inc.	OZRK	Bank of the Ozarks
STI	SunTrust Banks, Inc.	WTM	White Mountains Insurance Group, Ltd.
MCO	Moody's Corporation	SNV	Synovus Financial Corp.
PGR	Progressive Corporation (The)	ISBC	Investors Bancorp, Inc.
AMP	AMERIPRISE FINANCIAL SERVICES, INC.	MKTX	MarketAxess Holdings, Inc.
AMTD	TD Ameritrade Holding Corporation	LM	Legg Mason, Inc.
HIG	Hartford Financial Services Group, Inc. (The)	CBSH	Commerce Bancshares, Inc.
TROW	T. Rowe Price Group, Inc.	BOKF	BOK Financial Corporation
NTRS	Northern Trust Corporation	EEFT	Euronet Worldwide, Inc.
MTB	M&T Bank Corporation	DNB	Dun & Bradstreet Corporation (The)
FITB	Fifth Third Bancorp	WAL	Western Alliance Bancorporation
IVZ	Invesco Plc	EV	Eaton Vance Corporation
L	Loews Corporation	CFR	Cullen/Frost Bankers, Inc.
EFX	Equifax, Inc.	MORN	Morningstar, Inc.
PFG	Principal Financial Group Inc	THG	The Hanover Insurance Group, Inc.
RF	Regions Financial Corporation	UMPQ	Umpqua Holdings Corporation
MKL	Markel Corporation	CNO	CNO Financial Group, Inc.
LNC	Lincoln National Corporation	FHN	First Horizon National Corporation
CBG	CBRE Group, Inc.	WBS	Webster Financial Corporation
KEY	KeyCorp	PB	Prosperity Bancshares, Inc.
NDAQ	The NASDAQ OMX Group, Inc.	PVTB	PrivateBancorp, Inc.
CINF	Cincinnati Financial Corporation	SEB	Seaboard Corporation
CNA	CNA Financial Corporation	FCNCA	First Citizens BancShares, Inc.
HBAN	Huntington Bancshares Incorporated	MTG	MGIC Investment Corporation

Table 17.7 List of macro state variables

1	VIX
2	Daily change in the 3-month Treasury maturities
3	Change in the slope of the yield curve
4	Change in the credit spread
5	Daily Dow Jones U.S. Real Estate index returns
6	Daily S&P500 index returns

Acknowledgements Financial support from Deutsche Forschungsgemeinschaft via CRC "Economic Risk" and IRTG 1792 "High Dimensional Non Stationary Time Series", Humboldt-Universität zu Berlin, is gratefully acknowledged.

References

Adrian, T., & Brunnermeier, M. K. (2016). CoVaR. *American Economic Review, 106*, 1705–1741.

Belloni, A., & Chernozhukov, V. (2011). l_1-penalized quantile regression in high-dimensional sparse models. *The Annals of Statistics, 39*, 82–130.

Chand, S. (2012). On tuning parameter selection of lasso-type methods - a monte carlo study. In *Proceedings of 9th International Bhurban Conference on Applied Sciences & Technology* (pp. 120–129).

Chen, K., & Chan, K. S. (2011). Subset ARMA selection via the adaptive lasso. *Statistics and Its Interface, 4*, 197–205.

Diebold, F. X., & Yilmaz, K. (2014). On the network topology of variance decompositions: Measuring the connectedness of financial firms. *Journal of Econometrics, 182*, 119–134.

Efron, B., Hastie, T., Johnstone, I., & Tibshirani, R. (2004). Least angle regression. *The Annals of Statistics, 32*, 407–499.

Fan, J., & Li, R. (2001). Variable selection via nonconcave penalized likelihood and its Oracle properties. *Journal of the American Statistical Association, 96*, 1348–1360.

Giglio, S., Kelly, B., & Pruitt, S. (2016). Systemic risk and the macroeconomy: An empirical evaluation. *Journal of Financial Economics, 119*, 457–471.

Hastie, T., Tibshirani, R., & Friedman, J. (2009). *The elements of statistical learning: Data mining, inference and prediction* (2nd ed.). New York: Springer.

Hautsch, N., Schaumburg, J., & Schienle, M. (2015). Financial network systemic risk contributions. *Review of Finance, 19*, 685–738.

Härdle, W. K., Wang, W., & Yu, L. (2016). TENET: Tail-event driven NETwork risk. *Journal of Econometrics, 192*, 499–513.

Hsu, N. J., Hung, H. L., & Chang, Y. M. (2008). Subset selection for vector autoregressive processes using lasso. *Computational Statistics and Data Analysis, 52*, 3645–3657.

Holló, D., Kremer, M., & Lo Duca, M. (2012). CISS - a composite indicator of systemic stress in the financial system, ECB Working Paper Series No. 1426.

Johansen, S. (1991). Estimation and hypothesis testing of cointegration vectors in gaussian vector autoregressive models. *Econometrica, 59*, 1551–1580.

Koenker, R., & Basset, G. (1978). Regression quantiles. *Econometrica, 46*, 33–50.

Koenker, R., Ng, P., & Portnoy, S. (1994). Quantile smoothing splines. *Biometrica, 81*, 673–680.

Kritzman, M., & Li, Y. (2010). Skulls. *Financial Turbulence, ans Risk Management, Financial Analysts Journal, 66*, 30–41.

Lee, E. R., Noh, H., & Park, B. U. (2014). Model selection via Bayesian information criterion for quantile regression models. *Journal of the American Statistical Association, 109*, 216–229.
Leng, C., Lin, Y., & Wahba, G. (2006). A note on the lasso and related procedures in model selection. *Statistica Sinica, 16*, 1273–1284.
Li, Y., & Zhu, J. (2008). L_1-norm quantile regression. *Journal of Computational and Graphical Statistics, 17*, 1–23.
Nardi, Y., & Rinaldo, A. (2011). Autoregressive process modeling via the lasso procedure. *Journal of Multivariate Analysis, 102*, 528–549.
Osborne, M. R. (1985). *Finite algorithms in optimization and data analysis*, Wiley series in probability and mathematical statistics. Chichester: Wiley.
Osborne, M. R., Presnell, B., & Turlach, B. A. (2000). On the LASSO and its dual. *Journal of Computational and Graphical Statistics, 9*, 319–337.
R Core Team (2014). R: A language and environment for statistical computing. R Foundation for Statistical Computing, Vienna. http://www.R-project.org/. Accessed 15 April 2015.
Tibshirani, R. (1996). Regression shrinkage and selection via the lasso. *Journal of the Royal Statistical Society Series B, 58*, 267–288.
Wang, H., & Leng, C. (2007). Unified LASSO estimation by least squares approximation. *Journal of the American Statistical Association, 102*, 1039–1048.
Wang, H., Li, G., & Tsai, C. L. (2007). Regression coefficient and autoregressive order shrinkage and selection via the lasso. *Journal of the Royal Statistical Society: Series B, 69*, 63–78.
Wang, H., Li, B., & Leng, C. (2009). Shrinkage tuning parameter selection with a diverging number of parameters. *Journal of the Royal Statistical Society: Series B, 71*, 671–683.
Yuan, M. (2006). GACV for quantile smoothing splines. *Computational Statistics & Data Analysis, 50*, 813–829.
Zhao, P., & Yu, B. (2006). On model selection consistency of lasso. *Journal of Machine Learning Research, 7*, 2541–2563.
Zou, H. (2006). The adaptive lasso and its Oracle properties. *Journal of the American Statistical Association, 101*, 1418–1429.
Zou, H., Hastie, T., & Tibshirani, R. (2007). On the degrees of freedom of the lasso. *The Annals of Statistics, 35*, 2173–2192.

Chapter 18
Dynamic Topic Modelling for Cryptocurrency Community Forums

M. Linton, E.G.S. Teo, E. Bommes, C.Y. Chen and Wolfgang Karl Härdle

Abstract Cryptocurrencies are more and more used in official cash flows and exchange of goods. Bitcoin and the underlying blockchain technology have been looked at by big companies that are adopting and investing in this technology. The CRIX Index of cryptocurrencies http://hu.berlin/CRIX indicates a wider acceptance of cryptos. One reason for its prosperity certainly being a security aspect, since the underlying network of cryptos is decentralized. It is also unregulated and highly volatile, making the risk assessment at any given moment difficult. In message boards one finds a huge source of information in the form of unstructured text written by e.g. Bitcoin developers and investors. We collect from a popular crypto currency message board texts, user information and associated time stamps. We then provide an indicator for fraudulent schemes. This indicator is constructed using dynamic topic modelling, text mining and unsupervised machine learning. We study how opinions and the evolution of topics are connected with big events in the cryptocurrency universe. Furthermore, the predictive power of these techniques are investigated, comparing the results to known events in the cryptocurrency space. We also test hypothesis of self-fulling prophecies and herding behaviour using the results.

M. Linton (✉)
University of York, Unter den Linden 6, Heslington York YO10 5DD, UK
e-mail: msl508@york.ac.uk

E.G.S. Teo
School of Business, Singapore Management University, 50 Stamford Road, Singapore 178899, Singapore
e-mail: elisabeth.bommes@googlemail.com

E. Bommes · C.Y. Chen
Ladislaus von Bortkiewicz Chair of Statistics, C.A.S.E. - Center for Applied Statistics and Economics, Humboldt-Universität zu Berlin, Unter den Linden 6, 10099 Berlin, Germany
e-mail: elisabeth.bommes@googlemail.com

C.Y. Chen
e-mail: chencath@hu-berlin.de

W.K. Härdle
C.A.S.E.-Center of Applied Statistics and Economics, Humboldt-Universität zu Berlin, Unter den Linden 6, 10099 Berlin, Germany
e-mail: haerdle@hu-berlin.de

W.K. Härdle et al. (eds.), *Applied Quantitative Finance*, Statistics and Computing,
DOI 10.1007/978-3-662-54486-0_18

18.1 Introduction

Cryptocurrencies such as Bitcoin have become more mainstream over the years with big companies adopting and investing in the technology. Once seen to be the domain of technophiles and radicals, cryptocurrencies are now widely traded on many exchanges throughout the world. Governments have also discussed the possibilities of adopting cryptocurrencies as a means to offer digital currency. The underlying network (called the blockchain) of cryptocurrency is decentralised, unregulated and highly volatile, making its situation at any given moment difficult to assess. On the other hand, an almost bottomless source of information can be found in the form of unstructured text written by cryptocurrency users on the internet. Crowd wisdom found in such networks can be a powerful indicator of major events affecting cryptocurrencies. We attempt to take advantage of this to analyse and assign quantitative meaning to such resources.

Early academic statistical analysis of Bitcoin includes Cheah and Fry (2015) and Cheung et al. (2015), both looked at speculative bubbles using Bitcoin price data. More related to this paper are works that looked at social media information and search engine data such as Kristoufek (2013), Mai et al. (2015) and Matta et al. (2015).

Utilizing techniques from dynamic topic modelling (DTM), text mining and machine learning, we pull data from a popular cryptocurrency forum and attempt to detect events such as new trends in currencies, fraudulent schemes or legal and economic issues. The DTM technique, as a type of unsupervised learning, is demanded when the taxonomy is unclear. Some important topics may be left out if one does a subjective judgement for taxonomy. The DTM is designed for summarizing the unknown but important features in the world. In addition to "discover" and "quantify" the hidden topics, the DTM is able to characterize the evolution of the hidden topics, which may be useful for evaluating the importance and persistence. Specifically, we collect user information and text associated with time stamps and apply unsupervised dynamic topic modelling, studying how opinions and the evolution of topics are connected with big events in the cryptocurrency universe. Furthermore, the predictive power of these techniques are investigated, comparing the results to known events in the cryptocurrency space. We also test hypothesis of self-fulfilling prophecies and herding behaviour using the results. For example, Smailović et al. (2013) were able to improve predictive power for stock markets by using sentiment derived from Twitter feeds. Cryptocurrency discussion forums tend to be very responsive and sensitive to events; this makes it a suitable candidate to test the predictive ability of dynamic topic modelling.

18.2 Data

A good, consistent and representative source of information regarding the cryptocurrency community can be found on talk forums such as http://bitcointalk.org. Acquiring the data from this platform requires deploying a web scraper to download the relevant html pages from the server and extract the embedded information. Good practices of web scraping were used to ensure there was no risk of overloading servers such as waiting fifteen seconds between each request and respect for the robots.txt protocol. Information regarding thread ids, post ids, usernames, time stamps, post titles, post texts, quotes of other posts and links were collected and stored in a database. There are three main discussion boards which were used in this study, they are "Bitcoin", "Economy" and "Alternative Cryptocurrencies". The two remaining discussion boards were "Other" which was discarded as it mainly deal with non-related topics and "Local" which is also discarded as discussions are in local languages. Each of the main discussion boards were divided into subforums such as "Trading Discussions" and "Scam Accusations". In total there were little under 200 subforums, half a million different threads with over 15 million posts (including local discussion). For the purpose of our study, we concentrate on the Bitcoin discussion subforum.

Knowledge is power so the more information we have, the better. Aside from this, the main motivations behind collecting these bits of information are as follows: Thread ids and post ids are used to uniquely identify posts and the thread they come from; usernames are used to associate each post with an agent in order to create a graph for herding and social network analysis; time stamps are used to classify posts into time slices for the dynamic topic model; post titles and post texts are used in conjunction to form a document for the dynamic topic model; links and quotes are used in order to analyse how posts relate to each other and other websites which is useful for herding and social network analysis.

18.3 Topic Modelling

We apply topic modelling to these forums in order to model trends in the community and to see how real life events effect the topics discussed and vice versa. The most commonly used model to model topics in machine learning is LDA (Latent Dirichlet Allocation) by Blei et al. (2003).

This model, however, makes the assumption that all documents modelled are exchangeable and therefore the aspect of time is completely lost and the idea of detecting events becomes pointless. Therefore, the model we use is the dynamic topic model proposed by Blei and Lafferty (2006), which is a variant of LDA that analyses documents in a set of predetermined discrete time slices and assumes topics evolve smoothly from slice to slice with Gaussian noise.

LDA is a generative probabilistic model for text, however it has also been applied successfully to other types of discrete data sets such as images. This model differs from most as it is completely unsupervised, therefore removing the bottleneck of having to acquire a trained model, and the problem it tries to solve is not classification into topics, but rather assigning topic distributions to documents. These properties mean that it is ideal to apply to large quantities of unstructured text where it would be impossible to obtain reliable training data to produce a model and simply classifying documents into topics would produce confusing and unrealistic results. Bao and Datta (2014) apply the LDA method to extract the risk types (meaningful topics) in Security Exchange Commission 10-K forms, and find many plausible and meaningful risk types that have been left out in a supervised learning scheme proposed by Huang and Li (2011). The inferred topics from a supervised learning only cover 78% of topic pools.

The Dirichlet distribution is defined on a $(k-1)$ dimensional simplex

$$\Delta_k = \left\{ q \in \mathbb{R}^k : \sum_{i=1}^{k} q_i = 1, q_i \geq o, i - 1, 2, \ldots, k \right\}. \tag{18.1}$$

It can be thought of as a distribution of random probability mass/density functions (pdf). An excellent example based introduction can be found in Frigyik et al. (2010).

Definition 18.1 Let Q be a real value in Δ_k and suppose that $\alpha \in \mathbb{R}^k, \alpha_i > 0$ and define $\alpha_0 \stackrel{\text{def}}{=} \alpha^T 1$. Then Q has a $Dir(\alpha)$ distribution with pdf $f(q;\alpha) = \frac{\Gamma(\alpha_0)}{\prod_{i=1} \Gamma(\alpha_i)} \prod_{i=1}^{k} q_i^{\alpha_i - 1}$.

Density plots are given in Fig. 18.1 for different α. Given a document with a certain word distribution, the task is obviously to determine α from the set of documents.

The gamma function is a generalization of the factorial function, $\Gamma(s) = s\Gamma(s-1)$ with $\Gamma(1) = 1$. The mean of a $Dir(\alpha)$ random variable is $EQ = \alpha/\alpha_0$. Note that α determines the "location" of words in documents, a "small" α creates sharp peaks on defined locations. You may think of the document that has been written by the poet in the flim "Shining", in the described $Dir(\alpha)$ framework, there is just one "big" peak of the words at "all work and no play makes Jack a dull boy". With just $k = 2$ words in a document the $Dir(\alpha)$ reduces to the Beta distribution with pdf

$$f(x; a, b) = \frac{\Gamma(a+b)}{\Gamma(a) + \Gamma(b)} x^{a-1} (1-x)^{b-1}. \tag{18.2}$$

For $\alpha = (a, b)^T$ with $Q = (X, 1-X) \sim Dir(\alpha)$ for $X \sim Beta(a, b)$.

In a Bayesian context, employed here entirely for numerical and computational reasons, one finds that the multinomial distribution with pdf

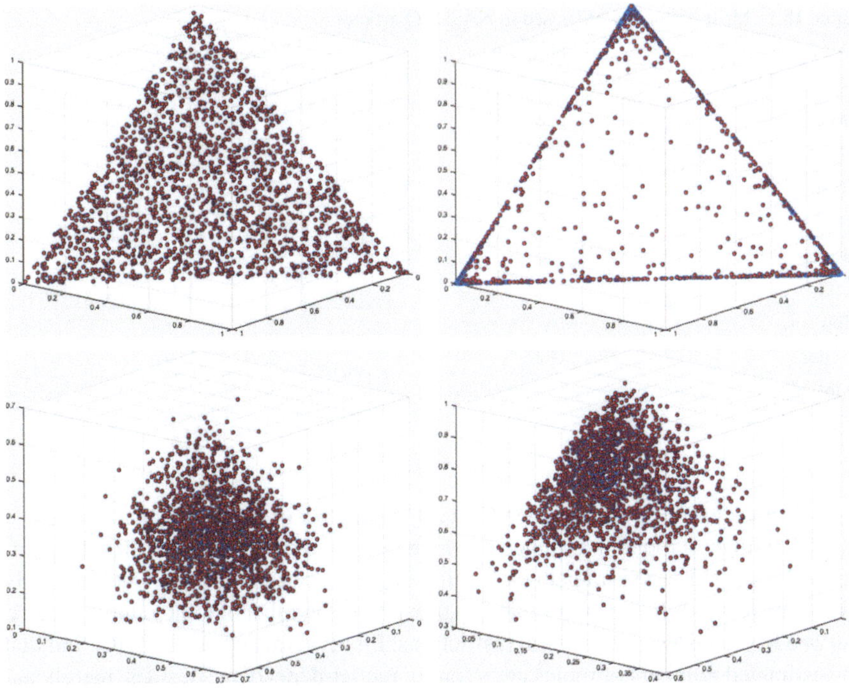

Fig. 18.1 Plots of sample pmfs drawn from Dirichlet distributions for various values of α. XFGtdmDirichlet

$$f(x; n, q) = \frac{n!}{\prod_{i=1}^{k} x_i!} \prod_{i=1}^{k} q_i^{x_i}, x, q \in \mathbb{R}^k \tag{18.3}$$

is a so called conjugate prior.

As the binomial distribution (for $k = 2$) is the conjugate prior for the Beta distribution, one finds that if $(X \mid q) \sim Mult_R(n, q)$ and $Q \sim Dir(\alpha)$, then $(Q \mid X = x) \sim Dir(\alpha + x)$. Again we refer for a proof of this to Frigyik et al. (2010).

The basic idea of a static Topic Model (TM) is to take a document as a sample of words generated by a $Dir(\theta)$ distribution, where θ represents the topic. More precisely it is assumed that a document is generated via the following imaginary random process:

1. For each topic k, draw a distribution over words $\vec{\beta}_k \sim Dir_v(\eta)$

2a. For each document d, draw topic proportions θ_d from over the $(k-1)$ simplex

2b. For each word $W_{d,n}$ within the document:

i. Draw a topic assignment $Z_{d,n} \sim Mult(\vec{\theta}_d)$, $Z_{d,n} \in \{1, \dots, k\}$

ii. Draw a word $W_{d,n} \sim Mult(\vec{\beta}_{z_{d,n}})$, $W_{d,n} \in \{1, \dots, V\}$

Table 18.1 Most frequent words used in NASDAQ articles

Word	Freq. (in k)	Freq. for top 5 sectors
Free	649	10
Well	238	9
Gold	235	1
Best	207	9
Fool	200	5
Strong	196	5
Like	172	5
Top	167	3
Better	162	0
Motley	152	2

β_z is a vector of β, one for each topic. β is a matrix of word|topic parameters.

The number of topics is assumed known beforehand though determining the number of topics (clusters) is rather challenging in unsupervised learning. One can easily find some methods being proposed for estimating the number of topics automatically, but one has to be aware of several restrictions. Firstly, Wallach et al. (2010) find that the estimated numbers of topics are strongly model-dependent. Besides, merely using fit statistics such as perplexity may be problematic due to a negative relation between the best fitted model and the substantive fit (Chang et al. 2009). To balance the substantive fit and statistical fit, Bao and Datta (2014) propose strategic procedures - Firstly, employing statistical fit to reduce the set of candidate models with different numbers of topics. Relying on the predefined perplexity, one can optimize the predictive power of model. In their case, the numbers can be chosen as 30, 40 and 50 in terms of perplexity and a converge in the range [30, 50] is shown. Secondly, the substantive fit for semantic coherence is compared among the competing models. To be specific, the model precision in word intrusion task is evaluated. It's so called "semantic validation". The semantic coherence of topics perhaps is the most useful indicator w.r.t the quality of topics, reflecting to how well the topic matches a human concept through a list of keywords. The number, 30, is therefore chosen due to its best semantic coherence performance.

Let us provide an example that sheds some light on this generation mechanism. Suppose that the "word universe" corresponds to the most frequent words in the NASDAQ analysis study by Zhang et al. (2016) and Bommes et al. (2017), as given in Table 18.1.

The idea is now that different topics have different word distribution as given by $Mult(\beta_z)$. Suppose there were $k = 2$ topics/sectors, corresponding to "finance" and "IT" and further suppose that the distribution of words over topics are generated by $Dir(\theta)$. To be precise, for $k = 2$, the Dirichlet distribution boils down to a $Beta(\theta)$ distribution. It could be the case that for the topic "finance", the third most frequent word "gold" is more concentrated. Whereas, for the topic "IT", concentration would be more around the words "fool" and "motley". See Fig. 18.2 below for an illustration

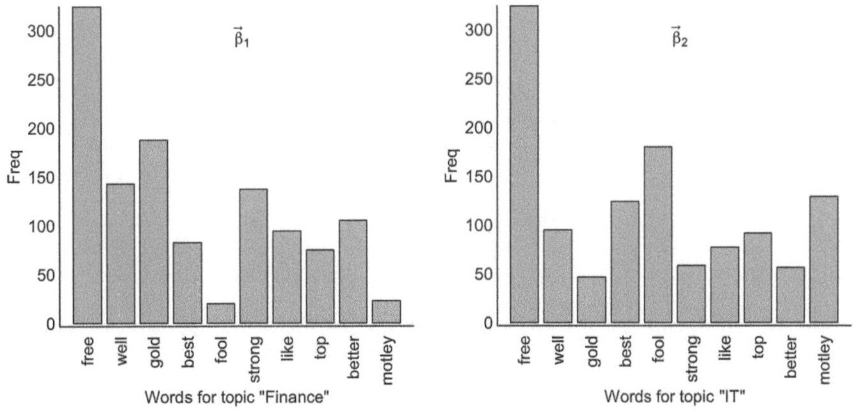

Fig. 18.2 Distribution of words by topic ($\vec{\beta}_1$ and $\vec{\beta}_2$). XFGdtmWDistr

that shows the random outcomes $\vec{\beta}_1$ and $\vec{\beta}_2$. In such as scenario, we would prefer a different word distribution for each these topics.

Step 2bi. now refers to the random mechanism that a word to be written down is drawn from $\vec{\beta}_1$ or $\vec{\beta}_2$. Suppose that the first has to be drawn from $\vec{\beta}_1$ since $Z_{1,1} = 1$, for $d = 1$ (1st document) and $n = 1$ (first word). So a random outcome as described in Step 2bii. could be the word $W_{1,1} =$ "gold" (the word with the second highest frequency in $\vec{\beta}_1$. For the next word ($n = 2$), $Z_{1,2}$ could take the value 1 again and now $W_{1,2} =$ "strong" could be the outcome. A third word could be via $Z_{1,3} = 2$, $W_{1,3} =$ "free", and so on. The task of TM is now to invert this mechanism and calibrate the observed documents to the parameters of the Dir and $Mult$ distributions.

The problem of static TM though is that there is no timeline, an issue that is of course necessary for the questions we would like to study here. The dynamic topic model, on the other hand models each time slice with LDA, but its parameters β and α are chained together in a state space model which evolves with Gaussian noise:

$$\beta_{t,k}|\beta_{t-1,k} \sim N(\beta_{t-1,k}, \sigma 2I) \tag{18.4}$$

$$\alpha_{t,k}|\alpha_{t-1,k} \sim N(\alpha_{t-1,k}, \delta 2I) \tag{18.5}$$

Like this we get a smooth evolution of topics from slice to slice. The state space diagram describes the model well.

Due to the nonconjugacy of the Gaussian and multinomial distributions, exact inference is intractable so the authors present two methods for approximate inference using variational methods: variational Kalman filtering and variational wavelet regression (Fig. 18.3).

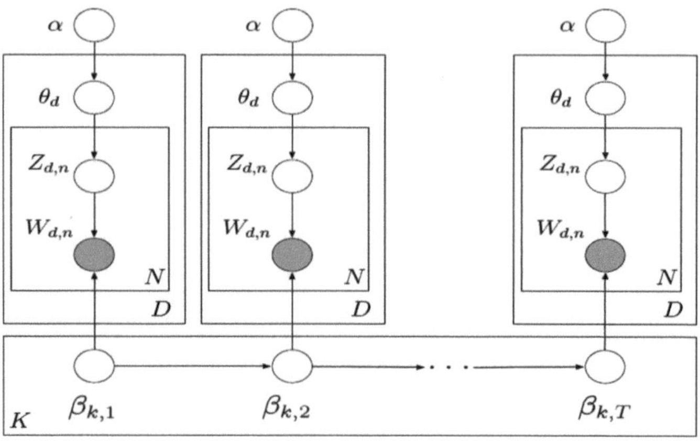

Fig. 18.3 State space diagram of the dynamic topic model

18.4 Preprocessing

Preprocessing steps make a big difference to the outcome of topic models. Especially when working in the domain of a forum where thousands of users post everyday, most likely without looking words up in the dictionary or worrying about the correctness of their grammar, we will find many spelling mistakes, slang and proper names that aren't going to be simple to handle. Therefore, a natural approach to preparing the data appropriately would be to use a POS tagging algorithm coupled with a tokeniser to infer from context what words have which function. Stop words will appear multiple times in each sentence without conveying any meaning and therefore are removed and so are functional words, verbs, adjectives and adverbs leaving us only with nouns, proper nouns and foreign words. In this way we have all the most important information from each post without losing out on non-standard vocabularies that arise in the community. To combat typos, the words occurring in fewer than 10 documents were removed and to get rid of generic words, the words appearing in more than 10% of the documents were also removed. In the end, from a dictionary of 500,000 words, we obtained one of 10,000 meaningful words. Once we had the cleaned text, the preparation for the dynamic topic model (code by Sean M. Gerrish) consisted of converting the corpus to a sparse matrix representation whereby each line represented a document and was in the following form:

$N\_unique\_words\ word\_id : word\_count\ word\_id : word\_count.....$

Also a file containing information about the time slices was prepared of the following format:

N_time_slices
$N\_docs\_slice\_1$
$N\_docs\_slice\_2$
...

Where N denotes number of documents in the corresponding slice. On top of these necessary files, for each corpus a file containing metadata, a dictionary file and a vocabulary file were also produced. The metadata file contains a header describing the fields and then each line represents a document with the following pieces of information: thread id, post id, date time, username, post text, post quotes and post links. This will come in handy for information retrieval and herding analysis. The dictionary file is a python dictionary object which maps ids to words and contains word count information. The vocabulary file is a human readable file where each line is a word from the dictionary and its position maps to its key.

18.5 Trends

As mentioned in the introduction, the data acquired from the forum was divided into subforums. The main subforums by posting volume are: 'Economics', 'Bitcoin Discussion', 'Altcoin Discussion' and 'Speculation'. The dynamic topic model was run on these subforums and in addition also with the subforum 'Scam Accusations'. The commonly used $50/k$ heuristic by Griffiths and Steyvers (2004) for the alpha parameter was chosen and a varying number of topics were modelled. All models were run with weekly data over the 2009/11/22 (when the forum was created) to 2016/08/06 period.

Each topic in the hidden structure is represented as a distribution over words and therefore the most human interpretable way of understanding what a topic is about is to look at the most probable words in each distribution. An example representation can be found in Table 18.2 in which some topics are shown for the last time slice in the Bitcoin Discussion subboard. Each time slice will have it's own similar representation. While the words may change over time as new trends emerge and fall, the topic will intuitively remain the same. For example, in the table shown we can see that topic 50 is about Bitcoin mining, but the top words in the first time slice are rather different even though we would still assign the same topic label to it; cpu, difficulty, proof, mining, adjustment, proof-of-work, power, attack were the top words in 2009 in topic 50, demonstrating how Bitcoin mining has evolved to cope with the increasing mining difficulty. In fact we can directly compare different mining hardware and how they were relevant over different periods of time in Fig. 18.4.

As we can see, in topic 50 the word CPU was very prominent initially and all the others were non-existent. Then when the network grew to an extent that the quantity of Bitcoins produced by CPU mining were worth less than what it cost to operate, GPU mining came into play. Another stride in mining hardware was the usage of application specific integrated circuits (asic). The first asic mining hardware project called the 'Avalon Project' was announced in 2012 on the forum and the peak in the third plot in January 2013 corresponds to the release of their first chip. In the fourth plot we see the timeline of Antminer, a brand of asics considered to be the current top of the line. As expected we can see a positive trend over the last years with peaks in discussion around releases of new models.

Table 18.2 Notable topics from 50 topic model on Bitcoin Discussion subforum from 2016/07/31 to 2016/08/06

Topic number	Most probable words
1	Value, gold, bar, dollar, rate, demand, interest, asset
2	Business, casino, house, trust, gambling, run, strategy, player
5	Government, control, criminal, law, study, regulation, state, rule
7	Use, service, option, cash, good, spend, fiat, convert
12	Account, payment, fund, card, paypal, party, merchant, credit
18	Score, online, pay, shop, bill, product, purchase, phone
20	Wallet, key, paper, computer, storage, code, data, secure
23	Price, trade, market, trader, drop, volume, sell, stock
24	Trading, term, hold, buy, pump, dump, earn, gamble
30	Exchange, bitfinex, lesson, cryptocurrency, crash, platform, altcoins, popularity
32	Investment, risk, invest, aim, impact, salary, making, way
33	Year, altcoins, end, today, adoption, prediction, happen, trend
35	Transaction, block, fee, chain, confirmation, hour, minute, hardfork
38	Altcoin, company, loss, hack, scam, hacker, scammer, road
42	Bank, system, security, fiat, banking, role, function, institution
45	Ethereum, split, advantage, issue, side, change, fork, core
48	Forum, post, topic, member, bitcointalk, thread, index, php
50	Mining, miner, network, power, pool, cost, reward, electricity

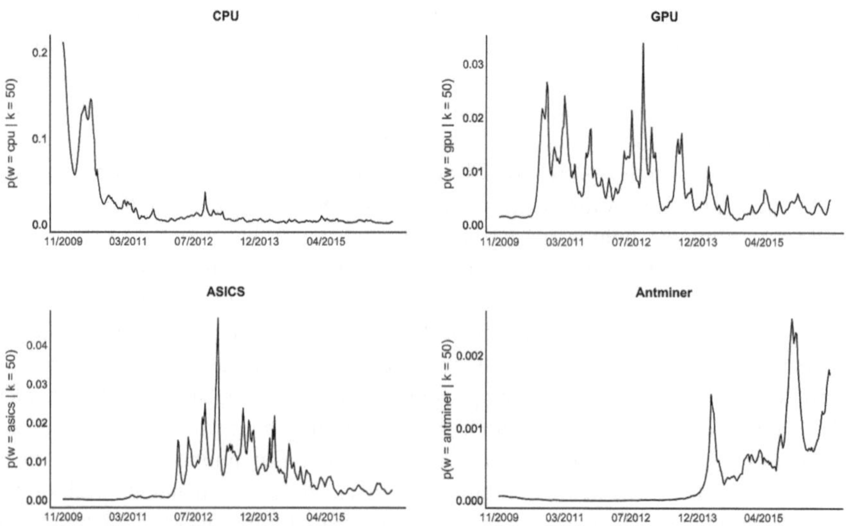

Fig. 18.4 Comparison of word evolution for different mining technologies 22/11/2009–06/08/2016. XFGdtmMining

As an up and coming and fast growing technology, Bitcoin has had its fair share of issues. In fact, due to its unregulated nature and uncertainty of legality or legitimacy as currency in most corners of the world, the cryptocurrency history is laden with high profile hacks, ponzi schemes and scam websites. Many of these go undetected for months until a certain point where gradually complaints start to stack up and a realisation or confirmation of the events takes place.

Probably the biggest example of such an event in Bitcoin history is the insolvency of the MtGox Bitcoin exchange in 2014. MtGox originally started off in 2007 as a platform for trading Magic: The Gathering Online trading cards which is where it got its name (Magic: The Gathering eXchange). In 2010, however, it was rebranded as one of the first exchanges where people could buy and sell Bitcoins. The exchange grew gradually and watched the price of Bitcoin go from less than 0.1 USD in 2010 to parity with the US dollar in 2011. At this point however, the owner of MtGox decided to sell the exchange in order to dedicate himself to 'other projects'. An internal email dating back from after the sale of the exchange revealed that already 80,000 Bitcoins (worth over $60,000 at the time) had already been missing before any of the public fiascos had occurred and had never been recovered. However, it was only three months later that a major event occurred. 60,000 accounts were exposed publicly and a compromised MtGox auditors account was used to create huge sell orders and crash the Bitcoin price from $17.51 to $0.01. As a result of this event the site was down for a week and many of the exposed accounts were used to steal coins from other bitcoin services due to password reuse. However, unlike many other Bitcoin services, MtGox managed to recover its reputation and became the largest Bitcoin exchange, handling 70% of all trades worldwide. Fast forwarding to 2013, when their real problems began, in June withdrawals of US dollars were suspended and even though a couple of weeks later in July it had been announced that withdrawals had fully resumed, as of September few withdrawals had successfully been completed. Complaints piled up over the next few months and on 7 February 2014 all Bitcoin withdrawals had been suspended for good. On the 24th of February all activities had halted, the website went offline and a leaked internal crisis management document claimed that 744,408 Bitcoins (worth almost half a billion dollars) had been lost and the company was insolvent.

As we can see, MtGox has had a roller coaster of a past with repeated security issues and poor management and has therefore been a major topic of discussion among users of the main Bitcoin forum. The main topics in which MtGox arises are predictively topic 23 about Bitcoin trading and markets and topic 38 about scams and hacks. Naturally the word/topic probability plot in Fig. 18.5 reflects this and we can see peaks corresponding to the main events. In topic 38 there is a clear peak in mid 2011 during the first hack and in February 2014 also. Meanwhile in topic 23 there is a gradual peak starting in mid 2013 when the transaction issues first occurred and trailing off at the same time MtGox starts to gain momentum in topic 38.

MtGox is only one example of the many scams and hacks resulting in huge losses that have occurred over the years and it is because of this that cryptocurrencies get a bad rap. Many services have come and gone, but none quite so spectacularly as MtGox.

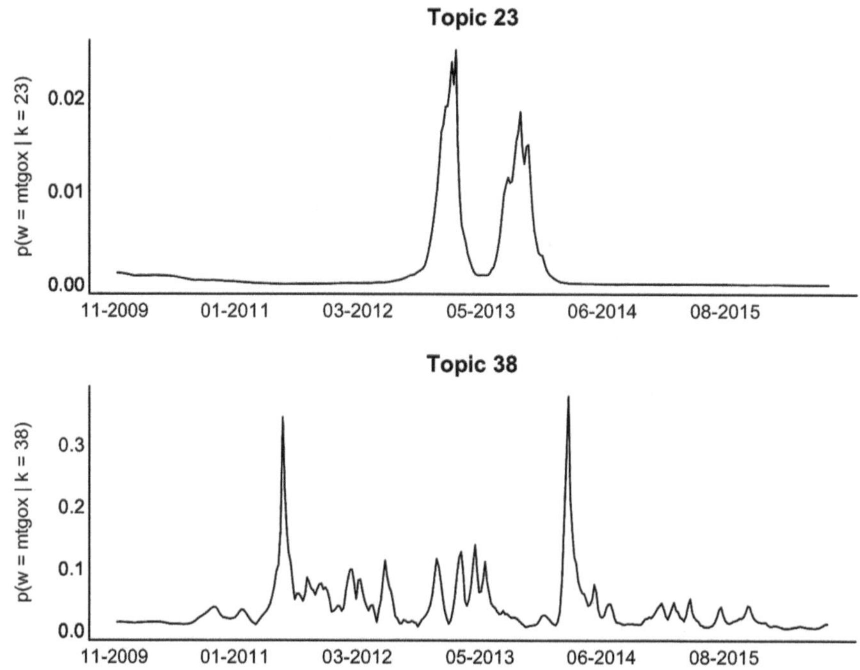

Fig. 18.5 MtGox word evolution 22/11/2009–06/08/2016. XFGdtmMtGox

Currency exchanges, mining hardware manufacturers, technology startups, mining pools and many other cryptocurrency related services have almost infallibly been victims of hacks and inside jobs, revealed as ponzi schemes, virus promoters etc. As soon as such events occur or are discovered, we would expect there to be gradual buildups or sudden explosions of discussion on the forum depending on the situation. In general, we would expect any event in the Bitcoin universe to be discussed on the forum and therefore be a part of the inferred generative process of the topic structure.

We want to evaluate the effectiveness of topic models in discerning these types of events. In our MtGox example, the word probabilities over time are characterised by relatively flat probabilities in general and spikes at the time of events. We can take advantage of this structure and hypothesise that it extends to other events. First we must validate this against other events. A curated list of Bitcoin services which have been victims of hacks or perpetrators of scams have been compiled over the years in a thread on http://bitcointalk.org (https://bitcointalk.org/index.php?topic=576337.0). This list will form our basis for event discovery validation. This could be done for other types of events however the most complete information can be found regarding scam/hack events since they are of relevance and interest to all involved with Bitcoin. We look at the topic *prominence* for this set of words and see if the model correctly partitions them in a scam/hack topic.

18.6 Choosing K and Analysis

The choice of the number of topics has been an issue ever since topic models were first introduced in 2003. For this particular study, we used the Umass coherence metric by Mimno et al. (2011) to evaluate which number of topics was optimal. This method involves taking the top N words for each topic and taking measures of their occurrences and co-occurrences in the corpus. Formally it is defined as:

$$\sum_{i=1}^{N-1}\sum_{j=i+1}^{N}\frac{D(w_i, w_j)+\epsilon}{D(w_i)} \tag{18.6}$$

where w_i and w_j are the ith and jth ranked words in a given topic respectively and $D(w)$ is the number of documents in which word w occurs. We set $N = 20$.

It has been shown to correlate well with human interpretations of what constitutes a coherent topic. In addition, the metric does not require external validation, simplifying the procedure and making it more versatile. To make the repeated training of models viable, we calculated Umass coherence on a subsample of 100 weeks of data. In Table 18.3 we can see the results of the coherence evaluation. We have taken the arithmetic mean and standard deviation of the output values over the 100 chained LDA models; higher values mean more human understandable topics. Clearly our model is optimal when we choose 30 for the k parameter since on average the topics are more coherent and stable over time. We also observe that lower numbers of k are more coherent than higher values, but are also less stable over time. While this method does a good job at finding the number of topics more attuned to human intuition, we would also like to study how this effects event detection.

The generative process described now gives us a multi-layer interpretation of the data. We have K topics with D documents and W words. Each topic can be described by a vector of length W of word/topic probabilities. Each document can be described by a vector of length K of topic/document probabilities. Each topic changes over each of the T time slices and therefore each topic/document distribution acquires a different meaning depending on where it is in the timeline.

Say we have a particular word w in our vocabulary we would like to learn something about. The best way to do this is to look at the word probabilities over a certain

Table 18.3 Topic coherence statistics

Number of topics k	μ	σ
10	–185.74	66.62
20	–204.28	65.57
30	–176.46	52.80
40	–202.10	68.99
50	–205.83	63.17

time slice in the topics. We can call this concept the word *prominence* and we would like to maximize this in order to find the most relevant topic.

$$\arg\max_{k} \frac{1}{t_j - t_i} \sum_{t=t_i}^{t_j} p(w|k, t_i) \tag{18.7}$$

Once we have found this topic (or topics if we want to find several), looking at the topics top words will allow us to discover in which context this term is discussed the most. We can also plot the evolution of the probability over time of this particular word in this topic and see when it was most used, when it came into use or passed out of use. Quite often words with same spelling but different meaning (homonyms) occur or words that can be discussed in different contexts (for example *price* could be present in a stock market topic or in a groceries topic). Whereas usually it wouldn't be a simple task to discern these words, topic models account for them very nicely and provide a useful perspective.

In addition to analysing the word/topic distribution we can also take a look at the topic/document distributions and determine in which time slice which topics were 'hotter' and which were 'colder' and identify trend starters. The hotter a topic k at time t, the more documents are going to exhibit higher mixtures of the topic. The inverse is true for colder topics. We can define the topic *temperature* as follows by Hall et al. (2008):

$$\sum_{d:t_d=t} p(k|d)p(d|t) = \frac{1}{D_t} \sum_{d:t_d=t} p(k|d) \tag{18.8}$$

where D_t is the number of documents in time slice t and t_d is the date document d was written.

18.7 Detection

From the list of events acquired from the forum, all those solely concerning individuals or causing losses of fewer than 1000 Bitcoins were removed. As a consequence of this procedure, we were left with 33 different Bitcoin services (and 37 different events). For each word we determine which topics the word achieves a topic *prominence* larger than a certain threshold. Typically, any given word will only appear in a handful of topics and most in just 1 or 2. Even though a certain topic may not have anything to do with a chosen word, topic models have the property that the probability of a word occurring in a topic is never 0, albeit negligible. Therefore we use a very low empirically tested threshold to determine which topics to test and discard the noisy ones. Then we analyse the topic *prominence* of the words conditioned on topics through time and determine an event occurring to be when its upper control limit is breached. I.e. when:

Table 18.4 Events in chronological order, an asterisk means undetected in the 50 topic model

Event	Dates	Topic
Ubitex* (1,138b)	2011-04 to 2011-07	None
Allinvain	2011-06-13	23
MtGox	2011-06-19	23
Mybitcoin	2011-06-20, 2011-07	23
Bitomat	2011-07-26	23
Mooncoin	2011-09-11	23
Bitscalper	2012-01 to 2012-03	23
Linode	2012-03-01	23
Betcoin* (3,171b)	2012-04-11	None
Bitcoinica	2012-04-12, 2012-07-13	23
Btc-e	2012-07-13	12
Kronos	2012-08	23
Bitcoin Savings and Trusts	2012-08-28	23
Bitfloor	2012-09-04	23
Btcguild* (1,254b)	2013-03-10	None
OkPay (main victim of 2013 Fork)	2013-03-11	30
Ziggap* (1,708b)	2013-02 to 2013-04	None
Just-Dice	2013-07-15	23
Basic-Mining* (2,131b)	2013-10	None
Silkroad2	2013-10-02	23
Vircurex* (1,454b)	2013-10-05	None
GBL	2013-10-26	12
Bips* (1,294b)	2013-11-17	None
Picostocks* (5,896b)	2013-11-29	None
MtGox	2014-02-24	23
Flexcoin	2014-03-02	23
Cryptorush	2014-03-11	23
Mintpal	2014-10-14	23
Silkroad2	2014-11-06	23
Bitstamp	2015-01-04	23, 25
Bter	2015-02-14	23
Cryptsy	2016-01-01	23
Shapeshift	2016-04	23
Gatecoin*	2016-05-13	None
Bitfinex	2016-08-03	12

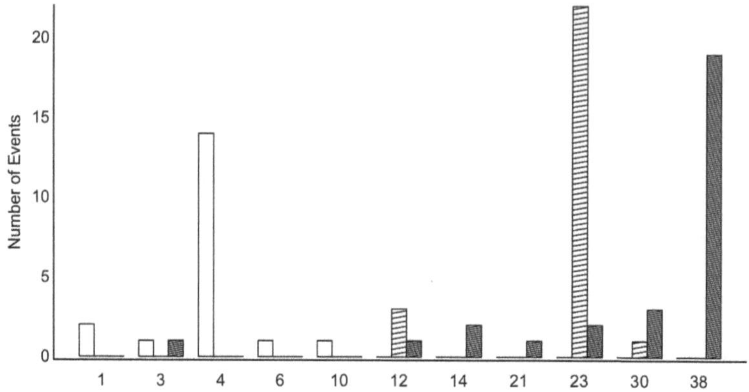

Fig. 18.6 Event partitioning over varying k parameters, 10 topics (*no filling*), 30 topics (*dashed filling*), 50 topics model (*densely dashed filling*). XFGdtmEvents

$$p(w|k, t_{i+1}) > \mu(p(w|k, t_{1:i})) + 3\,\sigma(p(w|k, t_{1:i})) \tag{18.9}$$

Table 18.4 contains the information regarding our events and the dates they occurred. We compared these events against those detected in our model using the method described and have marked with an asterisk those that went undetected.

Most of the events causing losses of circa 2000 Bitcoins and under (indicated) went undetected and almost all of those causing larger losses were identified. As hypothesized in the previous section, the large majority of these events were found to be in a single topic (topic 38), demonstrating the effectiveness of topic models in discriminating event types and providing an indicator for future such events.

This event detection algorithm was also run on our 10, 30 and 50 topic models. For the varying number k we can see what effect it has on our event distribution in Fig. 18.6. With the number of topics considered to be most coherent, our events are grouped mainly into a single topic. On the other hand, the less coherent topics are composed of many junk topics in the higher k case, or more general topics in the lower, therefore resulting in inconsistency in the experiment. A lower k results in fewer detections as our topics will each be less relevant and a higher k results in many junk topics and detections across more topics.

In addition, for each event we can observe the impact it has on the topic structure by measuring the deviation of the topic $temperature$ from the mean at the time in which it occurred. Since our timeline and number of time slices is large and we are using a symmetric Dirichlet prior, our topics are going to be rather general and fixed through time and the change in temperature between different times won't be significant. However, one can note in Fig. 18.7 that all values are positive at the times the events occurred and appreciate the event hierarchy that follows.

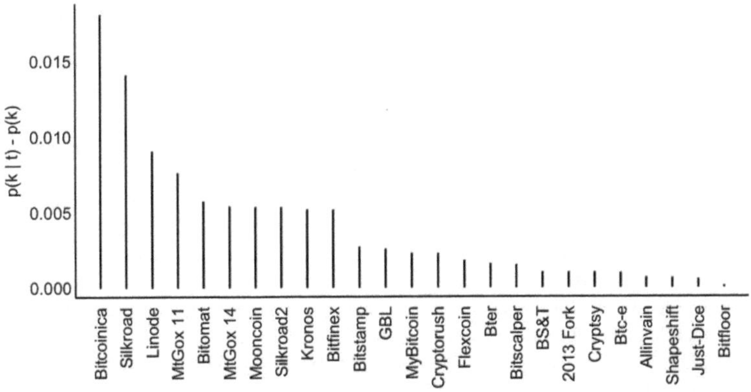

Fig. 18.7 Plot of ordered topic *temperatures* at time of event with k being the event topic and t being the time of the event XFGdtmTemperature

18.8 Conclusion

In the above piece of work we have introduced and explained topic models. A dataset has been created from user posts on http://bitcointalk.org by using web scraping; then text-mining techniques were used to prepare the data for dynamic topic modelling and consequently a walk through of all the steps for constructing such a model has been provided. We have presented a study and exploration of the popular cryptocurrency forum in this framework and employed an event detection technique to capture the effect of high profile scamming and hacking on the community. The number of topics parameter has been shown to be optimal for event detection when it accords with a measure of topic coherence. In addition, the constructed model partitions almost all of the events above a certain severity in a single topic.

References

Bao, Y., & Datta, A. (2014). Simultaneously discovering and quantifying risk types from textual risk disclosures. *Management Science*, *60*(6), 1371–1391.

Blei, D., Ng, A. Y., Jordan, M. I., & Lafferty, J. (2003). Latent Dirichlet allocation; *Journal of Machine Learning Research, 3*, 993–1022.

Blei, D., & Lafferty, J. (2006). Dynamic topic models. In *Proceedings of the 23rd international conference on Machine learning (AMC)*.

Bommes, E., Chen, C. Y., Härdle, W. K. (2017). Textual sentiment and sector-specific reaction. *Forthcoming*.

Chang, J., Boyd-Graber, J. L., Wang, C., & Blei, D. M. (2009). Reading tea leaves: How humans interpret topic models. *Advances in Neural Information Processing Systems*, 288–296.

Cheah, E. T., & Fry, J. (2015). Speculative bubbles in Bitcoin markets? An empirical investigation into the fundamental value of Bitcoin. *Economics Letters*, *130*, 32–36.

Cheung, A., Roca, E., & Su, J. J. (2015). Crypto-currency bubbles: An application of the Phillips-Shi-Yu (2013) methodology on Mt. Gox bitcoin prices. *Applied Economics*, *47*(23), 2348–2358.
Frigyik, B. A., Kapila, A., & Gupta, M. R. (2010). Introduction to the Dirichlet distribution and related processes. *Technical Report*, Department of Electrical Engineering, University of Washington.
Griffiths, T., & Steyvers, M. (2004). Finding Scientific Topics. *Proceedings of the National Academy of Sciences of the United States of America*, *101*(Suppl1), 5228–5235.
Hall, D., Jurafsky, D., & Manning, C. (2008). Studying the history of ideas using topic models. *Proceedings of the Conference on Empirical Methods in Natural Language Processing*, 363–371.
Huang, K. W., & Li, Z. L. (2011). A multilable text classification algorithm for labeling risk factors in SEC form 10-K. *ACM Transactions on Management Information Systems (TMIS)*, *2*(3), 18.
Kristoufek, L. (2013). BitCoin meets Google Trends and Wikipedia: Quantifying the relationship between phenomena of the Internet era. *Scientific Reports*, *3*, 3415.
Mai, F., Bai, Q., Shan, Z., Wang, X. S., & Chiang, R. H. (2015). The impacts of social media on Bitcoin performance. In *Proceedings of the Thirty Sixth International Conference on Information Systems (ICIS 2015)*.
Matta, M., Lunesu, I., & Marchesi, M. (2015). Bitcoin spread prediction using social and web search media. *Proceedings of DeCAT*.
Mimno, D., Wallach, H. M., Talley E., Leenders, M., & McCallum, A. (2011). Optimizing semantic coherence in topic models. In *Proceedings of the Conference on Empirical Methods in Natural Language Processing*, 262–272.
Smailović, J., Grčar, M., Lavrač, N., & Žnidaršič, M. (2013). Predictive sentiment analysis of tweets: A stock market application. In *Human-Computer Interaction and Knowledge Discovery in Complex, Unstructured, Big Data* (pp. 77–88). Berlin: Springer.
Wallach, H. M., Jensen, S. T., Dicker, L. H., & Heller, K. A. (2010). An alternative prior process for nonparametric Bayesian clustering. *Proceedings of the Thirteenth International Conference on Artificial Intelligence and Statistics (AISTATS)*, *9*, 892–899.
Zhang, J. L., Härdle, W. K., Chen, C. Y., & Bommes, E. (2016). Distillation of news flow into analysis of stock reactions. *Journal of Business and Economic Statistics*, *34*, 547–563.

Erratum to: Copulae in High Dimensions: An Introduction

Ostap Okhrin, Alexander Ristig and Ya-Fei Xu

Erratum to: Chapter 13 in: W.K. Härdle et al. (eds.), *Applied Quantitative Finance*, Statistics and Computing, https://doi.org/10.1007/978-3-662-54486-0_13

In the original version of the book, the following links have to be newly included in the caption of Fig. 13.4 in Chapter 13. The erratum chapter and the book have been updated with the change.

The updated online version of this chapter can be found at https://doi.org/10.1007/978-3-662-54486-0_13

W.K. Härdle et al. (eds.), *Applied Quantitative Finance*, Statistics and Computing,
DOI 10.1007/978-3-662-54486-0_19

MIX
Papier aus verantwortungsvollen Quellen
Paper from responsible sources
FSC® C105338

If you have any concerns about our products,
you can contact us on
ProductSafety@springernature.com

In case Publisher is established outside the EU,
the EU authorized representative is:
Springer Nature Customer Service Center GmbH
Europaplatz 3, 69115 Heidelberg, Germany

Printed by Libri Plureos GmbH
in Hamburg, Germany